102 THINGS TO DO WITH A HOLE IN THE GROUND

First published in Great Britain
by Eden Project Books 2024.

The Eden Project
Bodelva
Cornwall
PL24 2SG
edenproject.com

Printed and bound in the UK
by Short Run Press Ltd.

ISBN: 978-0-95622-135-3

10 9 8 7 6 5 4 3 2 1

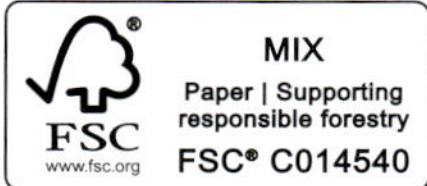

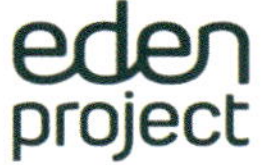
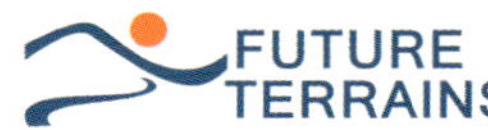
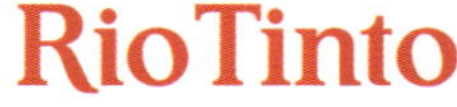

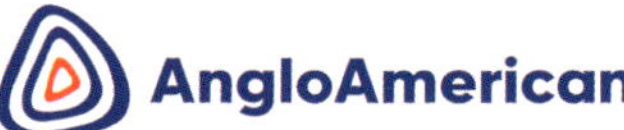
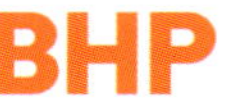
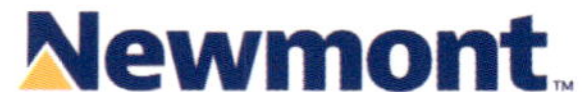

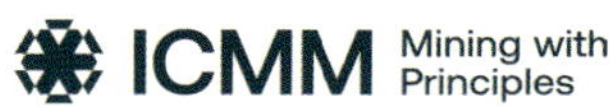
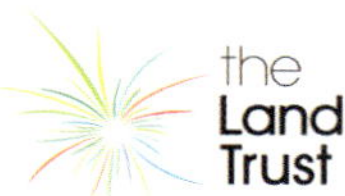
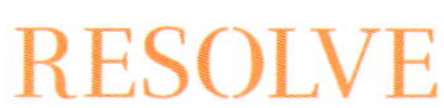

Foreword

It was an absolute honour to be asked to write the foreword to this amazing and important book that was years in the making. The original *101 Things to Do with a Hole in the Ground* book single-handedly changed the global perception of former mine sites. To this day, people with copies of this book, first published in 2009, hold on to them like a treasure. It not only showed what is possible but also how to reuse our natural resources in unique ways to create new natural, social and economic value for communities that once relied on mining operations for their livelihoods. The Eden Project continues to be a shining example of this.

Minerals and metals are essential to the products and technologies of modern society. As we transition to a low-carbon future, the need for and dependence on minerals and metals will grow exponentially. Critical minerals such as cobalt, copper, lithium, nickel, and rare earth elements are needed in large quantities for clean energy technologies, from solar panels and wind turbines to electric vehicles.

The market size of critical minerals alone has doubled over the last five years. If all current projects that are in the pipeline succeed and achieve production, it is likely that only two-thirds of the critical minerals needed for the energy transition will be covered.

What can be learned from the past to ensure responsible mining in the future? The mining industry has been associated with a host of negative environmental, social and governance (ESG) impacts, including environmental pollution, biodiversity loss, displacement of communities, and loss of land and livelihoods. A trust deficit exists between communities and the mining industry and the failure to effectively manage these impacts will have profound implications for clean energy transitions as well as continue to cause damage to the environment and communities in the vicinity of mining deposits.

For mining to be more acceptable, the industry must use and manage these natural resources without compromising the integrity of the environment or the well-being of local communities in the future. In my work I advocate that well-planned mine closure and post-mining transition can help achieve this goal.

That is why I love this book so much. It takes us on a journey of the possible. A well-managed transition from mining to new and productive post-mining land use can help build trust in the sector. We have learned a lot from dealing with the legacy of historical, abandoned, and certain declining mining industries. As a result, the design and permitting of a modern mine should be driven by planning for the future of altered underground and surface landscapes, and the communities that remain after closure.

This book, *102 Things to Do with a Hole in the Ground*, promotes the idea that beneficial uses can be derived from repurposing old mines and infrastructure. It updates and expands the messages of its predecessor. The case histories presented in this 'Second Edition' show that new life and purposing after mine closure are realistic objectives.

This book showcases examples of human resourcefulness when confronted by challenges from mining – past and present. The numerous projects that have generously allowed their work to be captured in this book will motivate us all.

Greg Radford, M.Sc.
Director, Intergovernmental Forum on Mining, Minerals, Metals and Sustainable Development (IGF)

Introduction

We all rely on mining for raw materials to fuel our lives and build our societies. In the coming decades, growing populations, expanding wealth and, crucially, the green industrial revolution will drive a huge expansion of mining and, inevitably, environmental and social change. Simultaneously, coal mines will close on an unprecedented scale as burning coal for power declines.

Mines have finite lives, but their legacies – good and bad – may persist for generations. *102 Things to Do with a Hole in the Ground* showcases many ingenious ways that old mine sites have been reinvented to rebuild ecological integrity and enhance the lives and livelihoods of the people connected with them.

Recognising the above, and mining's chequered reputation, investors and society are pushing the mining sector towards more responsible performance, including how mines end. As a result, the language of closure is changing; phrases like social transition, just transition, nature positive and nature-based solutions describe progressive approaches away from 'traditional closure' and recouping costs at the end of a mine's life to 'transition' – where a mine doesn't technically 'close', but catalyses the availability of 'new' assets and resources (e.g. mine water, repurposing buildings and infrastructure, new mineral resources in old wastes, cheap energy, repaired land) to stimulate new socio-economic, environmental and cultural values after mineral extraction has ended.

102 Things, like its 2009 predecessor *101 Things to Do with a Hole in the Ground*, features this paradigm shift in action. It follows the same philosophy but delves deeper into what makes specific projects tick. Since

the first book, a host of new exemplars have emerged, and we have done our best to capture the best of these and update accounts of some of *101 Things'* original projects. We have been generous in interpreting what constitutes 'a hole in the ground'. The mine typically forms an important part of a post-mining story, but it is not the whole deal. Alongside the reuse of mining voids, such as Australia's Kidston pumped storage hydro project, the breadth of creativity and imagination at work is demonstrated in stories about re-mining mineral wastes, repurposing mining features, infrastructure and services, and reusing mine water. There are also inspiring stories about regenerating mining communities and reimagining mining landscapes.

We've scoured the world for this compelling collection of post-mining tales to offer tastes of the possible from the wealth of flavours out there and tried hard to include the breadth of geography, types of mine and after-uses. We relate the stories of projects, programmes and places as well as myriad individual uses. It was challenging to sift these from the hundreds we uncovered and summarise often complex, decades-long stories into a few hundred words.

Only a few regions are hotbeds of post-mining creativity and invention. These tend to occur where industrialised mining has happened the longest. In particular, Europe's long association with industrial-scale mining, its large population and land pressure, mean a deeper experience of dealing with the after-effects of mining and a greater imperative for regenerating its old mines. Places like Germany's coal regions offered so many eligible stories that they could fill this book on their own. Since *101 Things*, it has been encouraging to see how other regions are coming to the fore in the post-mining world, such as Appalachia, Australia, China and South Africa. In some mining regions though, gaining traction due to language or cultural issues or a lack of direct contacts was problematic.

102 Things is not intended as a menu of options to be applied to a specific place or problem nor as guidance on closing a mine or regenerating a community. It sits alongside the excellent library of technical guidance on the subject by others. Neither is it intended to justify mining developments under any circumstances. Its stories celebrate ingenuity in making the best from a bad lot, but the lesson is that the best should be the norm. And it's always better to avoid the damage in the first place rather than repair it afterwards. If mining must take place, then solutions to its aftermath need to germinate when the mine is conceived and not left to chance.

There are – arguably – no new problems in this field. *102 Things* attests that they have all been solved somewhere. The main challenges are of perspective, will and communication, from considering a mine site not as a problem to be fixed but as an opportunity to be nurtured and learn from existing successes and avoid reinventing the wheel. Each story exemplifies a successful response to an inactive mine's local context and, as such, challenges the conventions; collaboration, knowledge and choices are the currencies that buy progress. Common attributes signpost a way to advancement: science-based problem-solving, constructive legislation, available funding, leadership and a critical mass of the right people thinking positively in the same place at the same time, looking in the same direction and speaking the same language. A group of people with a good idea and the wherewithal to realise the opportunity in the drama of damaged land can drive astonishing progress.

How communities and stakeholders participate in these issues is fundamental. *102 Things* includes several examples, from immersive community workshopping in defunct German lignite pits to transparency and trust-building at Beenup in Western Australia, trialling new underground technology at Canada's NORCAT or potential new careers in the restoration economy for Banjima Traditional Owners in Western Australia. In some cases, a hands-off approach may even be appropriate to enable a gradual natural recovery as at Germany's Sielmann Natural Landscape. There is a valuable lesson here: innovation is often brought about by inclusivity and collaboration – where answers may be found by looking beyond the conventional closure cast of engineers and environmental managers.

On a cautionary note, we are sensitive to accusations of greenwashing. This book aims to inspire innovation and better practice in realising new values from mines at the end of their lives. The use of an example from an organisation is not an endorsement of that organisation, mining company or otherwise. Those who have sponsored and supported *102 Things* have all agreed with the fundamental principle of Eden's editorial independence.

Finally, any landscape tells a unique story of aeons of interactions between uncountable generations of people and their land. The landscape experienced today is just the current chapter in that long narrative. The opening of a new mine heralds another chapter in the landscape's tale, which ends when mining ceases. That instant should not be the end of the story but should be the prelude to a vibrant, new chapter.

Acknowledgements

Making this book was a four-year journey from the initial mid-Covid discussions between Professor Mike Johnson (independent adviser to Rio Tinto) and me about how we might update Georgina Pearman's 2009 Eden book, *101 Things to Do with a Hole in the Ground*. This led to conversations with Dan James and Robert Lowe at the Eden Project, Peter Toth and Nigel Kieser (Rio Tinto – the original sponsors of the first book) about how we could collaborate to make it happen. Founding sponsorship from Rio and commitment by Eden catalysed the project, which was ably overseen by an advisory group of Dan James, Robert Lowe (my co-author and editor) and Nigel Kieser, superseded by Alexandra Le Van and Virginia Alexander of Rio, and enthusiastically chaired by Mike Johnson.

Further support and generous sponsorship were provided by Anglo American, BHP, and Newmont, as well as intellectual and moral support from Resolve/ Regeneration, The Land Trust, and ICMM, among many others. The sponsorship has covered the book's production costs, enabling a not-for-profit approach to sales and keeping the price low, the primary aim being to facilitate as wide a distribution as possible. It also enabled the original vision for *102 Things* to evolve into the stand-alone tome it has become, although grounded in the perspective and style of the original. The Eden Project's editorial independence was guaranteed at the outset.

The years of research and travels to distant lands were made more manageable with help from the above individuals and organisations, among others, who let the light in on some truly superb examples of post-mining regeneration. EnviroMETS' Allan Morton, who has been a stalwart advocate for the book Down Under, went beyond the call of duty in organising my Australian itinerary.

The Eden Project's publishing team, which consists of Robert Lowe and graphic designer Sam Jarrold, professionally assisted by Eden Project publishing veteran Mike Petty, ensured the beautiful, informative product you hold in your hands. A big thank you, too, to Greg Radford for setting the scene in his fine foreword and Natalia Valderrama for researching and writing the Latin American stories.

Hundreds of other people around the world have provided contacts, project information, access to images and help with logistics. They are acknowledged at the back of this book, along with the photography credits.

Finally, the *102 Things* project was an impressive team effort. I hope *102 Things* does the team and all those involved in the project justice for their support.

Peter Whitbread-Abrutat
Future Terrains International

About the authors

Peter Whitbread-Abrutat

Pete is a mining-environmental consultant through his company, Future Terrains. Before this, he was deeply involved in the early days of Eden's creation and remained part of the Team for many years. He was instrumental in establishing and managing Eden's engagement with the mining industry, including Eden's former partnership with Rio Tinto, which – among other things – *101 Things to Do with a Hole in the Ground*.

Pete is an Eden Associate, a Chartered Environmentalist and an IEMA Lead Environmental Auditor with over 30 years of international experience in mining ESG and sustainability, mine closure and international development in developing and conflict-affected countries. He is a Churchill Fellow for his 2011 travelling research project, *Exploring World Class Landscape Restoration*. He earned a PhD from Camborne School of Mines (University of Exeter) in revegetating mine wastes and a Natural Sciences degree from Cambridge University.

Robert Lowe

Rob has worked for the Eden Project for over 15 years. As the Eden Project's in-house editor, his role is to bring depth and insight to the Eden visitor and membership experience through its publications, notably the guidebooks and quarterly Eden magazine. He project-manages these and other Eden publications from concept to print and provides copywriting and editorial services to the wider business. He was also involved in producing *101 Things to Do with a Hole in the Ground*. Prior to joining the Eden Project, Rob held editorial, marketing and digital development roles at Oxford University Press and Taylor & Francis. He holds a doctorate from the University of Oxford on contemporary medical science in James Joyce's *Ulysses*.

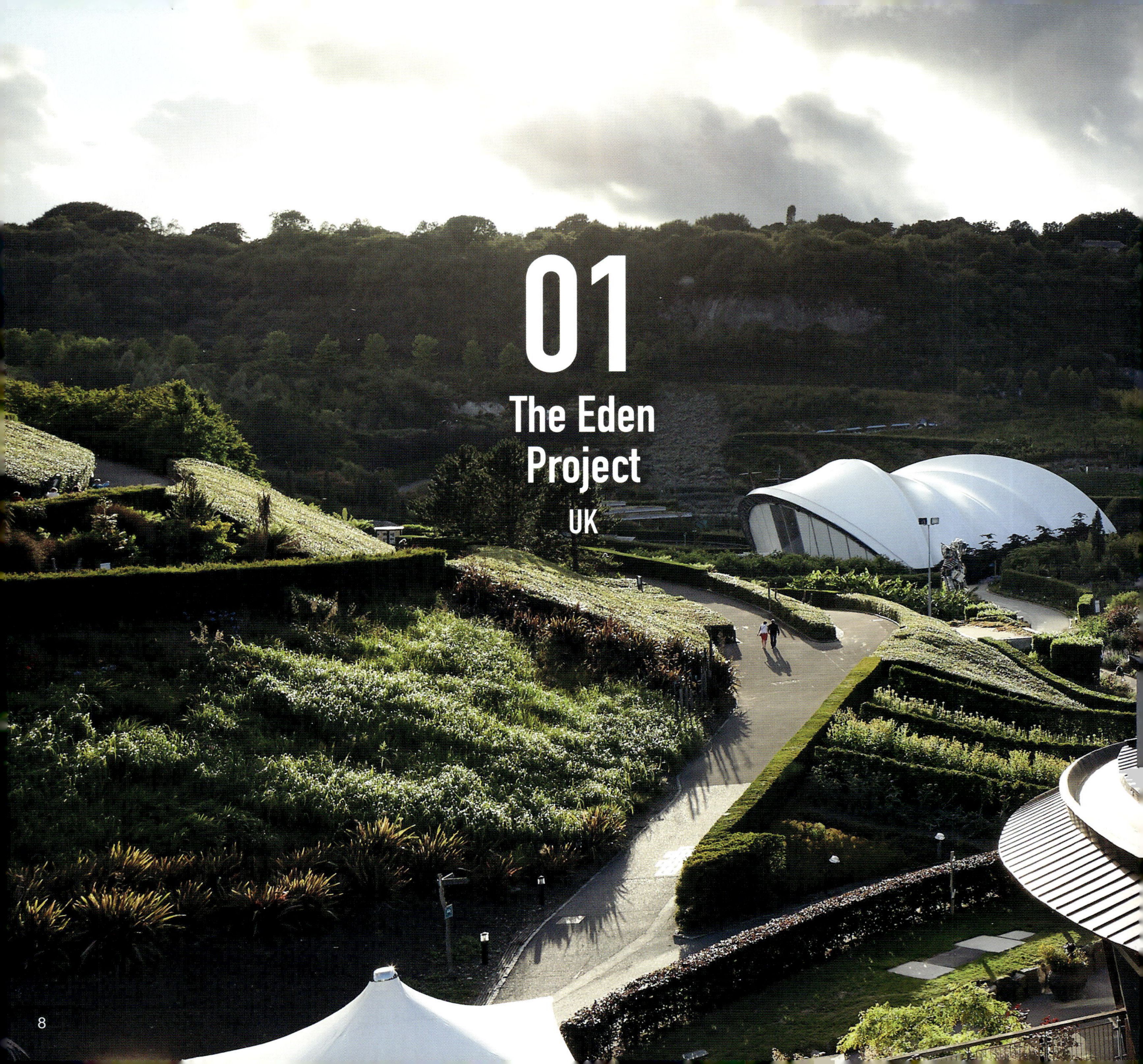

01
The Eden
Project
UK

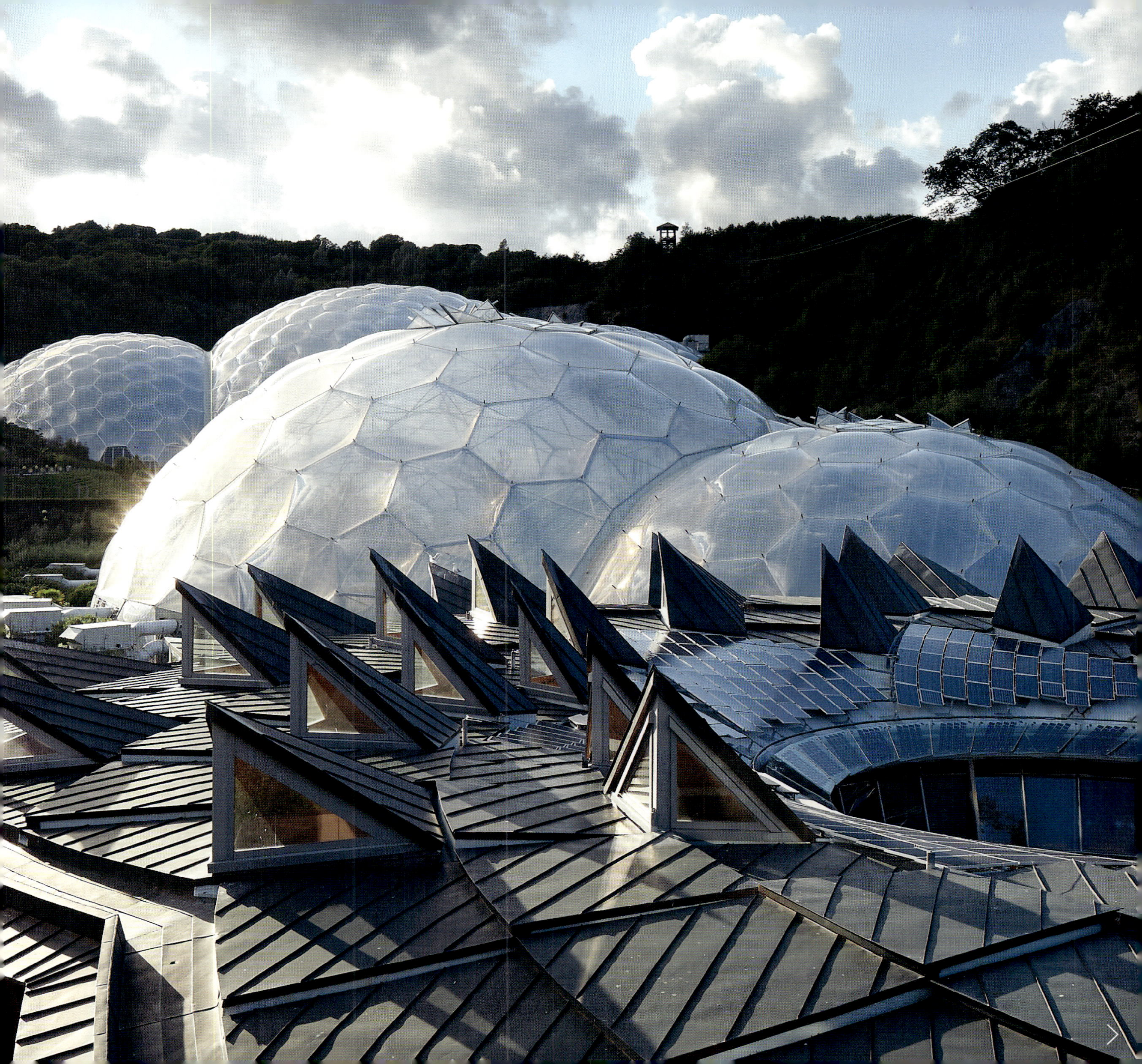

In the 1990s, Tim Smit (now Sir Tim), the founder of the Lost Gardens of Heligan, and a team of experts came up with the idea to create a global garden in Cornwall dedicated to the plants that had shaped human history. The Heligan site was swiftly deemed too small to accommodate the scale of this ambition, and he and his rapidly expanding team began to look further afield to the clay-mining country around St Austell. Already the focus had shifted from one of purely education to include economic and environmental regeneration.

Clay mining defined the region's economy, and its open pits and waste tips shaped the landscape. Its kaolin was used mainly for paper (clay from Bodelva, the site that was eventually chosen for the Eden Project, once gave the *Financial Times* its pinkish hue), but although there are still active mines in the region, the industry is in decline, and the area was widely recognised as deprived. Cornwall had long been a popular holiday destination, but few chose to stay in the clay country.

There were many contenders before Bodelva was settled on, and not all of them were pits, but once Smit saw it he knew it was the one. Yet it's hard to imagine a place less suited for the site of a botanic garden than a 160-year-old worked-out china clay mine. It was essentially a blank slate: kaolin mines are chemically benign compared to other types of mine site, so there was no contamination to remediate, but there wasn't any soil either. A team of scientists at the University of Reading and from the Eden Team blended silica sand waste from the china clay industry and composted bark and green waste into

six recipes to cover the different planting areas of the site. Over 83,000 tonnes of soil were created to bring life into the 18-metre-deep pit. Other challenges included unstable pit sides and incessant rain during construction on a site that was 15 metres below the water table.

Funding was also a challenge. Restormel, the borough council of the time, provided the first chunk of funding and set a precedent for others to follow. The Millennium Commission provided GBP 37.5 million of lottery funding; a further GBP 50 million came from the EU and regional funding bodies and topped up with GBP 28 million in loans and income generated during the construction phase. Many of these grants were regeneration funds. A considerable amount of the funding went into stabilising the slopes and installing water management systems.

Nothing like this had been attempted before – at least not in Cornwall – and recognising people's increasing curiosity about the project, the Visitor Centre opened a year before the site to allow them to watch the spectacle of the Biomes being constructed.

Over 1.8 million people visited in the first year – twice what was expected. Millions more have followed. Since it opened in 2001, it has brought over GBP 2.2 billion into the Cornish economy and supported hundreds of local suppliers over the years. It now employs 350 people and provides volunteering opportunities for a further 150.

Just over 11 hectares of diverse gardens now grow where little could. There are orchards and allotments, crop displays and sculptures, as well as pockets of wild landscapes from around

the world, including American prairies and South African veld – and beyond all this are the Biomes.

Both backdrop and star attraction, the two Biomes consist of a series of geodesic domes designed by Grimshaw Architects. Their steel frames hold 'pillows' of highly durable plastic that admits the ultraviolet light the inhabitants require. The Rainforest Biome is home to thousands of the world's rainforest plants – from the exotic to the workhorses that we take for granted, like chocolate, sugar and bananas – and exhibits that reveal the biodiversity rainforests support and their role in creating our weather. The Mediterranean Biome features the plants from the Mediterranean Basin that shaped civilisations, like olives and vines, as well as rare and resilient plants from four of the world's Mediterranean climates: California, Southwest Australia, and South Africa. Throughout, the Eden Project emphasises the positive stories of resilience and regeneration.

In 2007, Eden added an education building to the site for its Schools programme: the Core. The building's

Previous page: The Core's single-source virgin copper roof was donated by Rio Tinto and sourced from its Kennecott Utah Copper Company's Bingham Canyon Mine, a mine known for its high environmental standards. From Utah, the copper was shipped to Germany for fabrication and was passed through the furnace separately to ensure traceability. The resulting rolls of sheet copper were shipped to Eden and cut to shape by specialist contractors in situ.

Right: The Rainforest Biome covers an area of 1.6 ha with no internal supports. It contains over 1,000 different tropical species, crops and cultivars from West Africa, South America, Southeast Asia, and tropical islands.

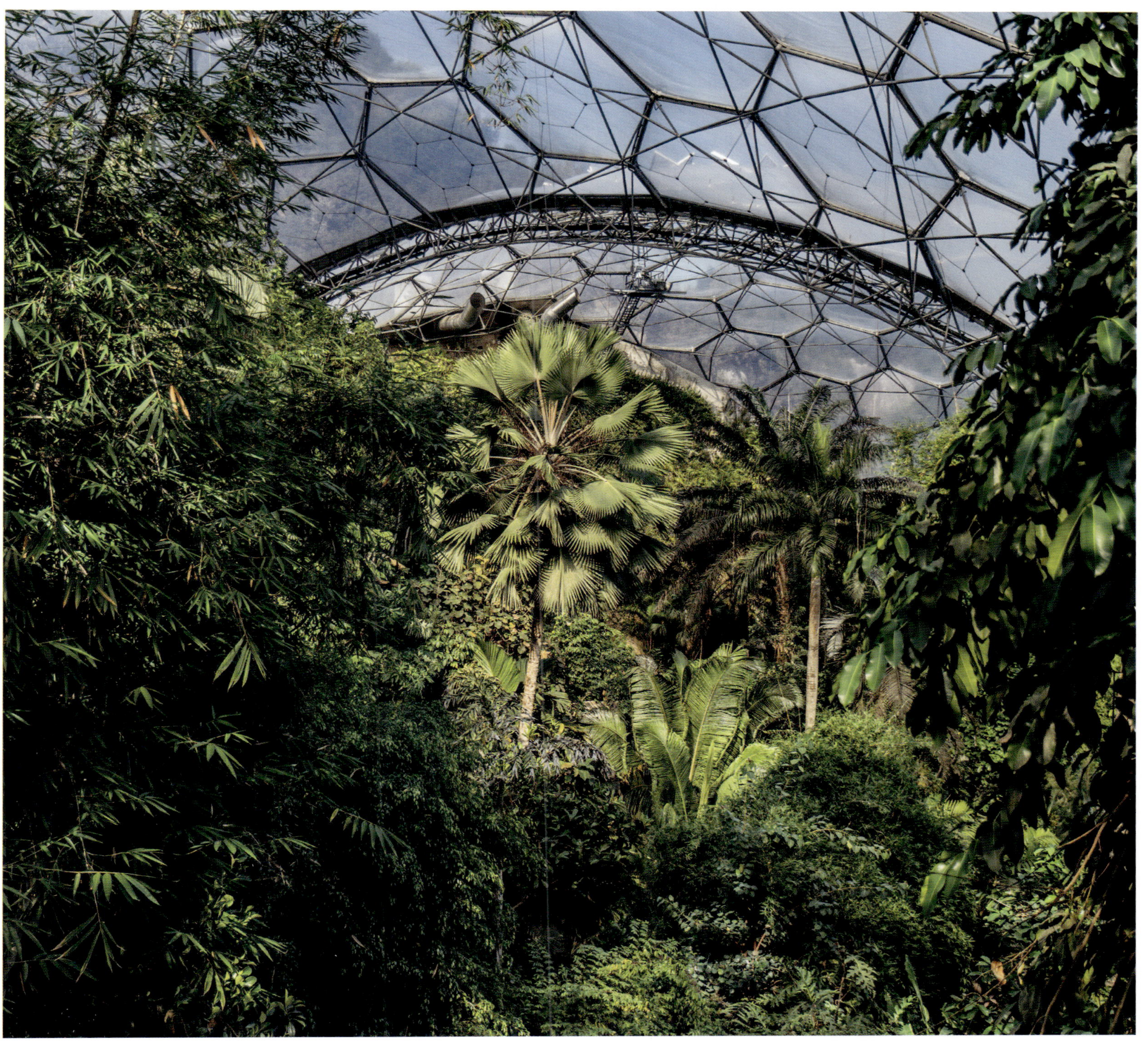

copper roof was the result of a pioneering 'rock to roof' sustainability project that traced the virgin single-source copper donated by Rio Tinto from Bingham Canyon mine in Utah, USA, through the smelting process to its installation on the Core.

A year-round programme of events helps bring the site to life, featuring science, nature and play activities for the general public and early years, and world-class live music at the Eden Sessions. It's not just for tourists; a winter programme runs every year that includes an indoor ice rink, and there are often weekend activities out of peak season.

But there is more to the Eden Project than its Biomes and gardens. Its outer estate of woodlands, orchards, wildflower meadows and kitchen gardens is used to host programmes that bring the therapeutic value of green spaces to those who need it most and give preschool children their first experience of 'wild' play in dedicated sessions.

A pioneer in waste recycling, local sourcing and sustainable construction, the Eden Project recently commissioned the UK's first operational deep geothermal plant in 37 years. This provides low-carbon heat to the Biomes, offices and a new on-site plant nursery,

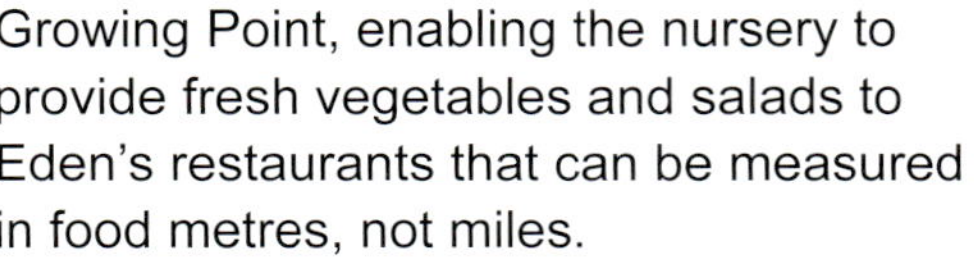

Growing Point, enabling the nursery to provide fresh vegetables and salads to Eden's restaurants that can be measured in food metres, not miles.

The Eden Project has established itself as one of the UK's most recognisable visitor attractions, but it is also a trailblazing educational charity that works with nature to respond to the planetary emergency. This mission starts with its visitors, but it also offers educational opportunities at every level, from preschool to postgraduate. Over 30,000 schoolchildren visit every year; there is a thriving apprenticeships programme, and Eden Project Learning provides degree and postgraduate qualifications in horticulture, event management, sustainability and land restoration in partnership with local HE providers.

The Eden Team is working on new Edens in the UK and overseas to regenerate neglected sites and overlooked communities. Eden also works with governments and NGOs on ecosystem restoration projects in Costa Rica, Mexico and elsewhere.

In 2018, the UK's National Wildflower Centre became part of the Eden Project. The NWC combines conservation, creativity, and colour to create new wildflower habitats that increase biodiversity and connect people with the natural world. The National Wildflower Centre also powers the Eden Project Wildflower Bank, a nature-positive biodiversity net gain company.

Almost 25 years since it opened, the Eden Project continues to offer a positive vision of the future, providing inspiration for others through its transformation of a sterile mining-scarred landscape.

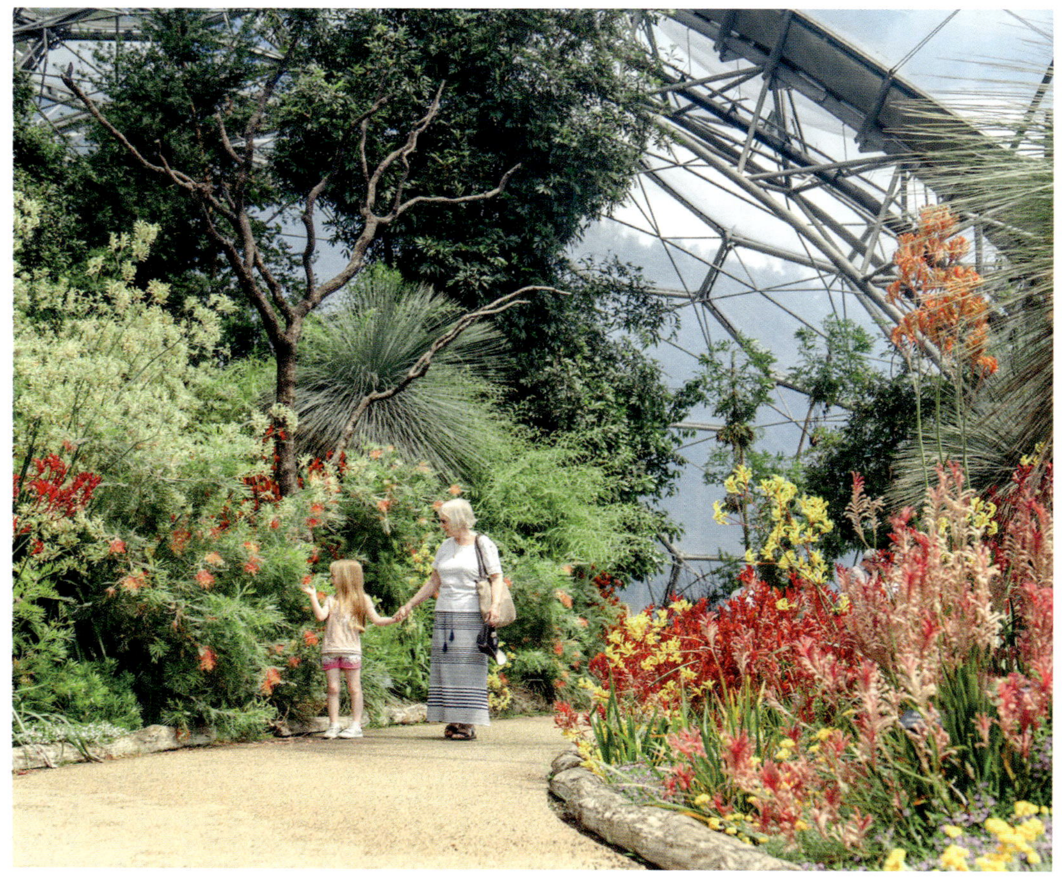

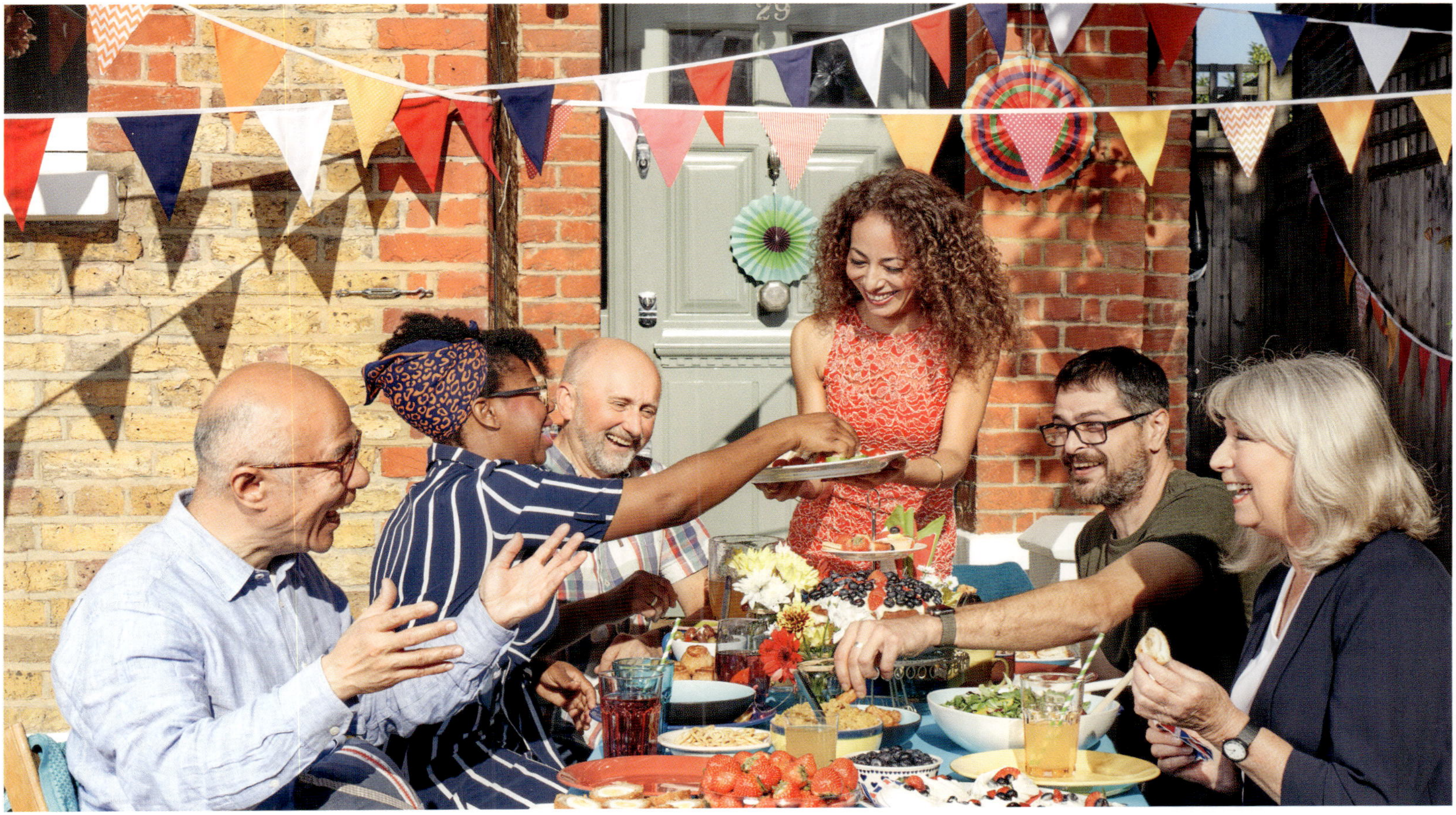

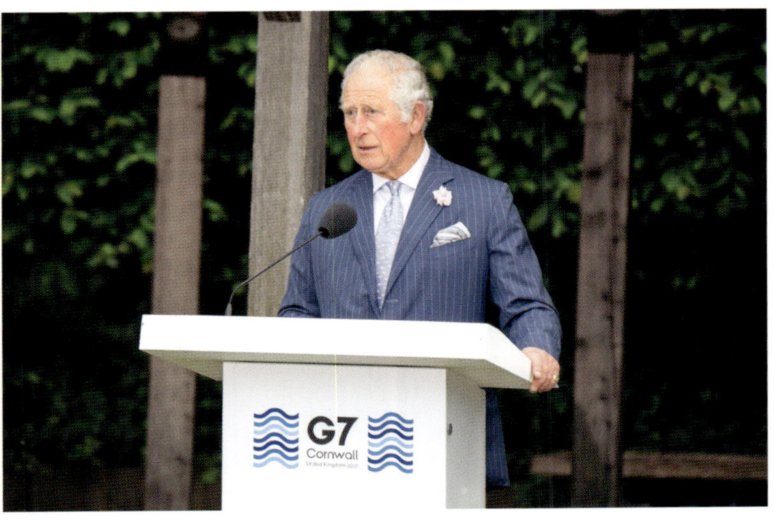

Opposite page: The Mediterranean Biome displays species and cultivars from places that share the region's climate and contains a restaurant that takes its inspiration from the Mediterranean diet.

Above: The Eden Project also invented The Big Lunch, a programme dedicated to bringing neighbours together for an annual street party that forges new relationships and can lead to real change. Millions of people take part every year, and £87m has been raised over the past decade, most of which has gone to local causes.

Left: The Eden Project hosted a royal visit and a G7 dinner for heads of state in 2021.

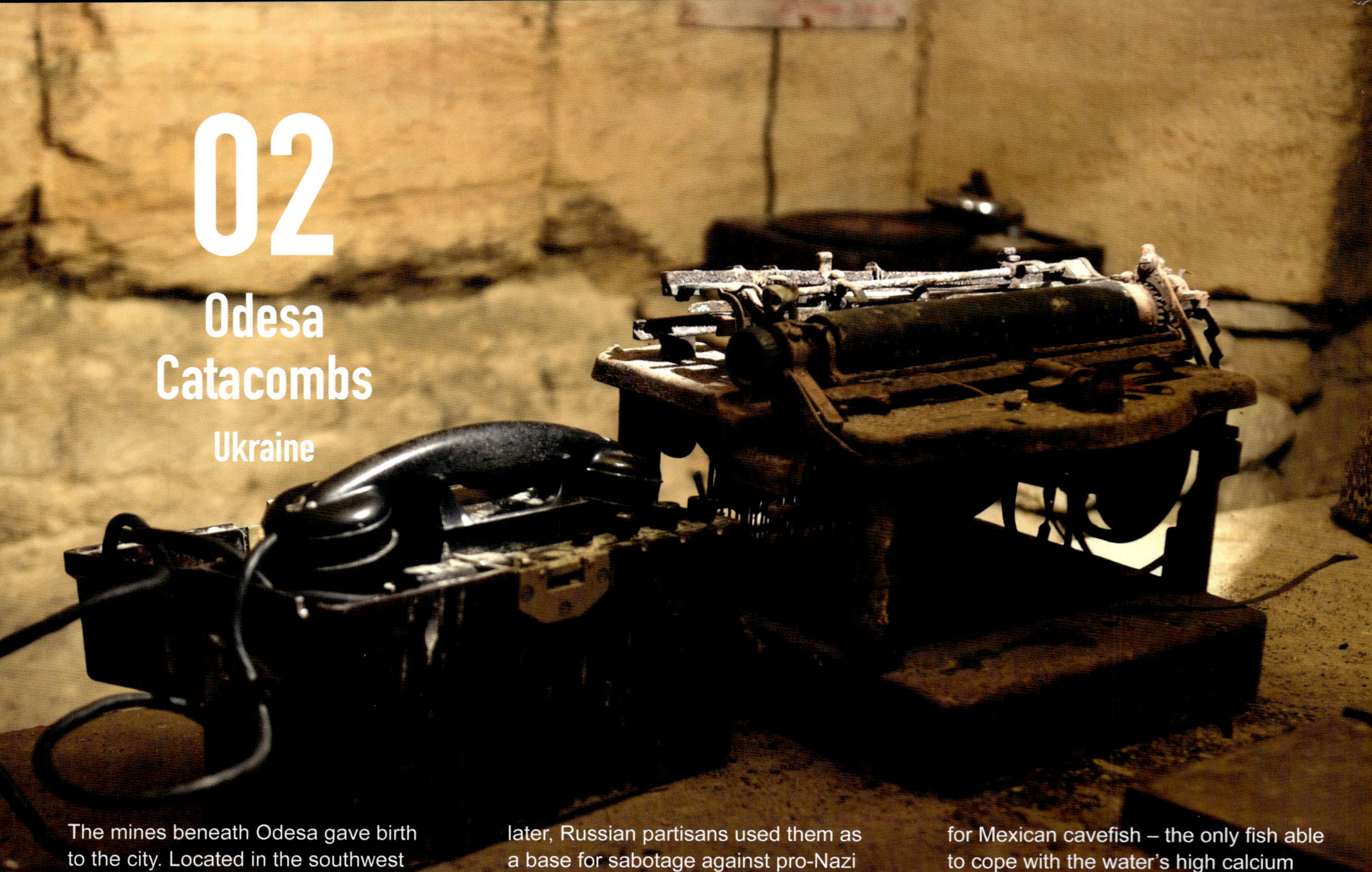

02

Odesa Catacombs

Ukraine

The mines beneath Odesa gave birth to the city. Located in the southwest of Ukraine, the seaport of Odesa sits over a labyrinth of passageways and stone quarries mined to create the city's grandest buildings.

Less than half of its estimated 2,500km tunnels have been mapped, making it one of the largest networks of tunnels in the world – significantly larger than the catacombs of Rome or Paris.

A network like this has obvious appeal during times of conflict. After the Russian Revolution, anti-communist Tsarists sheltered in its depths. Twenty-five years later, Russian partisans used them as a base for sabotage against pro-Nazi occupiers. Today, as Russia wages war on Ukraine, Odesans are once again taking refuge inside.

Until recently, the catacombs mainly drew tourists. Two museums provide limited access to catacombs that contain a Soviet-era nuclear bunker and an underground school with the remnants of an abacus and a blackboard. You can also take tours – one of the city's museums sits over one of the entrances. One enterprising tour guide has used the water that leaks in to create small ponds for Mexican cavefish – the only fish able to cope with the water's high calcium content. He's also built a small bar where visitors can sample local spirits.

Unlike its Roman and Parisian counterparts, the Odesa catacombs weren't used to bury the dead, but corpses have been found in their depths, often perfectly mummified. Rumours abound of a young woman who wandered away from a New Year's Eve party in the catacombs and whose body wasn't found for another two years…

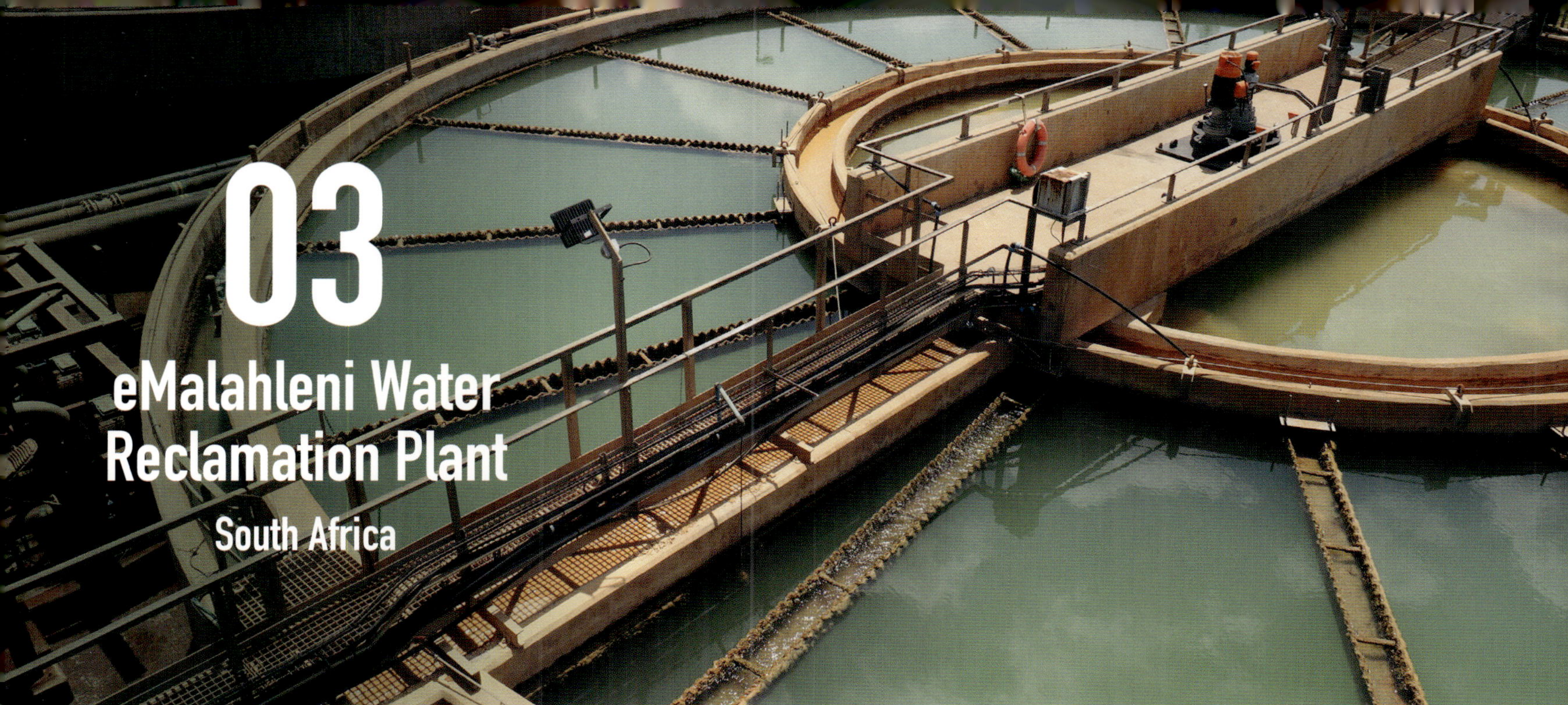

03

eMalahleni Water Reclamation Plant

South Africa

Increasing water scarcity in South Africa's Mpumalanga Province due to increased demand and declining rainfall and the availability of copious amounts of mine water have dovetailed into a neat answer to two problems: finding a long-term solution to the contaminated mine water and providing potable water for local communities.

Only some of the water treated at coal exporter Thungela's enormous eMalahleni Water Reclamation Plant (EWRP) is released into a receiving water body. Unusually, most of the now potable water enters the local municipality supply for eMalahleni's growing population of about 500,000 people. EWRP provides 20% of the municipality's needs, while some of the cleaned water is returned to the mines for their operational needs and drinking water.

Commissioned in 2007 and built on Greenside Colliery land, EWRP treats between 25 and 40 million litres per day from five opencast and underground coal mines – both operating and closed – the furthest being 26km away.

Treatment occurs in a conventional three-stage process, including a final ultrafiltration step. A turbulent spectrum of oranges and chocolate browns churns through angular concrete tanks and a tangle of pipework, with the dissonant accompaniment of hard-working pumps, to the circular blue and green calmness of the settling tanks. The gypsum by-product (from the main calcium and sulphate contaminants) is collected and used by farmers as a soil improver. Brine (sodium sulphate), however, is a benign waste product that is currently stored on-site, pending the development of reuse options.

Initially, the project had to satisfy the government's environmental concerns, notably that residents would refuse to drink 'reclaimed' mine water. The idea was eventually welcomed by the community, however, because of the poor quality and limited availability of the municipal supply. 'Taste testing' with local people was also carried out.

Collaborative research is underway to further minimise the waste streams. These include a final step to produce elemental sulphur as a possible future commercial product.

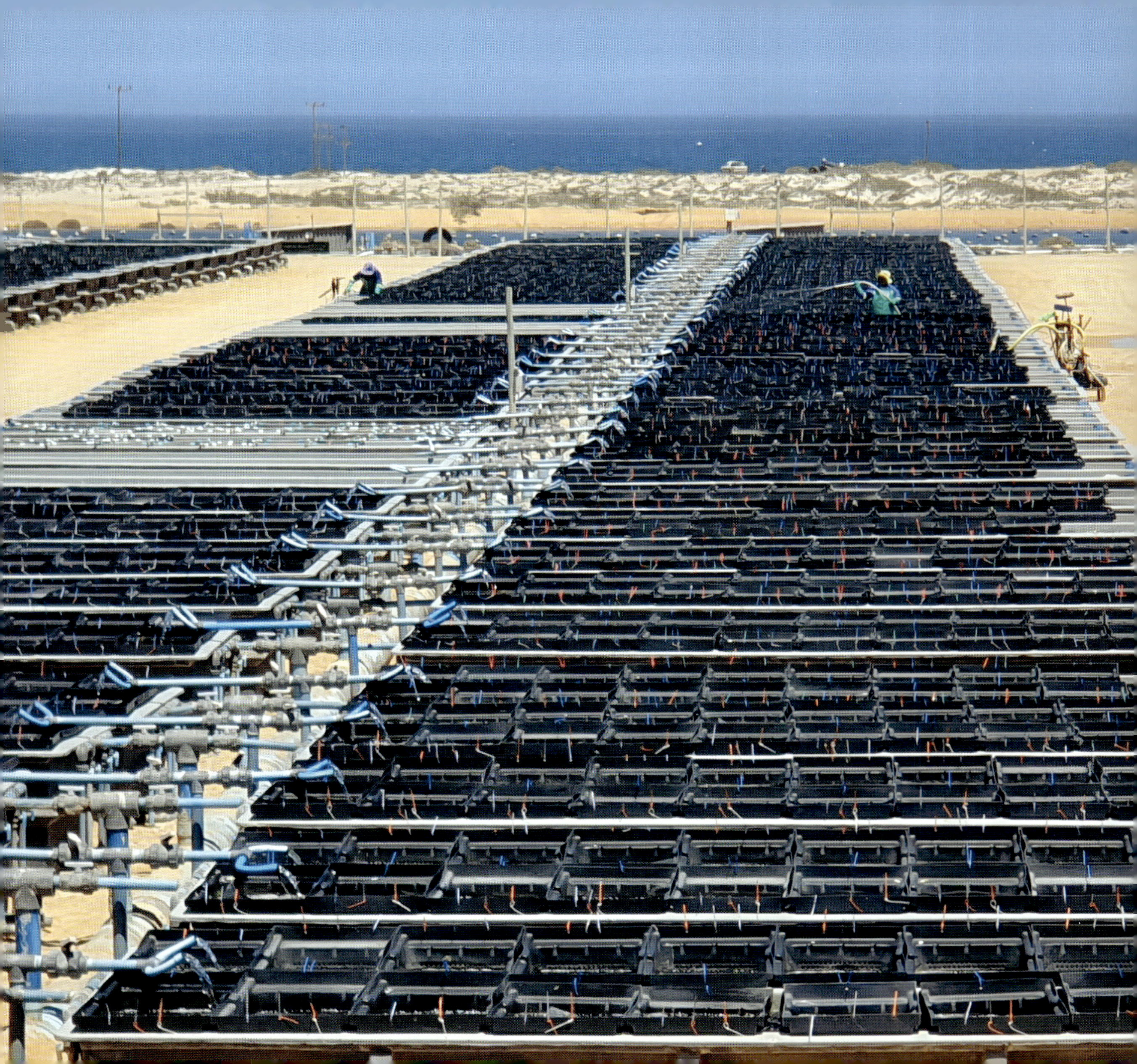

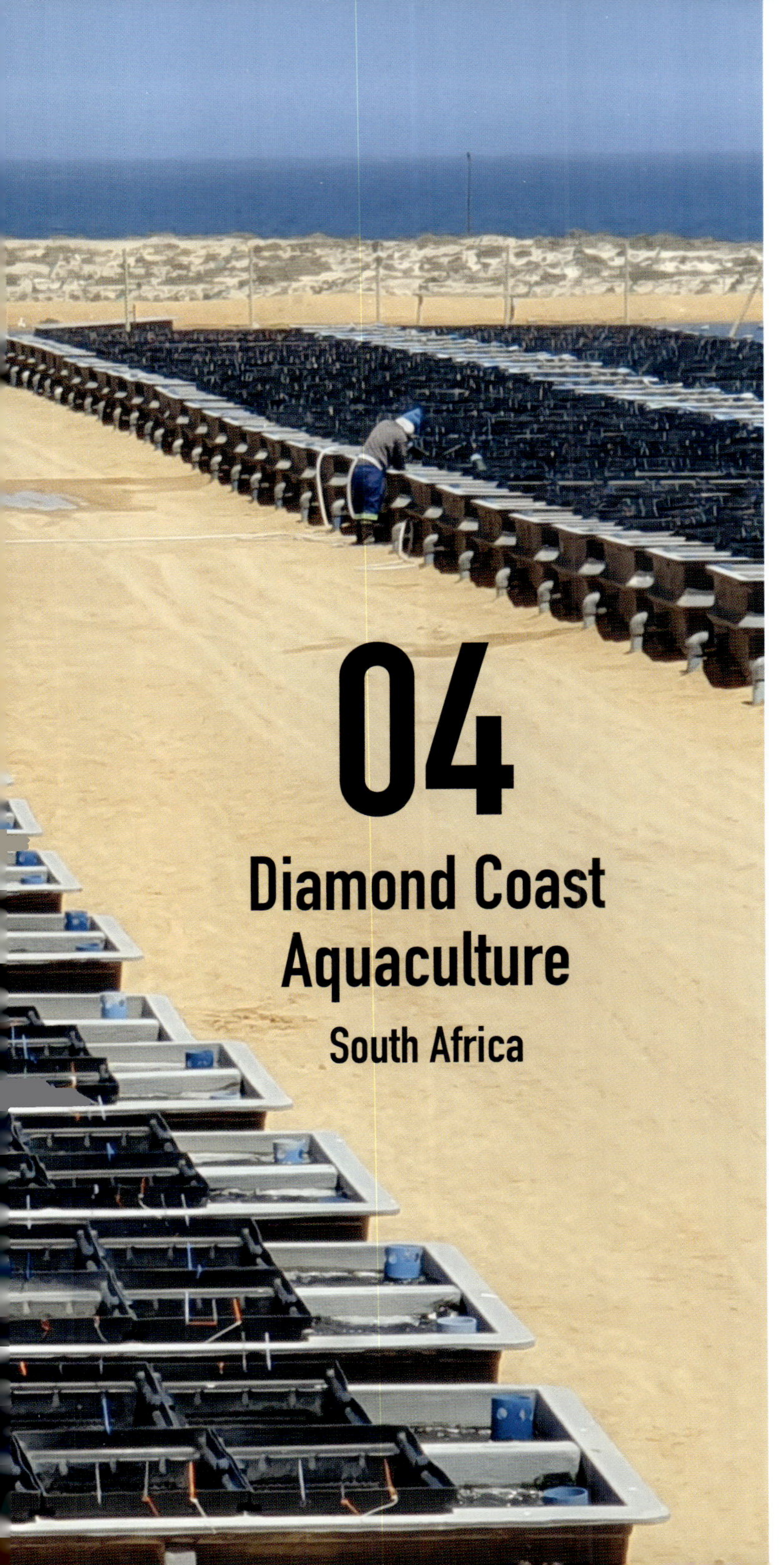

04

Diamond Coast Aquaculture

South Africa

Seawater bubbles in serried ranks of black tanks stretching to the Namaqualand blues of sky and ocean. An arid backdrop of derelict buildings and mine waste dumps shows there is still work to be done. Just offshore, small privately owned boats hoover diamonds from the shallow seabed with long pipes held on the seafloor by divers.

Abalone is a marine snail and a culinary delicacy. Seeing an opportunity in a South African operational diamond mining area, Diamond Coast Aquaculture began its abalone farm in 2010 after acquiring land and several buildings from De Beers, including the pumphouse – since refitted – and a former workshop now used for storage.

The large fibreglass tanks, housing thousands of abalone, are fabricated on-site. Fresh seawater is a prerequisite, so the proximity of the Atlantic Ocean is essential, although about 50% of the water used is recycled. *Ulva* seaweed, cultured in large circulating tanks, absorbs contaminants from the shellfish, and when broken down, it is fed back to them.

Abalone takes over four years to achieve a commercially valuable size. It's a high-value crop and easily transportable, so security is tight to deter poachers. Diamond Coast Aquaculture ships 100 tonnes of abalone per year to a packaging and export firm, which exports across the world – particularly to China and Taiwan.

The company's workforce of 72, plus contractors, hail mainly from surrounding communities – including the former diamond mining town of Kleinzee (a proclaimed town under the jurisdiction of the local municipality), which is suffering severe post-mining decline.

Originally excavated during mining operations, an artificial lagoon sits between the abalone farm and the sea. Kleinzee Mariculture leases this land from Diamond Coast Aquaculture to farm oysters on ropes suspended from regimented lines of floats.

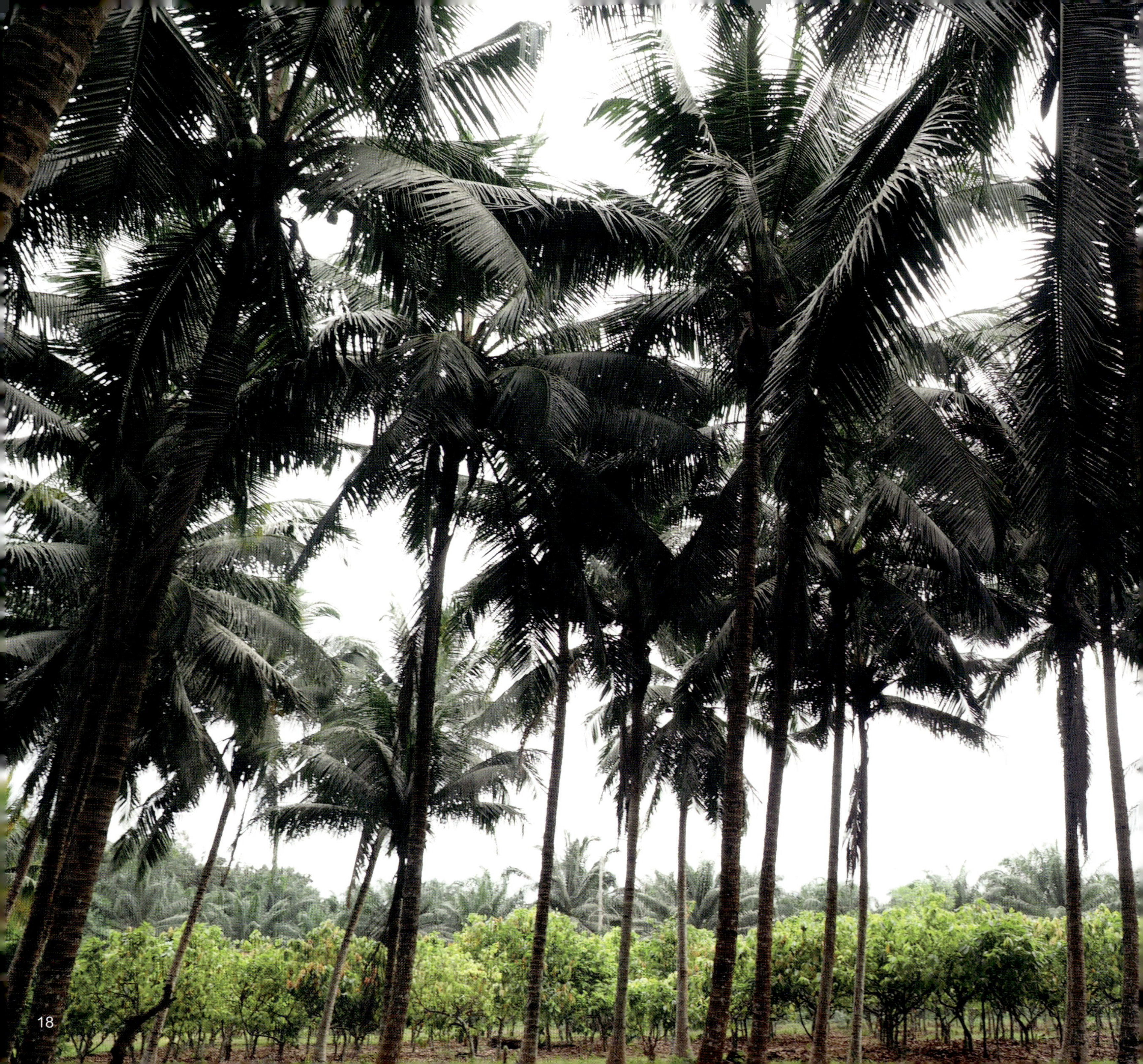

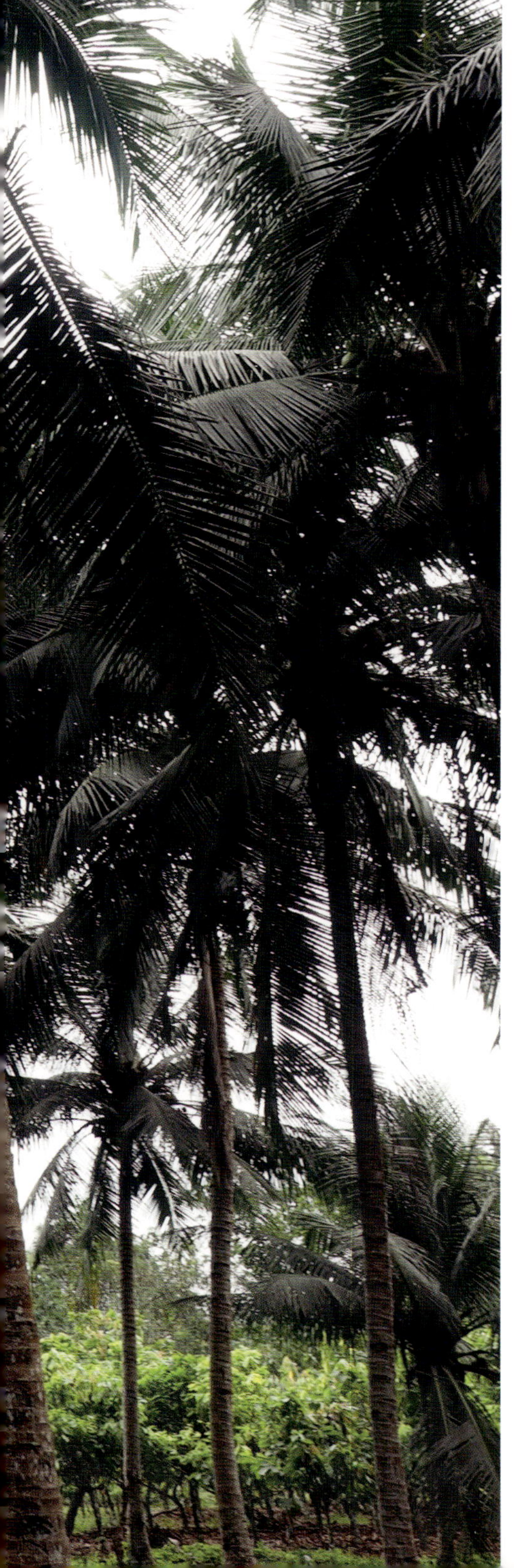

05
Agroforestry Systems, Damang Gold Mine

Ghana

Beneath the cool shade of the towering coconut palms, you have to pinch yourself that you are actually on top of a tailings (or mineral processing waste) site, which – not that long ago – was a white sand industrial wasteland. Today, the South Tailings Storage Facility (TSF) of Abosso Goldfields Ltd, Damang mine is a veritable garden of paradise – a mosaic of agroforestry plantations and a haven for wildlife.

The tailings' infertility was initially addressed by planting leguminous trees (which fix atmospheric nitrogen in their roots, thus increasing fertility) and herbaceous plants, which, over the years, enhanced soil nitrogen and organic material. Various food and cash crops, determined by consultations with local land users, were subsequently planted into these agroforestry systems, including oil palm, cacao, coconut, citrus, maize, cocoyam, plantain and vegetables. Laboratory analysis confirms that these crops are healthy and have not taken up any potentially toxic substances from the mineral substrate.

The tailings pond is now a self-sustaining wetland with fish and economic plants such as sugarcane, bamboo and raffia palm, and enriches the local wildlife too. Indeed, ecological indicators show biodiversity is recovering across the whole facility.

A broad range of technical experts and local stakeholders developed and agreed how the site would be used and collaborated throughout the restoration phase. In 2012, nine hectares of oil palm were given to selected communities by the government, which also enhanced the mine's social license to operate.

The model developed on this TSF has been extended to the mine's other TSFs, with new cash crops such as cashew and mango being trialled. The initiative has influenced the Environmental Protection Agency's thinking, becoming a model recommended by the Agency and a testament to socio-economic investment.

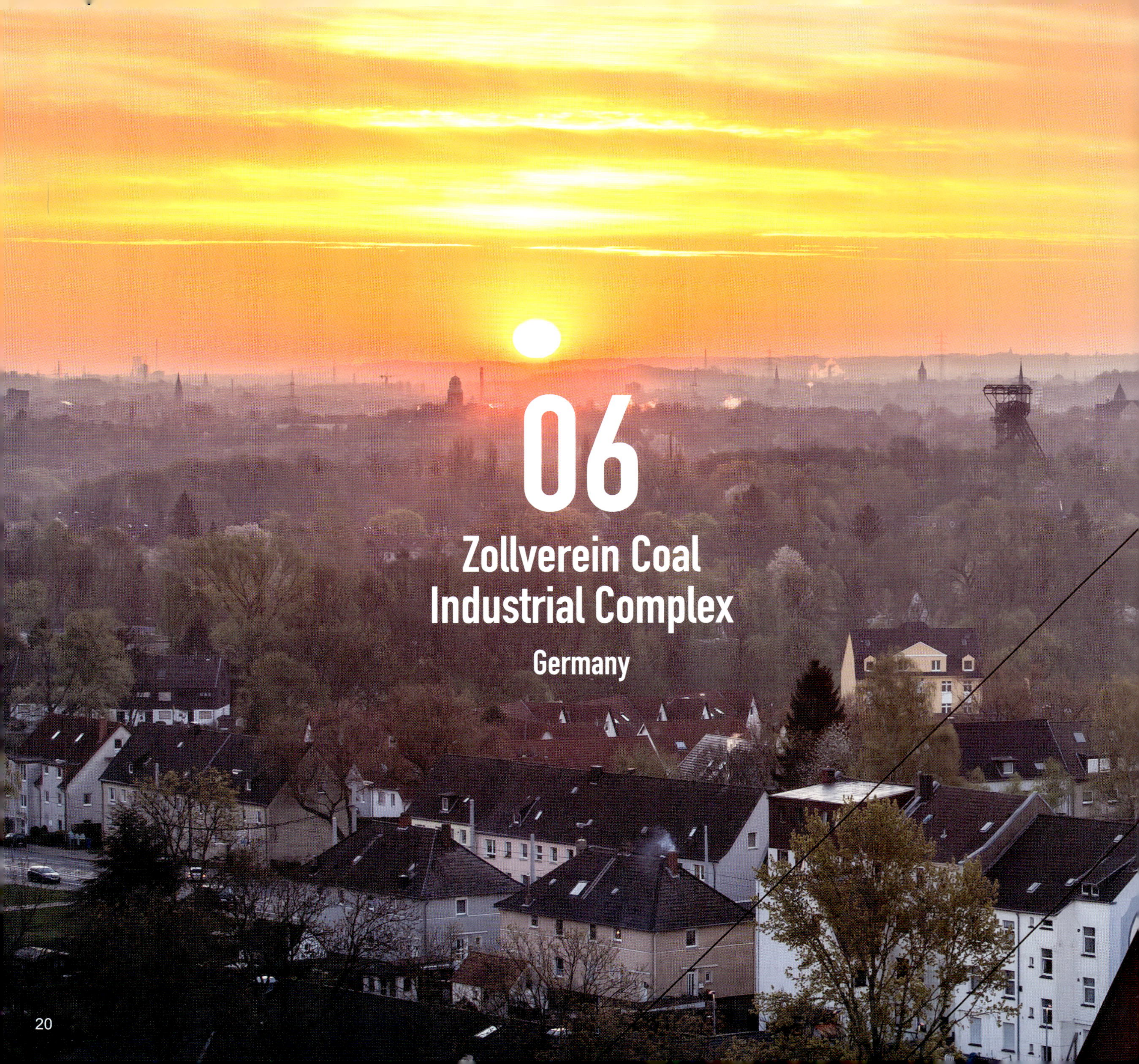

06
Zollverein Coal
Industrial Complex
Germany

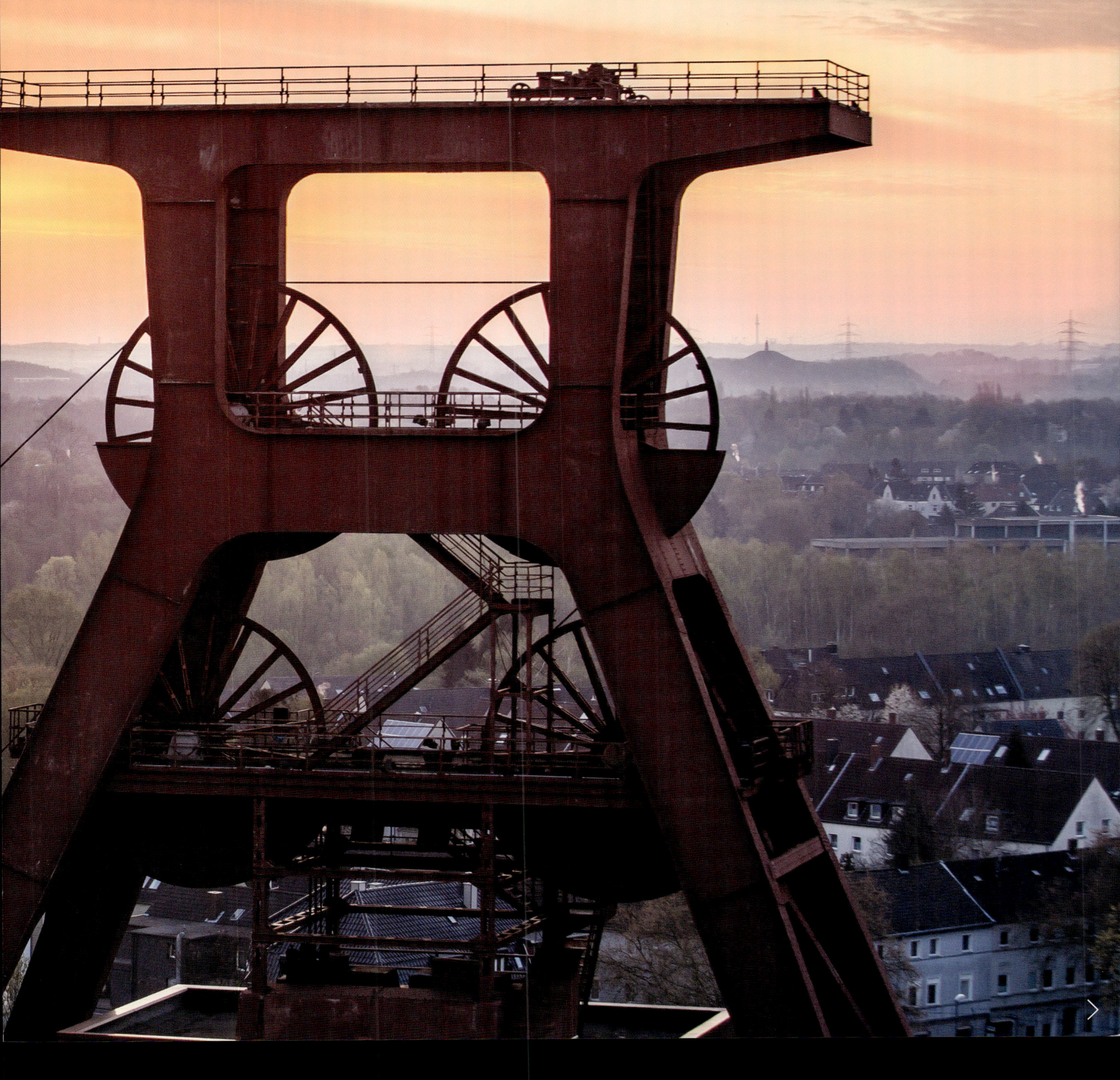

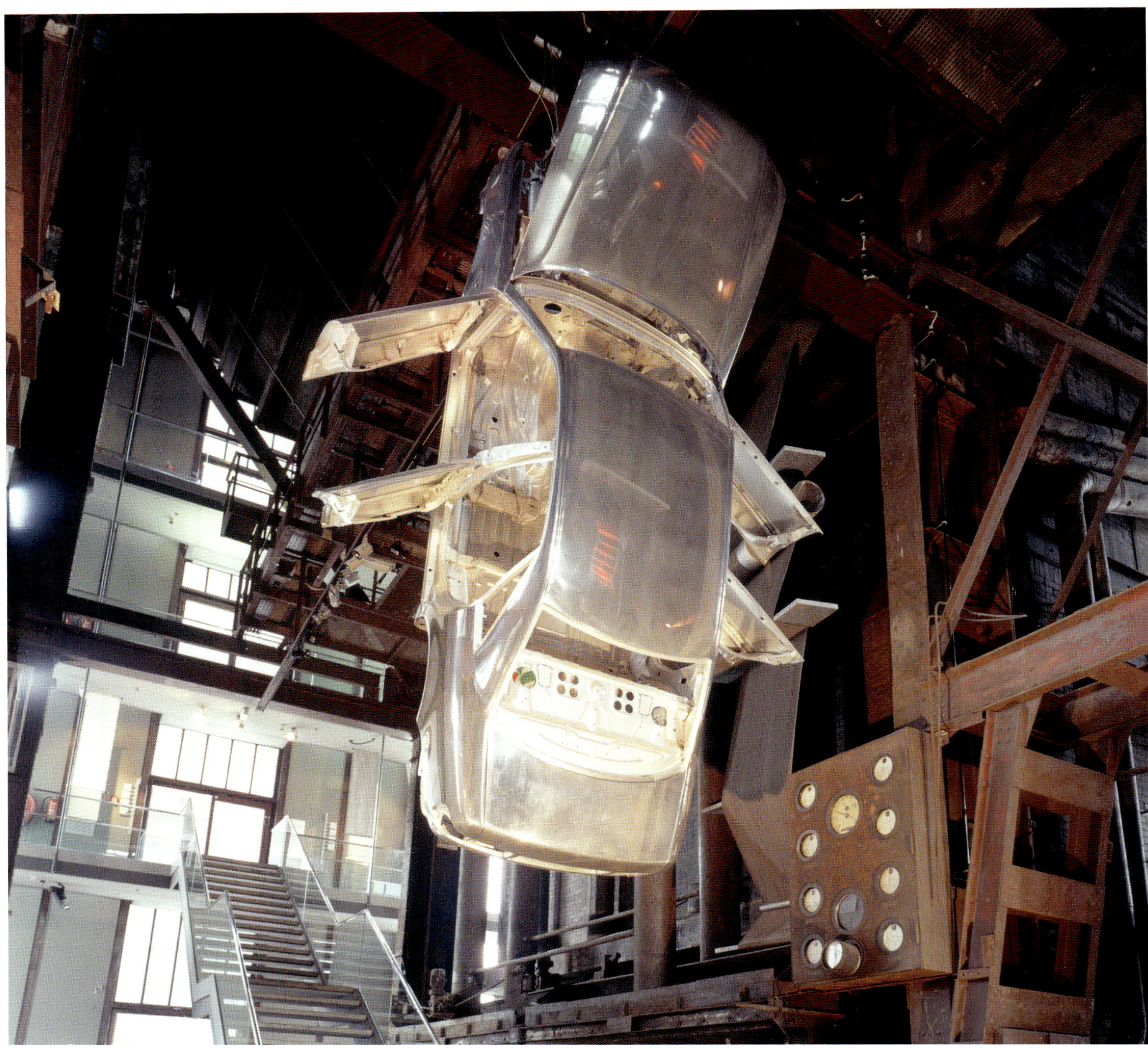

The iconic Zollverein coal and coking complex is monumental in every sense. Grounded in the Bauhaus design principle of 'form follows function' and the 1920s' cultural awakening of Germany's Ruhr region of heavy industry, Zollverein's size, symmetry and density purvey power, permanence, and importance – it was once proclaimed 'the most beautiful colliery in the world'. Its imposing architecture is visible from afar. The brown bricks and repainted red iron frames of its twenty buildings speak the same design language across the site, melding with the rusting metalwork of industrial decay.

Zollverein mined its first coal in 1851. Multiple shafts were sunk in the complex reaching a final depth of 1,000 metres. The emblematic Shaft XII and its associated buildings were constructed as a model colliery in 1932, the architecture and symbolism celebrating with pride the status of the mine and its workforce in local society and culture. It closed in 1986 – the last of the 290 collieries around Essen to do so – having produced an estimated 240 million tonnes of coal and employed 600,000 people. The colliery complex fed the adjoining Zollverein coking plant, which produced almost 9,000 tonnes of coke daily for the steel industry and employed 1,000 people. It closed in 1993.

On closure, the site was instantly protected as an industrial monument. The preceding decades of regional industrial decline had seen many historical and beautiful buildings destroyed. In the 1980s a cultural reawakening recognised that the Ruhr's historical roots were disappearing, which led – in 1989 – to the ten-year regenerative International Building Exhibition (IBA) Emscher Park. This focussed Ruhr minds on preserving industrial heritage and setting foundations for building new futures. In 2001 Zollverein was designated a UNESCO World Heritage Site, which led to further expansion of the site's capacities and offers. Its cavernous, blackened coal washing plant was converted into the stunning Museum of the Ruhr and and the Portal of Industrial Heritage, while the site's former boiler house is now the Red Dot Design Museum. It is the most popular leisure destination in the Ruhr, with two-thirds of its 1.5 million annual visitors emanating from beyond the region. The site is a crucible for the transition of a traditional working-class culture dating back to the industrial revolution to a new cultural trajectory. Zollverein rents space to 130 companies that include two museums, cafés and restaurants, concert venues, performance spaces, commercial offices, shops, art galleries, furniture showrooms, craft workshops, as well as offering a seasonal swimming pool and an ice rink.

The complex is arguably the most significant anchor point of the Ruhr's Industrial Heritage Route, which was launched in 1999 during the decade of the IBA Emscher Park, and subsequently agitated for the creation of the European Route of Industrial Heritage, which spans the continent. From the roof of Zollverein's Museum of the Ruhr, the low-lying landscape lays out visual connections to other iconic industrial landmarks. In the distance, the coking plant's forest of high, brick chimneys towers over the naturally recolonising real forest. The site sits within what is now a 70-hectare park, where full rein is given to the natural recovery of contaminated land and exploration of its multi-use trails is encouraged. The park was awarded the 2018 European Prize for Urban Public Space. Industrial ecology pervades throughout; fresh-leaved, young trees bind the dark, contaminated soil, while white birch bark brightens the drab brickwork. The trees gradually form areas of seclusion and tranquillity and birds sing between the buildings where once was industrial din.

Red Dot Design Museum
The Red Dot Design Museum (left) at Zollverein is the world's largest exhibition of contemporary design, housing around 2,000 international objects selected through its eponymous design awards.

The museum attracts 150,000 visitors per year and occupies Shaft XII's former boiler house. It was converted by renowned British architect Lord Norman Foster in 1997. The exterior retains the original brown bricks, red steel framework and clean lines and angles of the surrounding structures, while the modern interior architecture of glass and concrete merges with original pipework and remnant boilers.

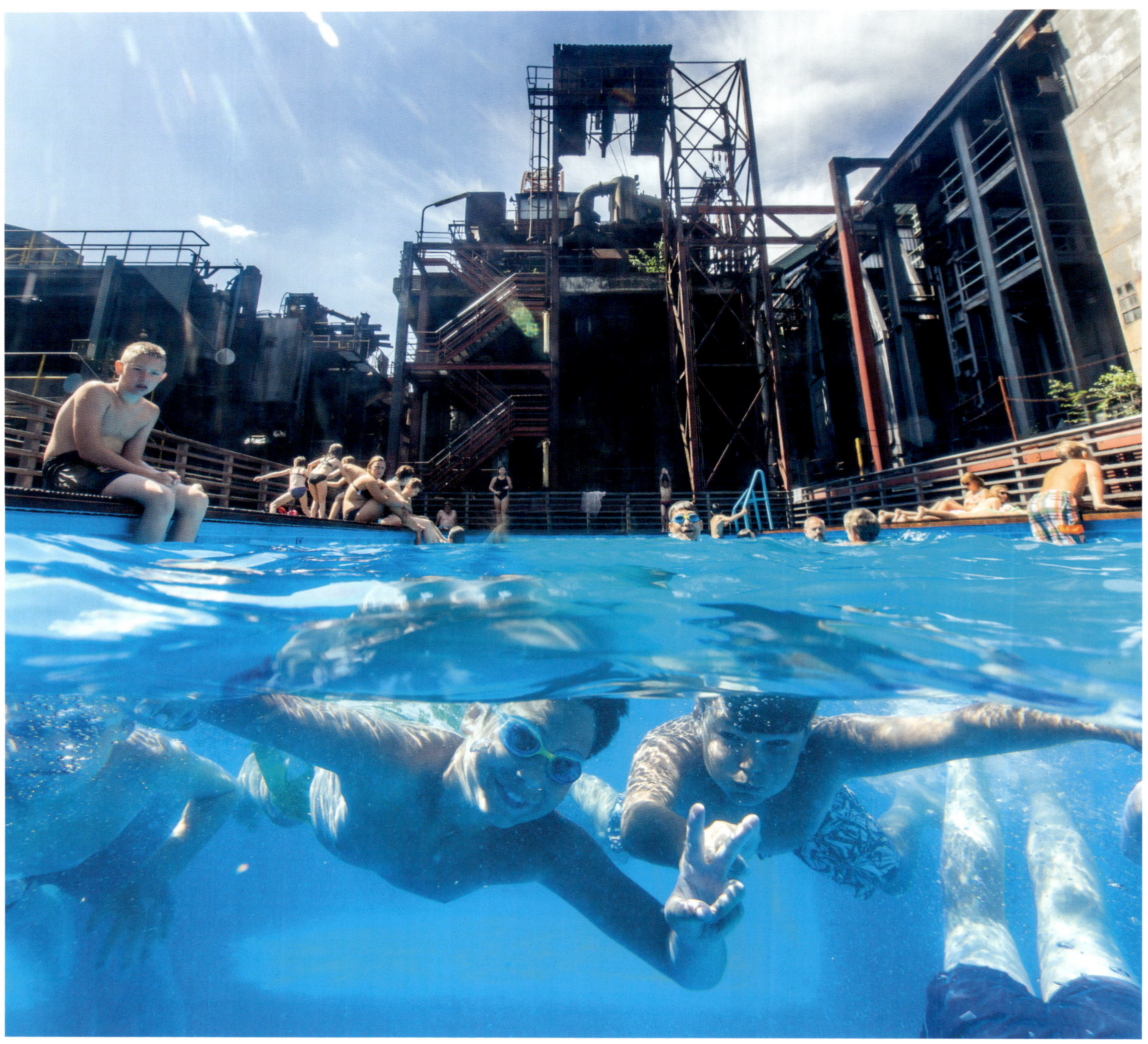

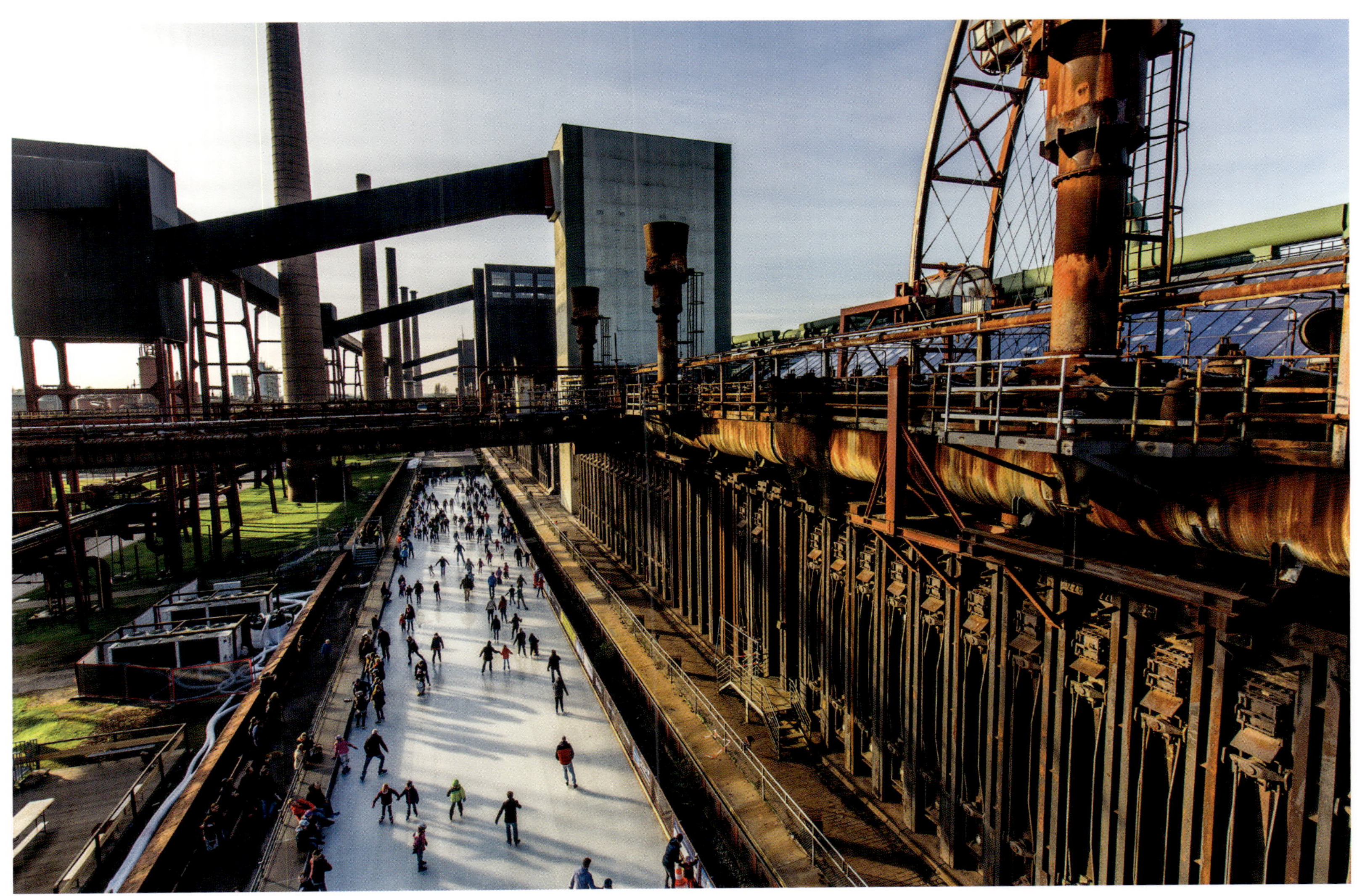

Above: Figures glide through the hulking industrial relic of Zollverein's old coking ovens. Open every winter, the 150-metre-long rink is dwarfed by the 800-metre-long plant.

Opposite page: Originally created 20 years ago as part of a contemporary art project and fabricated from repurposed shipping containers, Zollverein's Factory Swimming Pool sits amongst the relict architecture of the site's immense coking plant, in a place that was once filled with acrid fumes, deafening noise and scorching heat.

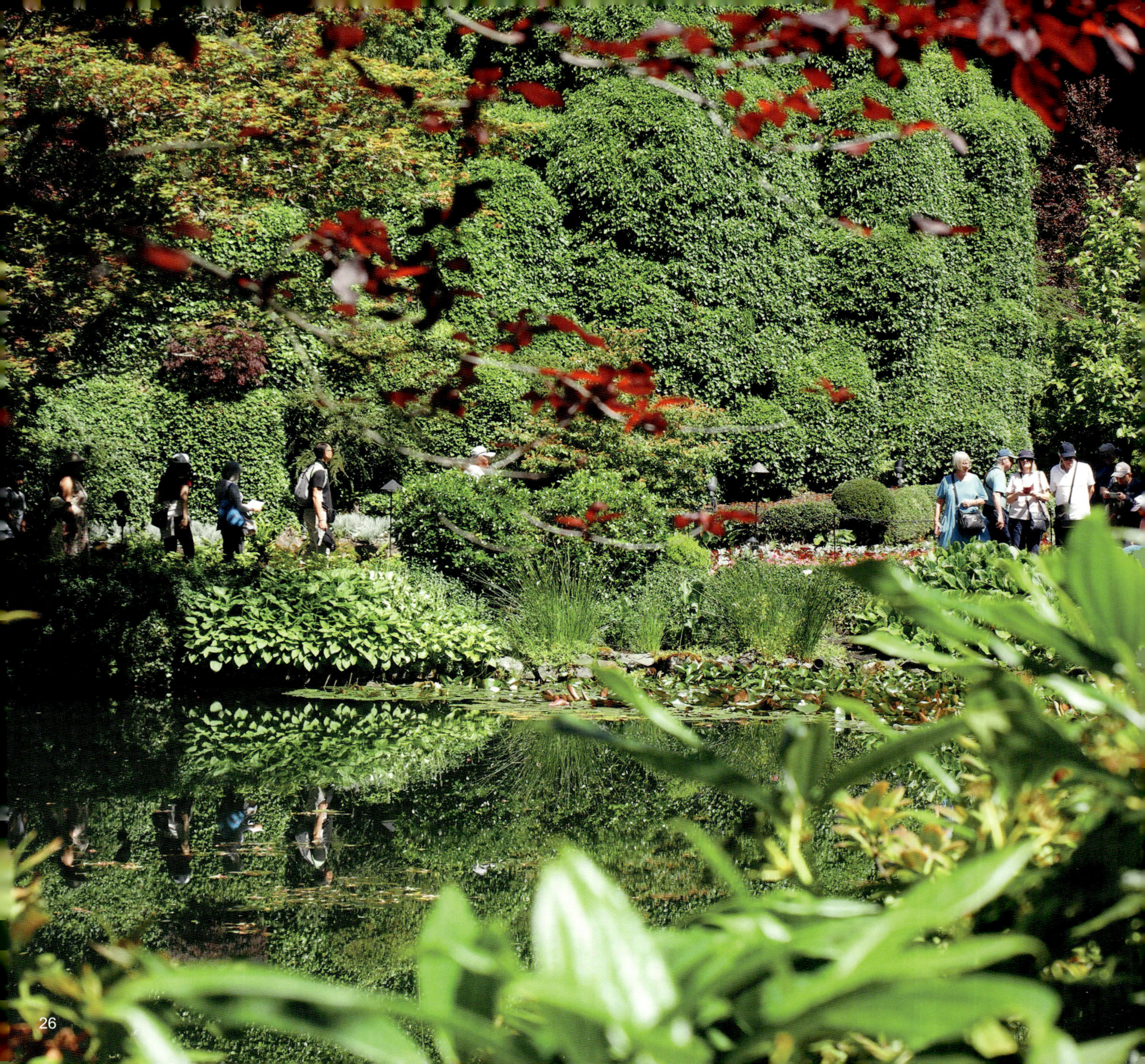

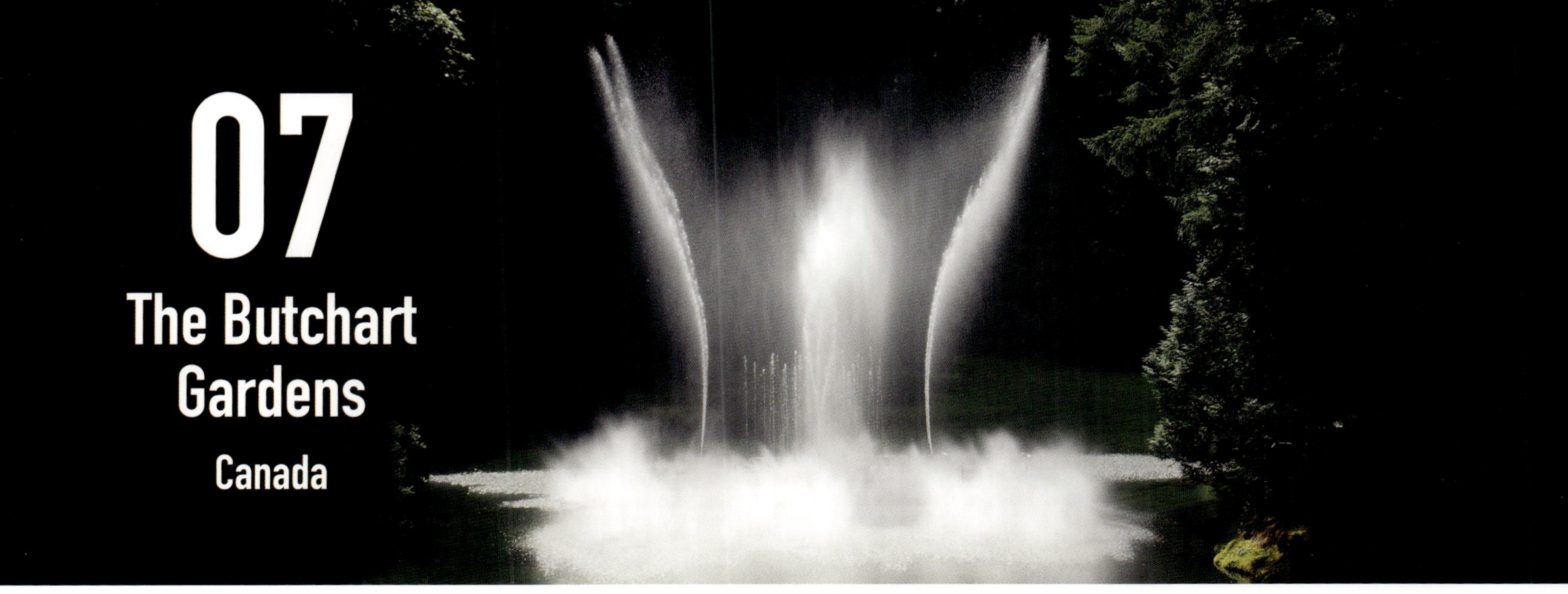

07
The Butchart Gardens
Canada

In the long history of mining, formal mine closure obligations and legislation are relatively new developments; but seeing the potential for beauty in post-mining wastelands is not new at all.

In the early 1900s, an idea germinated in the mind of Jennie Butchart for a 22-hectare garden in the limestone quarry developed by her husband Robert to produce cement for the burgeoning towns of Canada's west coast. By her own admission, she knew nothing about gardening but had a passion to learn, plant and create to beautify the land next to their residence.

The moist south of Vancouver Island, where the Butchart Gardens are located, is the country's mildest region, and the local microclimate is ideal for growing a wide range of sensitive garden plants. Once the limestone resources were exhausted, the transformation began. Hundreds of tonnes of rock debris were collected and redistributed around the site to create raised beds and edging, while topsoil was imported from a local farm by horse and cart. Tree-planting locations were planned, and the deepest parts of the quarry developed naturally as ponds. The depressing grey quarry walls were masked by ivy prodded into their crevices and pockets by Jennie, seated precariously on a bosun's chair.

The Japanese Garden, close to the coastline, was the first garden. Over time, the quarry was gradually transformed into today's tranquil Sunken Garden, followed by the Italian Garden, developed on a former tennis court, the Rose Garden and, most recently, the Mediterranean Garden.

In the late 1930s Ian Ross, the Butcharts' grandson, envisioned a more theatrical visitor experience, with light shows, concerts and a Christmas festival. The former quarry is still recognisable in some areas, particularly around the graceful and dynamic Ross Fountain, which he installed in a small pit lake to mark the garden's 60th anniversary in 1964.

To celebrate the 100th anniversary of the gardens in 2004 and to acknowledge their place in the traditional territory of the WSÁNEC people, two totem poles were carved by master carvers of the Tsartlip Nation and the Tsawout Band and are proudly displayed on-site.

The gardens are still owned and managed by the Butchart family, and their 'old quarry' is visited by over a million people annually. The gardens employ up to 600 staff who are renowned for their friendly welcome and willingness to help; indeed, as early as 1915, the site attracted 18,000 people who all received a cup of tea on arrival!

The Butchart Gardens are now formally recognised as a National Historic Site of Canada. The truth is that the land is far more valuable economically and culturally as a garden than it ever was as a quarry.

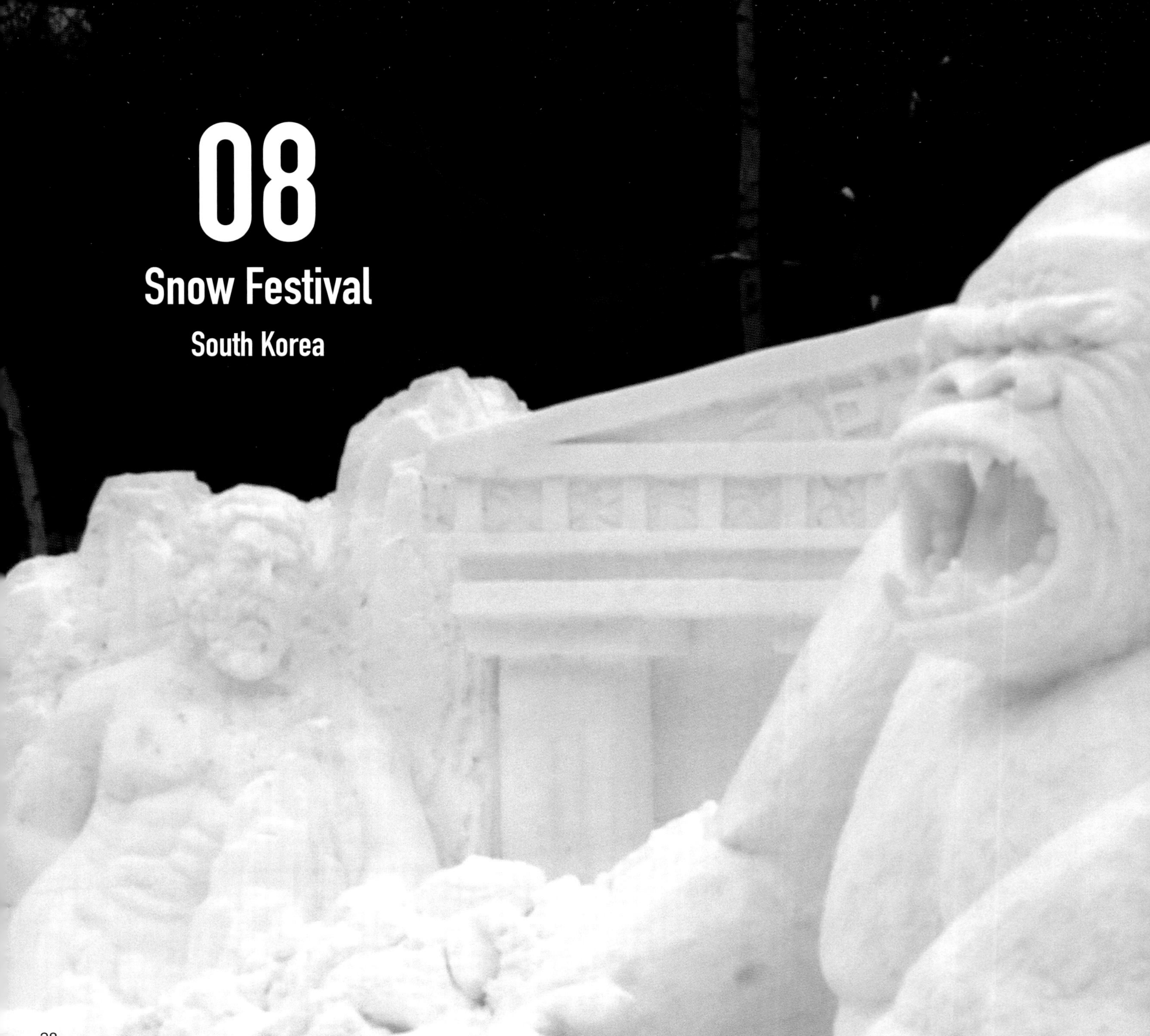

08

Snow Festival

South Korea

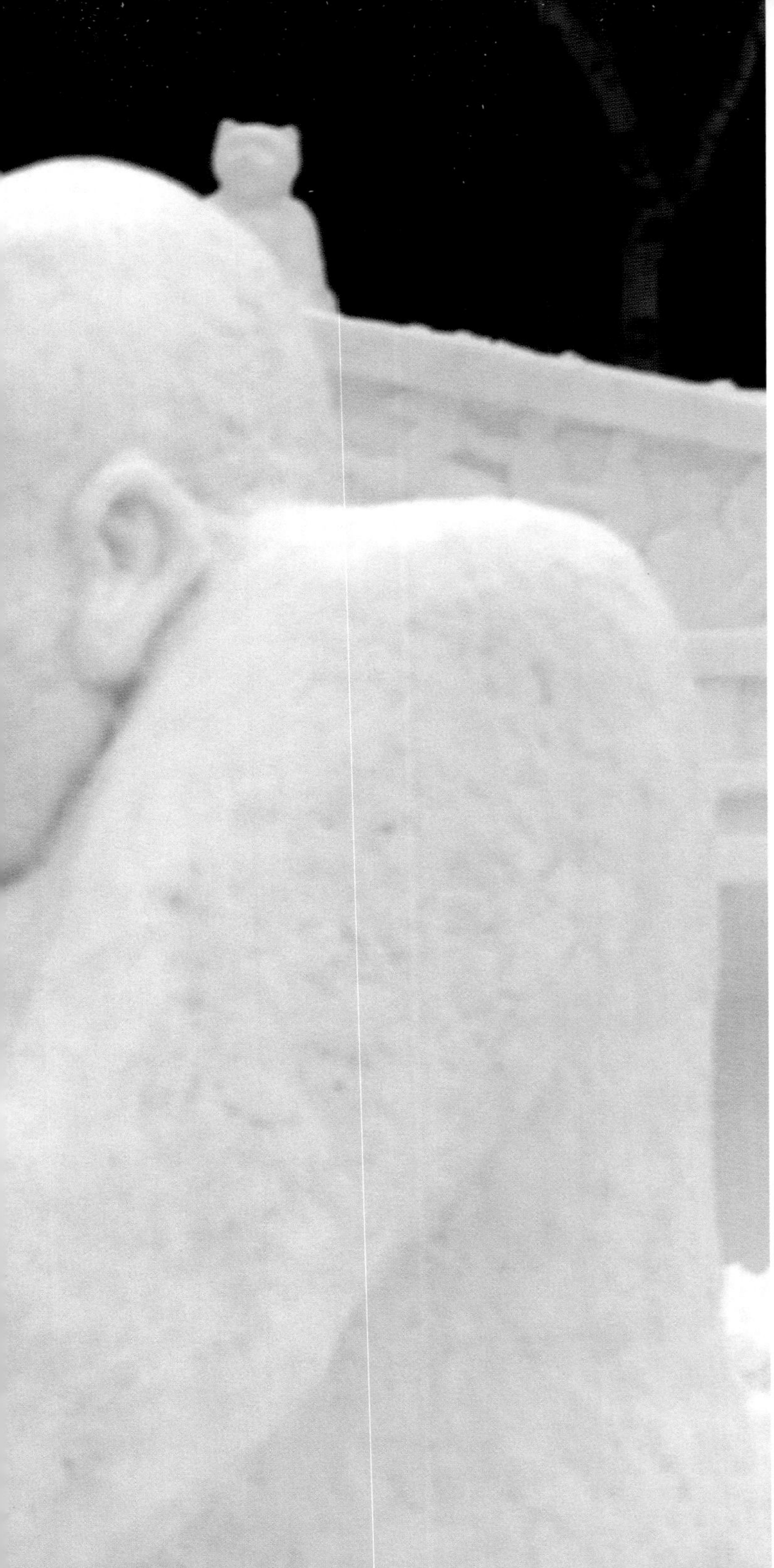

Until 2018, the mountainous Gangwon-do province was best known as a largely rural coal-mining district. One of the poorest provinces in South Korea, it is 80km from the Demilitarised Zone – a third of the original province still lies on the north side of the border.

Its fortunes began to change when the small town of Pyeongchang turned the region's long snowy winters to its advantage and won its bid to host the 2018 Winter Olympics. The province has since used this as a springboard to reinvent itself as a winter sports destination. It has also created a series of seven festivals designed to attract tourists, themed around ice-fishing, sledging and trekking, but the most popular feature is huge ice and snow sculptures. The oldest of these festivals, the Daegwallyeong Snowflake Festival, only dates back to the mid-1990s, but they have become an important source of income for the province.

One of the most popular ice sculpture festivals, The Taebaek Snow Festival, features an eclectic mix of characters: Buddha rubs shoulders with K-Pop stars, Snow White and Pinocchio. Tourists in bright puffa jackets queue for selfies next to their favourites, and singers in skimpy shiny dresses serenade the crowds from a small stage near pop-up stalls. Despite the event's popularity, the Taebaek economy is still heavily reliant on mining – 25% of its income is from coal. This can't last. The South Korean government is transitioning to renewable energy, and Korea Coal Corporation has agreed on a settlement to phase out coal mining in the province by 2025. Miners will receive financial compensation, but it remains to be seen whether its seasonal tourist economy will be enough to maintain its relatively limited prosperity, especially as climate change turns many of the world's seasonal snowfields permanently green.

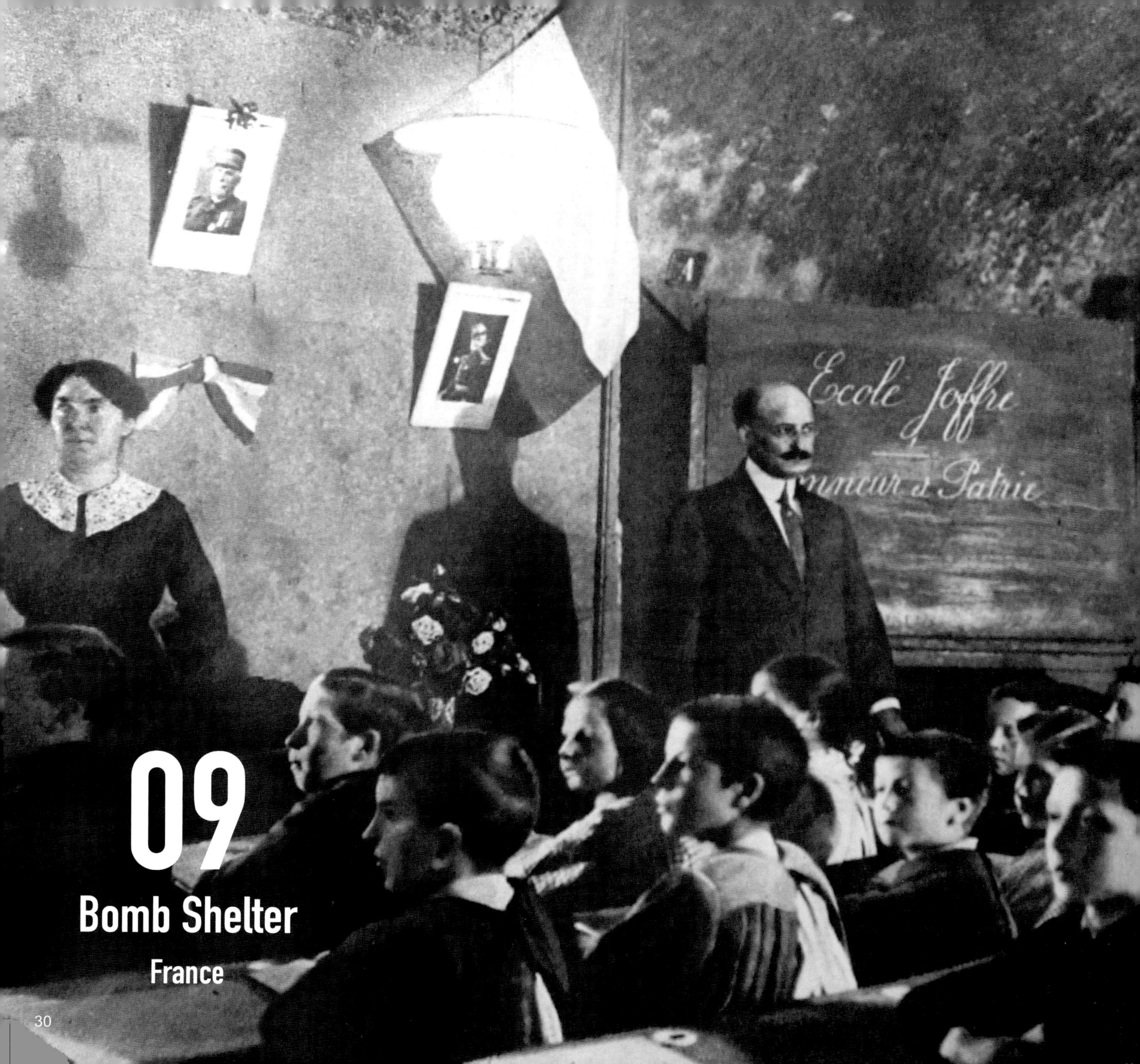

09

Bomb Shelter

France

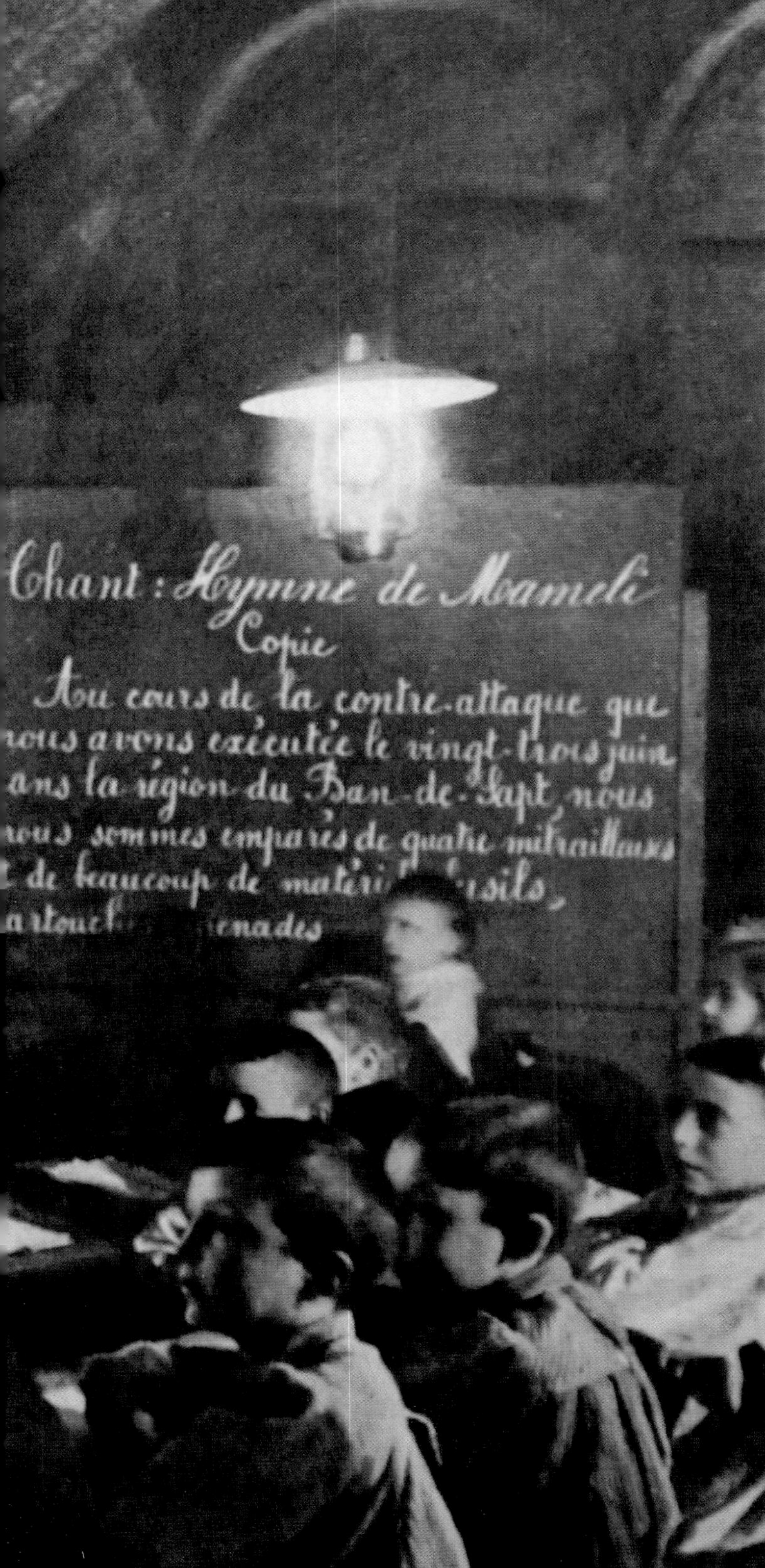

Over 200km of chalk caves and tunnels known as *Les Crayères* lie beneath Reims in north-eastern France. Begun by the Romans, who excavated the chalk for roads and buildings, *Les Crayères* were extended by the region's champagne producers in the 18th century. The chalk soil provides the ideal conditions for champagne vineyards, while the caverns below offer the perfect environment to store and age the resulting champagne. But in September 1914, when German forces approached Reims, *Les Crayères* provided something even more valuable to its citizens – sanctuary.

The lack of a natural frontier between Germany and Champagne had made Reims an obvious target, but when the German army failed to occupy the city, they resorted to shelling it flat. Over a thousand consecutive days of bombing followed. Most of the city's 120,000 population fled, but over 20,000 people retreated to the caves beneath the city to live and work by lamplight.

While the city above was being shattered, a new Reims was created underground that included a school (left), complete with playground and gymnasium, a church with packing crates for pews, as well as grocers and butchers to service the population. Life in the caves was hard, but over the previous century, the champagne firms had invested in ventilation shafts, drainage and water to protect their champagne and make *Les Crayères* a viable working environment.

Many of those who stayed were already familiar with the conditions underground because they worked for the champagne houses. These were mainly women, children and those men unable to fight.

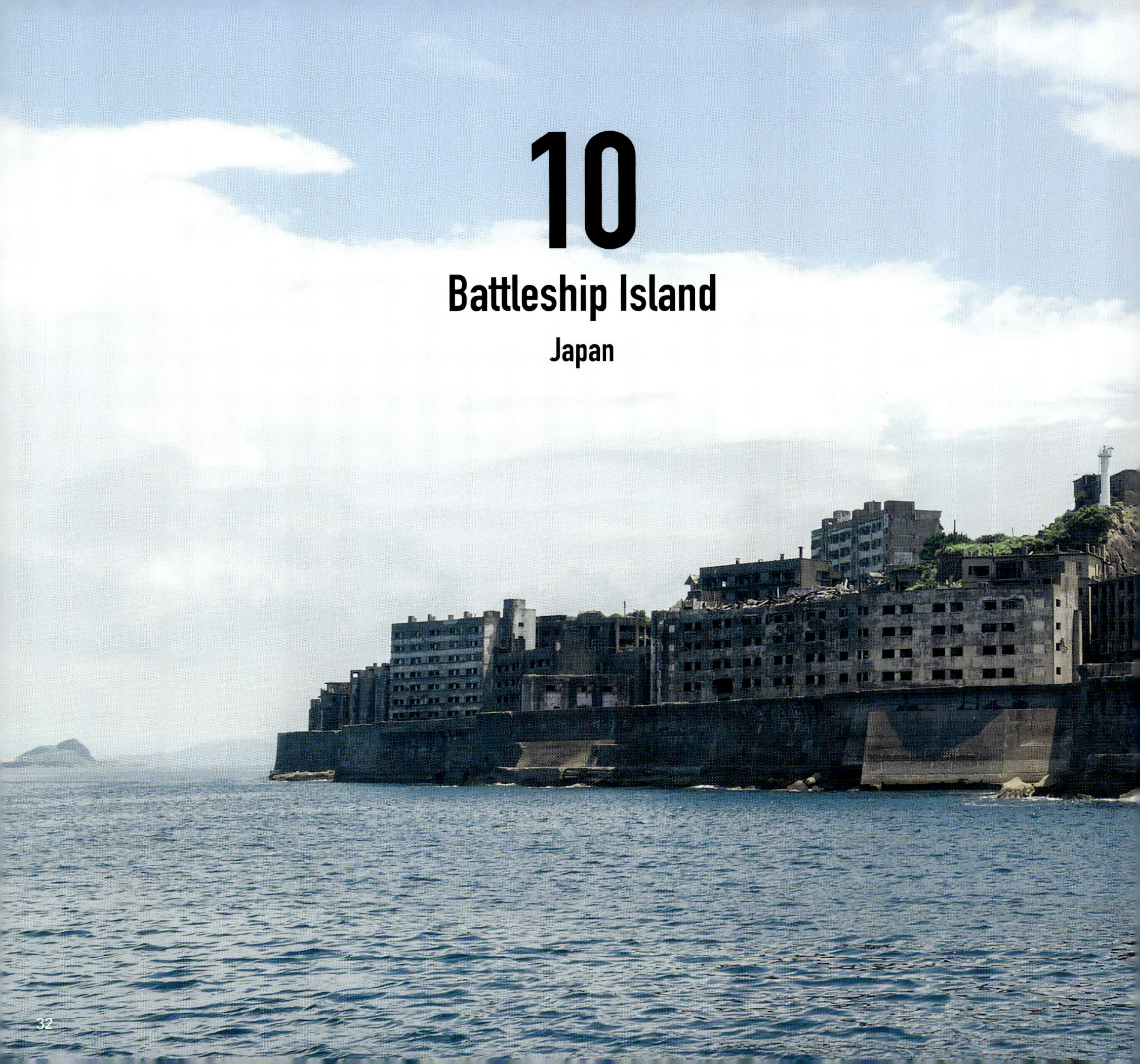

10
Battleship Island
Japan

It takes half an hour by boat to reach Hashima Island from Nagasaki. With its concrete towers and looming sea walls, it's easy to see why it was nicknamed *Gunkanjima* or Battleship Island. From 1900 until the early 1970s, the island was owned by the Mitsubishi Corporation mining the high-grade coal below the seabed. Mine waste was used to expand the footprint of the site and provide dormitories and flats for miners and their families, along with shops, a cinema, a school, a hospital, a *pachinko* parlour, and two swimming pools. Nonetheless, the island is tiny – it was said that you could walk the length of it in the time it took to smoke a cigarette – and at its peak, it was the most densely populated place on Earth, its six and a half hectares home to over five thousand people. Everything, including fresh water, was imported, and nothing grew until soil was brought over from the mainland for roof gardens in the early 1960s.

During World War II Hashima produced 410,000 tonnes of coal, but that yield had a human cost: over a thousand Chinese and Koreans were conscripted as forced labour, and many died.

After the war, Hashima's coal production became a crucial part of Japan's post-war economic recovery effort. Then in January 1974 Mitsubishi suddenly announced it was ceasing mining on the island, and by March all of its inhabitants had departed. Battleship Island had become a ghost ship, and for the next forty years the island was left to rot, battered by the region's frequent typhoons.

Tourists are permitted to visit Hashima, which is now owned by Nagasaki City and became a UNESCO World Heritage site in 2015. It's likely that much of the island's appeal relates more to its post-apocalyptic appearance – and to its brief cameo in the 2012 James Bond film *Skyfall* – than its industrial heritage.

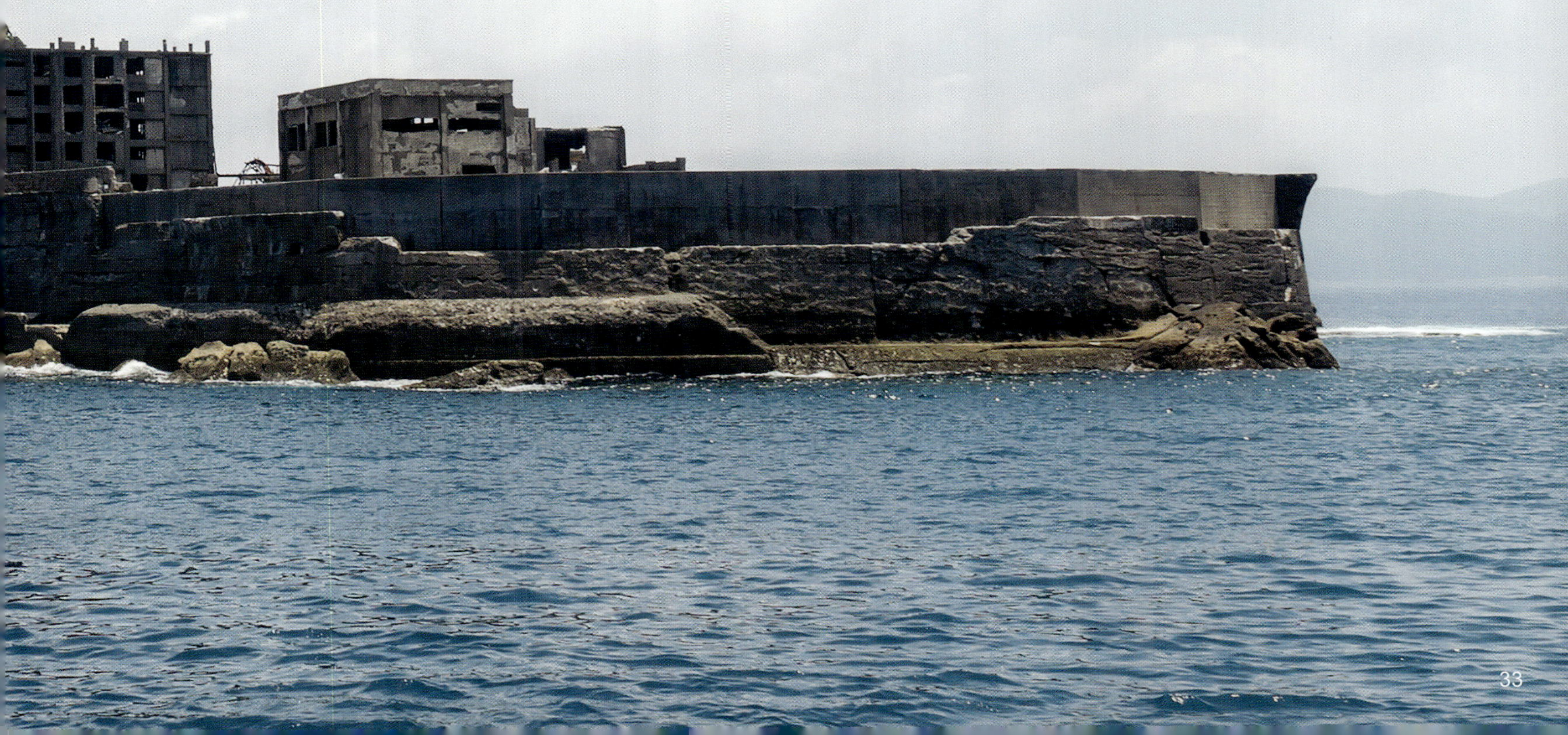

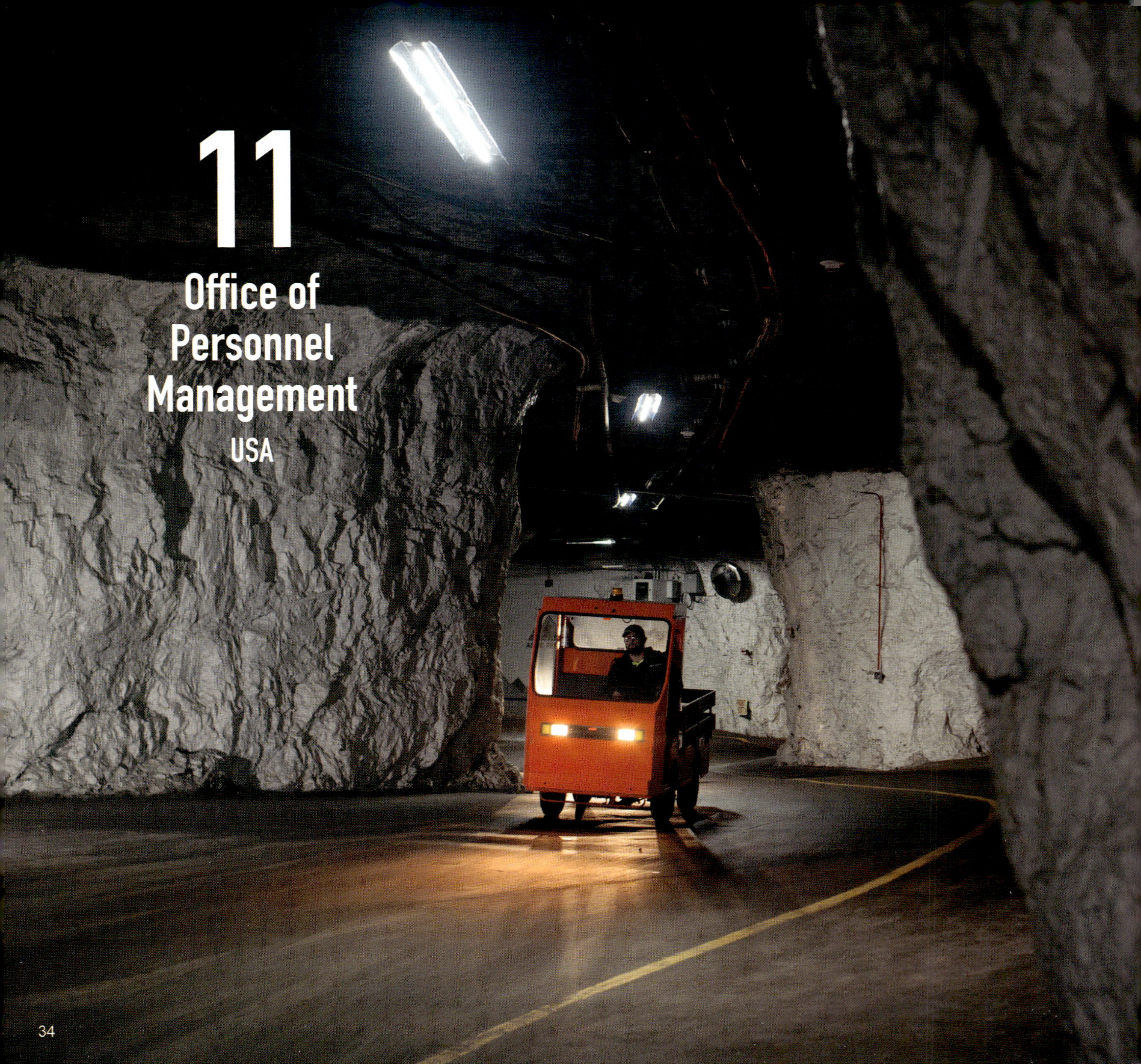
11
Office of
Personnel
Management
USA

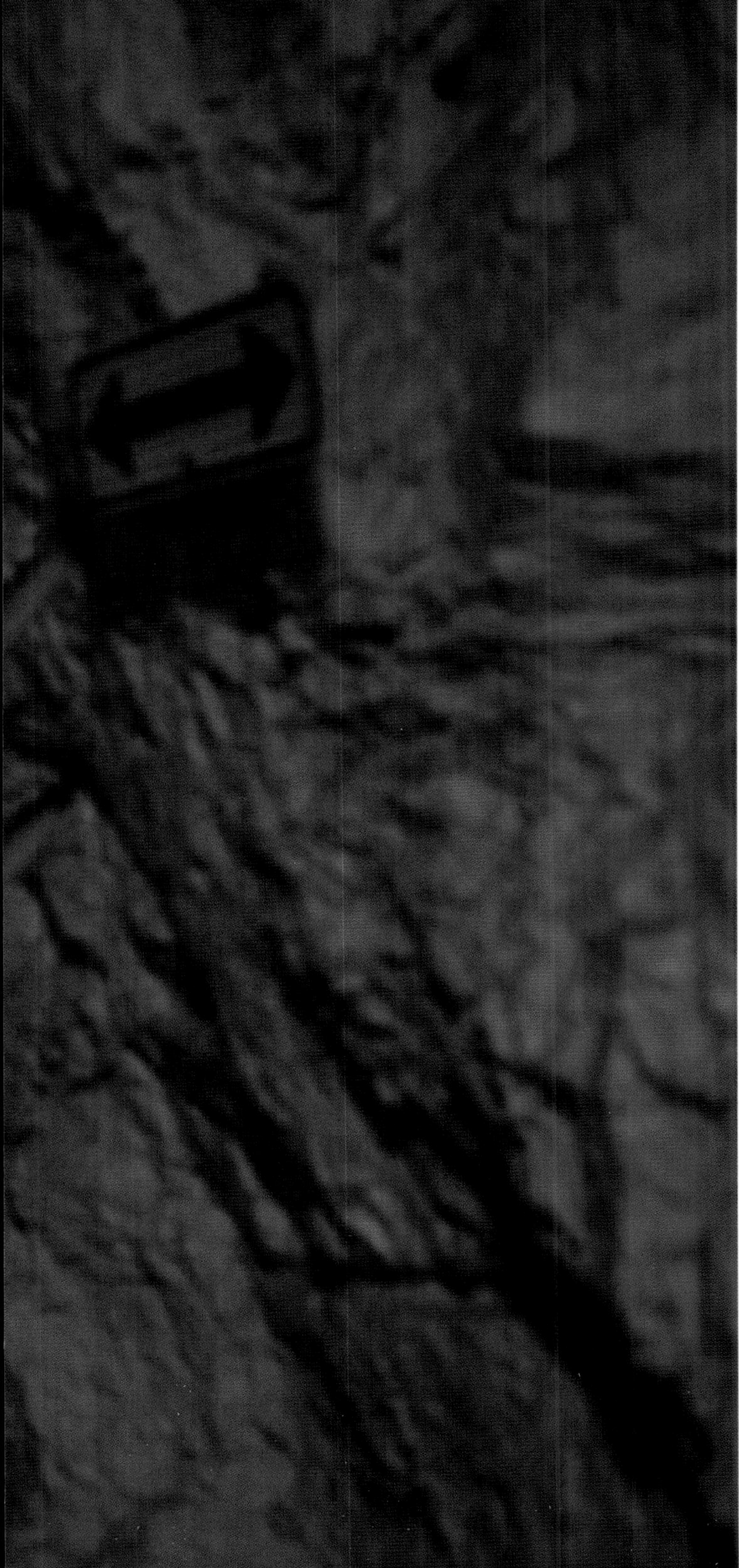

Since the 1960s, trucks full of documents have been arriving at the gate of a former underground limestone mine in Butler County, Pennsylvania, owned by US Steel and quarried for the nearby steel mills for over half a century. Mining ceased when more cost-effective opencast sites were found instead. Mines were highly valued assets during this period of the Cold War for their imperviousness to nuclear attack, and the first tenant was the government. A number of agencies occupied the site, including some clandestine organisations, but today the Office of Personnel Management (OPM), which processes the pension paperwork for federal employees, is the only government tenant.

While other government departments have transitioned to entirely digital systems, OPM still shuffles paper around its vast underground facility. Millions of dollars were spent attempting to digitise the arcane pensions procedures in the early 1980s and again in the 1990s, but its complexities mean that trolleys are still wheeled along the whitewashed limestone corridors.

The mine remains a valued local employer in a predominantly rural area; around 600 people work in the OPM archive, which is 230 metres underground. The documents are keyed in by hand, but some files have to be cross-referenced against existing paper records, and all of them end up being filed in one of the eight vast caverns that make up the OPM's part of the facility. The OPM archive is a mile and a half inside the mountain, which means that staff rarely get the chance to see daylight in the winter.

Only 5% of the former mine has been developed into storage space. It boasts its own roads, store and even a fire service, but there are no restaurants, as open flames are not permitted in the facility. The site is now owned by Iron Mountain, a data services company, which holds music, photography and film archives, as well as more conventional banks of servers and storage. The OPM remains its largest tenant and probably will be until someone cracks the digitisation problem. At least 100,000 federal employees retire every year, so if you had a Holy Grail to hide, this would be the place to do it.

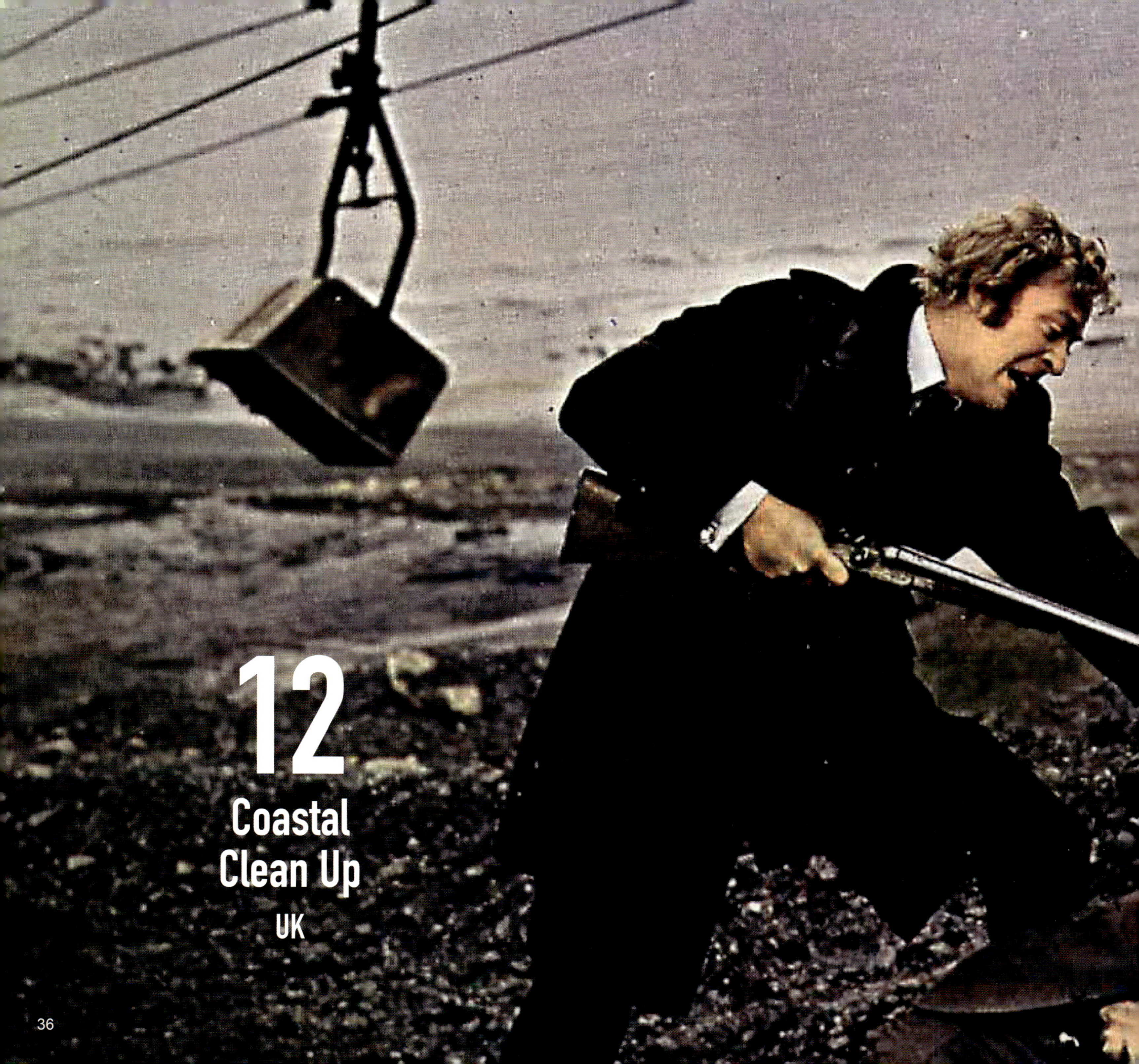

12
Coastal
Clean Up

UK

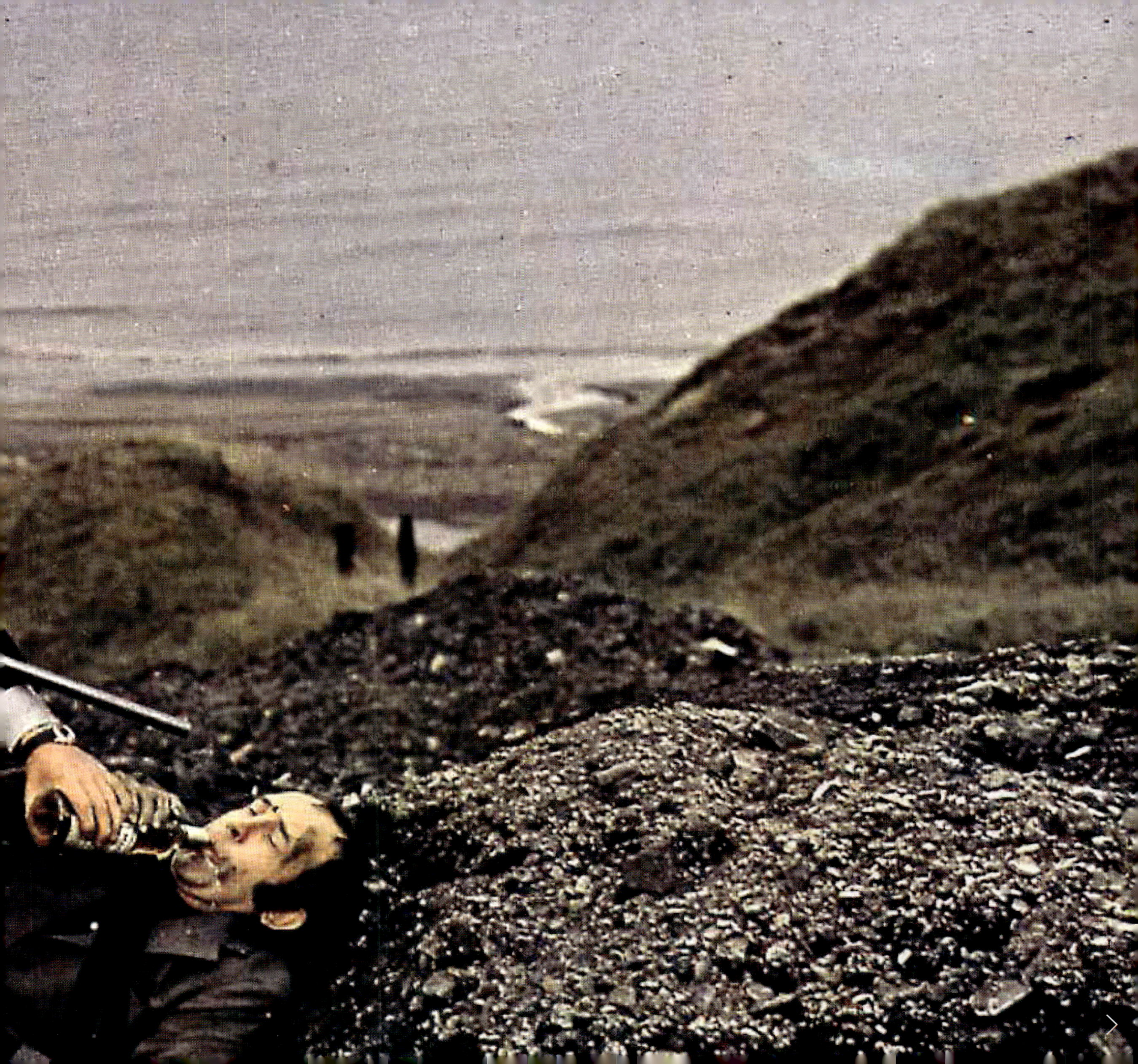

A single rifle shot under a depressing grey sky and a man falls to the ground. He dies on a black beach, his body lapped by black waves. The killer, gun in hand, walks nonchalantly away along a black cliff, his back to the camera.

The dramatic finale of *Get Carter*, the iconic British gangster film starring Michael Caine, was shot on the bleak Blackhall Colliery Beach, County Durham, in northeast England in 1970. This striking backdrop was the result of a century of indiscriminate dumping of spoil by coastal collieries of the East Durham coalfield; an estimated 40 million tonnes of waste smothered cliffs and beaches for over 12 kilometres and darkened the shallow sea several kilometres out from the coast. Bulldozers spread and compacted the waste, extending beaches seaward by up to 150 metres and raising them by six metres or more in places. The weight of overlying waste fractured sewage pipes, adding to the toxic cocktail. Widespread fly-tipping followed, and the beach became littered with burnt-out cars. By the time Easington Colliery closed in 1993 and its buildings were demolished, County Durham's devastated coast was regarded as a world-class environmental disaster and a lost cause.

Launched in 1997, the five-year Turning the Tide environmental regeneration programme swept clean the worst-affected 18 kilometres to encourage investment into this economically flat-lining region. Costing GBP 10.5 million, the programme was an effective partnership of 14 organisations led by Durham County Council. The work was funded by public and private bodies, including national and local government, the European Union and a water company; almost half came from the UK's Millennium Commission.

Firstly, skeletal cars and fly-tipped detritus were removed. Then the remnant cliff-top colliery sites were stripped of their topsoil, and 1.3 million tonnes of colliery waste covering the cliffs and beaches were moved onto these areas. These were capped and covered with alkaline soil to recreate the coastline's rare Magnesian limestone grasslands. The partnership also purchased and converted a further 225 hectares of adjacent fringe agricultural land to grasslands.

Since the clean-up, the area has received national and international recognition and is now a Site of Special Scientific Interest and a National Nature Reserve. Today, wildflowers gently nod in the breeze, accompanied by the lilting melodies of skylarks. Almost all vestiges of the mines have been removed, and it is hard to imagine just how bad it once was. Families, dog walkers, cyclists, artists and naturalists, not to mention people writing books on mining, now appreciate the wild, open spaces made accessible by new coastal paths. Subtle controls are in place to limit access in some areas and guide people along certain routes to avoid damaging these restored habitats. Over the long term, there is also much to do to connect local communities with their coast and restore the seabed.

The once infamous, devastated beaches are slowly recovering. The sheer extent of the damage done to the beach and seabed was too costly to clean up, and it has been left to the tides to wash away the residue. It's a natural work in progress that will take decades to recover. Today on Blast Beach, it feels peaceful but strange. A head-high, compacted shelf of waste at the rear of the beach is gradually being eroded. Bizarre clear orange pools, iron-stained, sea-worn cobbles and boulders, a lingering, ferrous odour and a remnant industrial archaeology of bricks, ironwork, pipes and old boots paint a picture of a place where the present is still coming to terms with its past.

Previous page: Blackhall Colliery Beach in 1970.
Opposite: Blast Beach, Seaham.

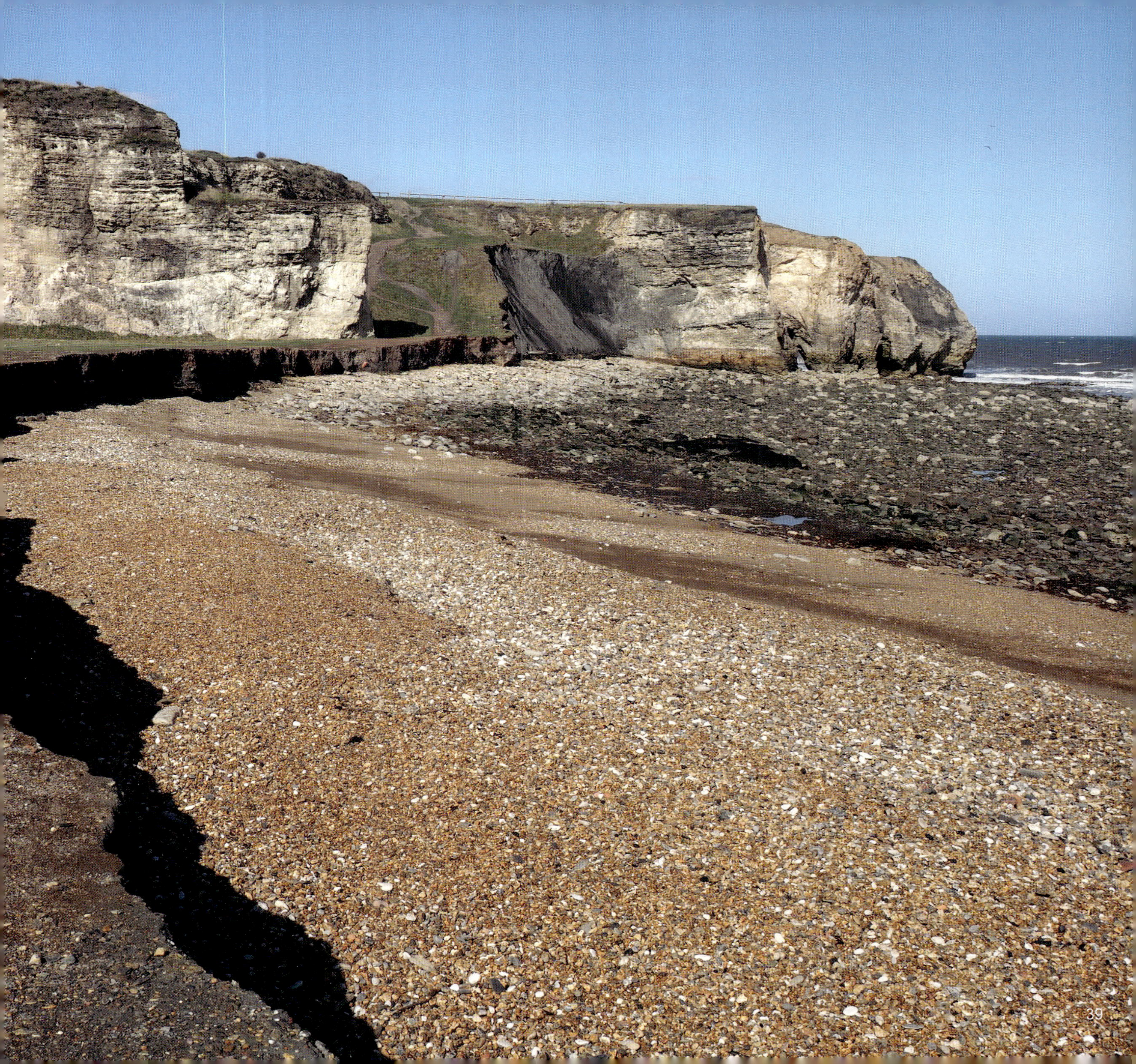

13

Kelp Renewal

UK

Decades of colliery spoil dumping onto England's County Durham coastline blackened both beaches and sea. Black tides smothered and poisoned life on the seabed, turning it into a submarine desert extending several kilometres from the shoreline.

Too expensive to clean up; time and nature are the healing forces here. However, the dense kelp forests that should occupy rocky outcrops are not recovering to the extent expected.

Kelp forests are one of the world's most productive habitats and provide valuable ecosystem services by aiding the sequestration of marine carbon (aka blue carbon), protecting coastlines against waves and storms, trapping sediments, stabilising the seabed and providing diverse habitats. Kelp requires stable rocky surfaces for its root-like holdfasts to grip, while its rigid leaf-like fronds extend upwards to intercept light for photosynthesis. Natural recolonisation is stymied by this underwater desert's enormous expanse, mine-waste contamination, and the high turbidity of the near-shore waters, which restricts light transmission. Low light conditions limit kelp development to shallower zones where wave-induced turbulence prevents young holdfasts from establishing.

Funded by the Durham Heritage Coast Partnership, scientists from the University of Newcastle are researching how to restore these critical marine habitats off the Durham coast. The main species, *Laminaria hyperborea*, has not been cultivated before, so commercial techniques developed for edible species are being adapted. Translocation experiments from donor sites will also be carried out, developing methods to transfer whole plants and – critically – fix them to the seabed long enough to allow establishment. Monthly seabed monitoring is carried out off Blast Beach and Easington Colliery in often challenging sea conditions – particularly during the winter.

This stretch of coastline is a key part of the 'Stronger Shores' initiative to strengthen a regional coastline against rising sea levels and more powerful storms. This UK-first is trialling nature-based solutions, including restoring kelp forests, oyster reefs and seagrass beds alongside more conventional engineering responses. Once established, such habitats will accelerate improvements in water quality, economic opportunities from coastal fisheries, carbon sequestration and seabed and coastal stability.

In a further nod towards a more optimistic future, the possibility of creating a marine national park is being explored. This would be a fitting epilogue to the story of the clean-up of what was once one of the most polluted coastlines in the world.

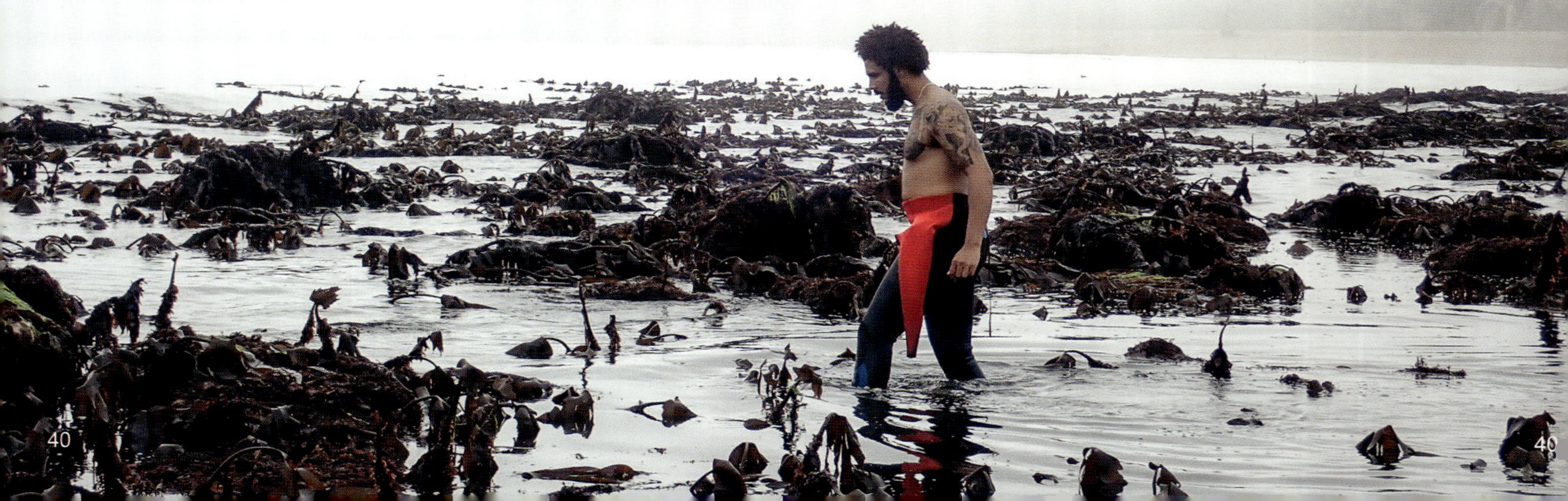

14
Sunway Lagoon Amusement Park

Malaysia

When the Malaysian tin mining industry collapsed in the 1980s, one mine operator saw an opportunity amid the vast wastelands created by over 100 years of operations. Jeffrey Cheah proposed turning part of the 323-hectare opencast mine in Selangor into Malaysia's first amusement park.

It opened in 1992 and has grown into a series of attractions. Sunway Lagoon Theme Park occupies a 35-hectare site, 46 metres below ground-level, offering 90 rides across seven themed areas, including water, wildlife, and the 'scream' park, attracting around a million visitors a year. Sunway Lagoon was just the first step in the regeneration of the mining site by Cheah's group. A 37-hectare Egyptian-themed mega-mall overlooks the park along with a hotel, and while its tourism and retail businesses are the most visible, the Sunway Group also invests in education and health.

Sunway University predates the theme park, opening on the same site in 1987. Melbourne's Monash University established an adjacent campus in 1998. Run by a non-profit foundation, Sunway University has partnerships with Harvard, Oxford and MIT, among others. The foundation endowed the Jeffrey Sachs Center on Sustainable Development to improve sustainability in the region. The Sunway Group has won environmental awards for its developments and sustainability remains one of its core values, guiding its expansion beyond its Selangor site.

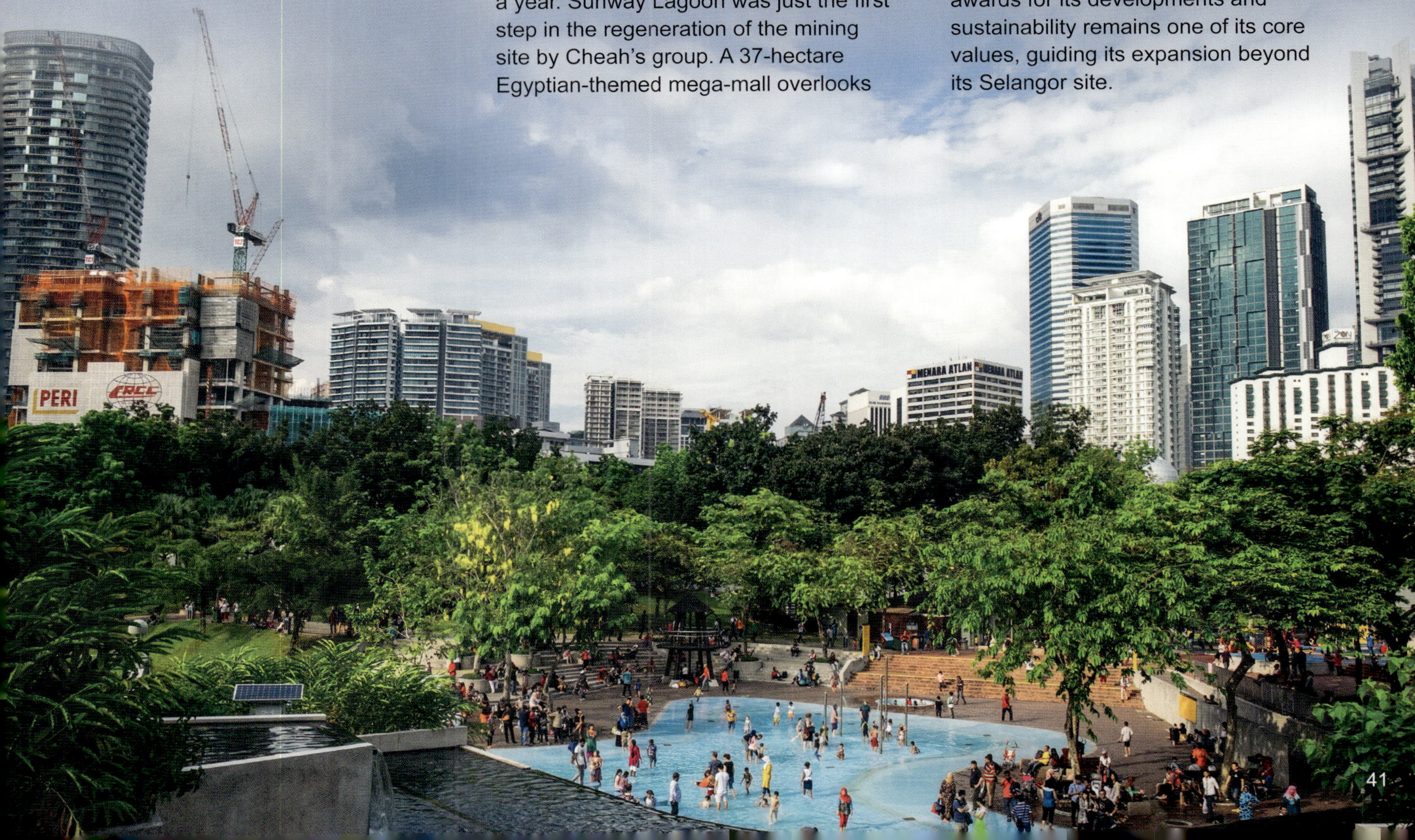

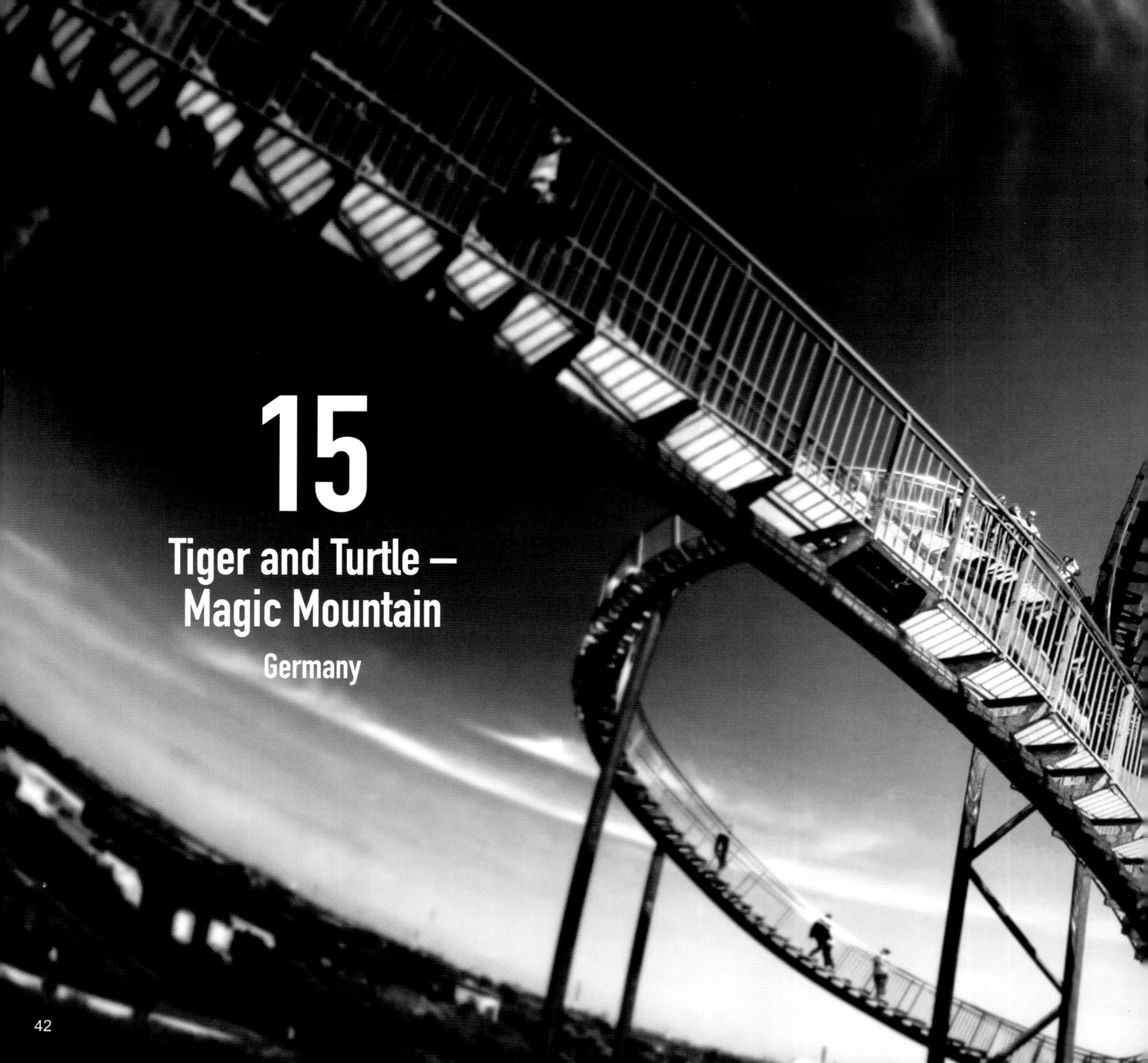
15
Tiger and Turtle –
Magic Mountain
Germany

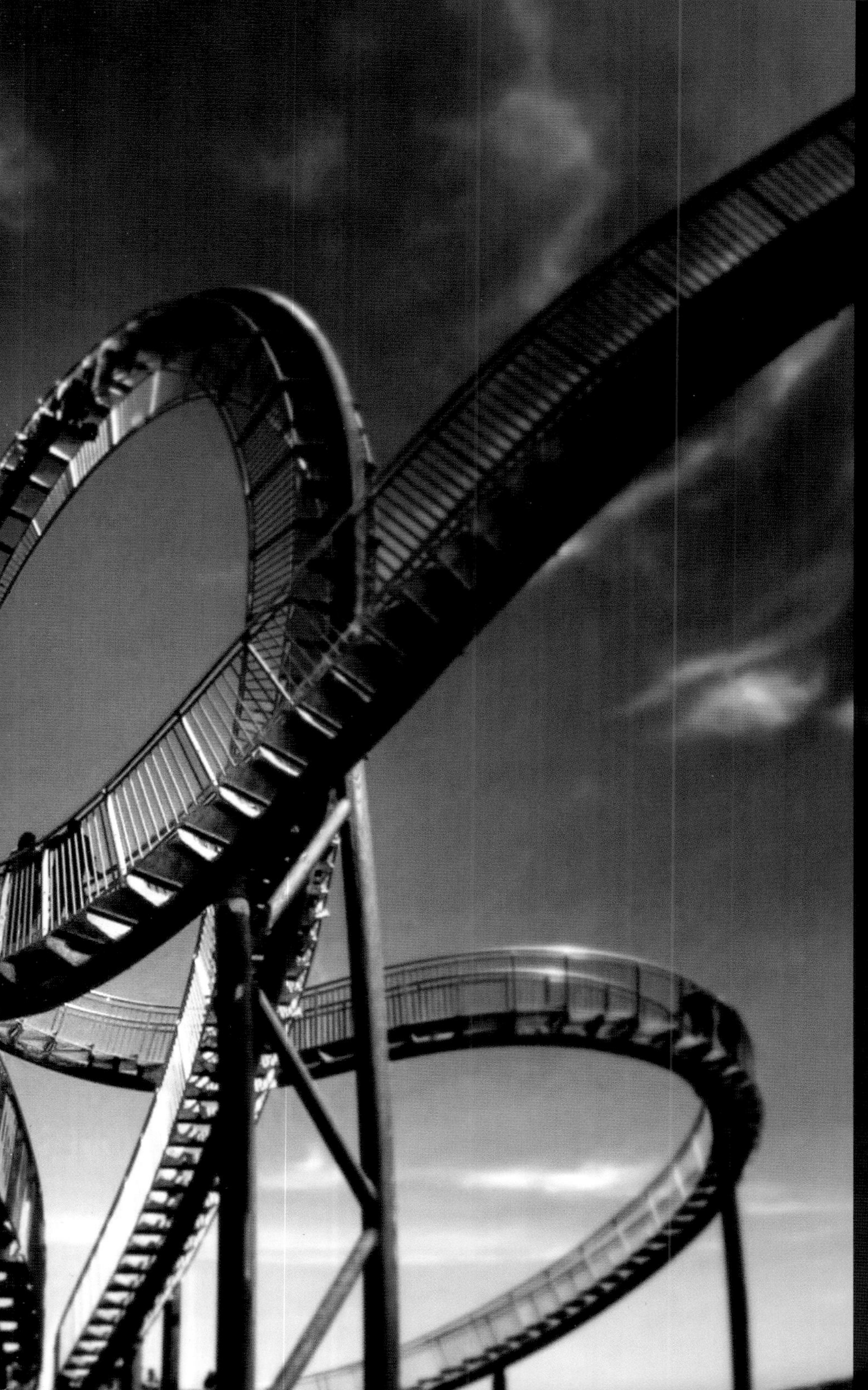

Frozen mid-writhe, the serpent shimmers in sunlight and sparkles diamantine at night. Its distinctive, dynamic twists and curves encourage exploration.

Tiger and Turtle – Magic Mountain is located in Duisburg's Angerpark, close to the Rhine. The installation's name derives from the fast tiger and slow turtle, representing the variable speeds at which you explore this pedestrian rollercoaster. Opened in 2011, it was the winning design in an international competition run by the city of Duisburg for a new landmark on the Heinrich-Hildebrand-Höhe hill.

Designed by Hamburg creative duo Ulrich Genth and Heike Mutter and constructed from zinc and steel in memory of its progenitor industries, the 13-hectare hill and its sculpture are the toxic legacies of MHD Sudamin's zinc smelter, which closed in 2005. The slag was covered with a geotextile membrane, then soil, into which 60,000 trees and shrubs were planted.

One emerges at the summit from the pathway that spirals upwards through the young, wooded slopes. A metallic odour on the breeze reinforces a sense of place. It pervades the 360-degree views over the Rhine and the low-lying industrial, cultural landscape of the western Ruhr – a view dominated by the neighbouring steelworks and punctuated by cooling towers and high chimneys, belching steam and fumes. Despite the modest elevation, on clear days the view extends all the way to Düsseldorf.

Visible for miles around, this dramatic, walkable sculpture attracts people even on a cool, grey Saturday afternoon. As a key element of the Ruhr's renowned Industrial Heritage Route, Tiger and Turtle – Magic Mountain contributes to the region's emerging new sense of identity.

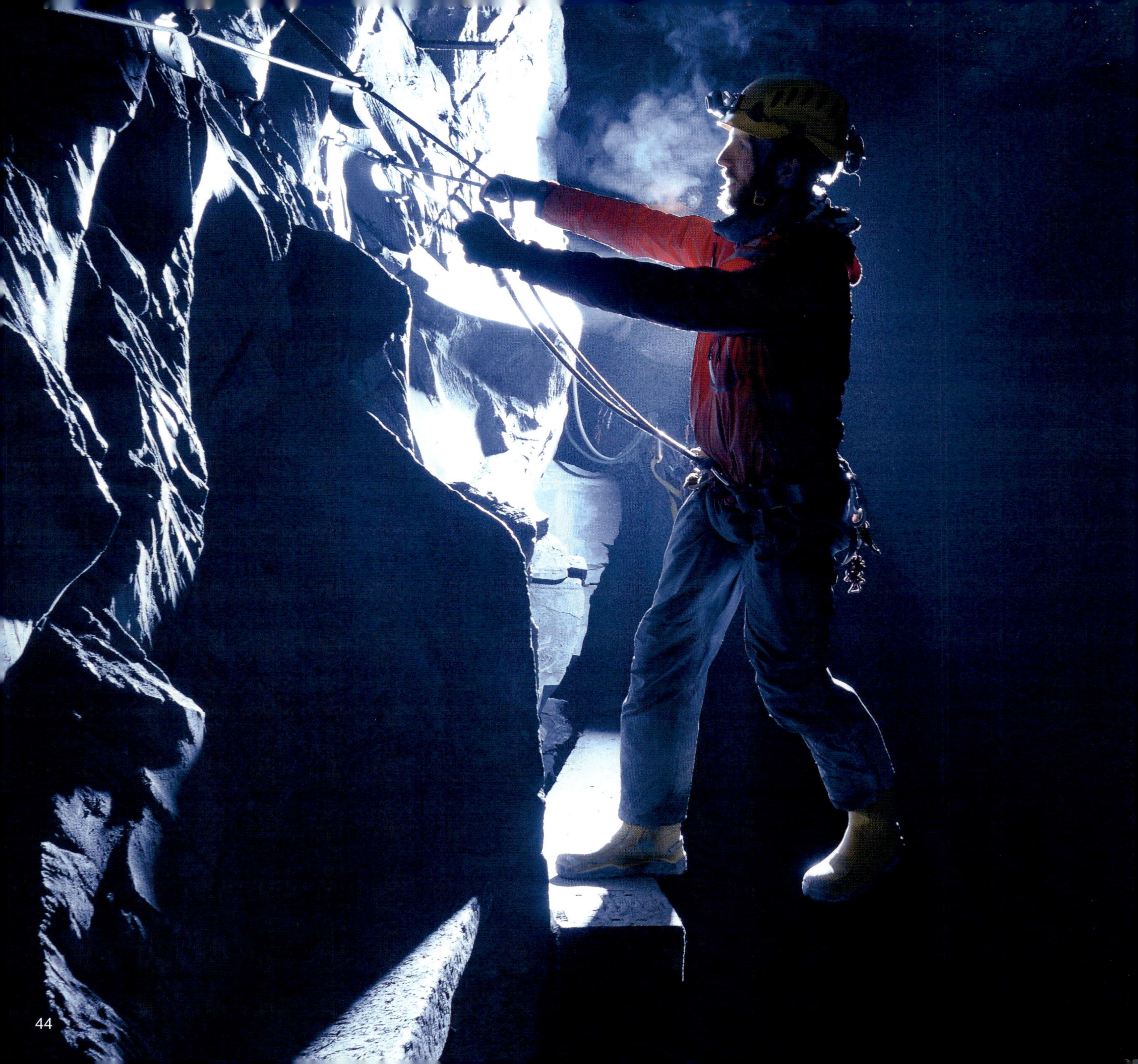

16

Go Below Underground Adventures

UK

Balancing with trembling knees on a slippery, narrow plank fastened with steel pegs to the vertical slate cavern wall, noses an inch from the rock and an infinite black void above and below, a small group shuffles tentatively sideways, their traverse is accompanied by the incessant drip of water on cold slate and the sound of it flowing through unseen tunnels and down waterfalls. There's a damp chill to the air and the smell of wet rock.

The group is exploring the dark, damp tunnels and caverns of North Wales's Cwmorthin Quarry with expert guides. For several hours they climb, traverse and abseil in the darkness, using the ladders, zip wires and narrow, wobbly bridges to explore.

It's an authentic underground experience that is a homage to the ingenuity of the Welsh slate miner, who moved around its labyrinthine confines with tonnes of slate. There is rusting machinery, old cart tracks, chains and an underground cabin, whose walls are covered in graffiti and the faint shadows of newspapers that announced the sinking of German battleships Bismarck and Graf Spee in WW2. All the time, two guides narrate the human stories of the hard lives underground.

Go Below was the brainchild of Miles Moulding. His childhood obsession was adventuring through underground mines in the honeycombed landscape of North Wales. He was discouraged from pursuing a mining career by his grandfather, whose father died in a mining accident, and became an IT specialist instead. He continued his underground passion as a hobby with like-minded mates before setting up the company in 2010.

Miles gravitated to Cwmorthin Quarry because it was 'degraded and forgotten', and he felt that its stories and the hidden world should be known again. Go Below made a deal with the local landowner, who also owns the underground workings. Self-funded, Miles and his team have spent years installing the infrastructure and have worked hard to create an authentic and naturally dramatic underground experience. All of Go Below's freelance trip leaders have qualifications from the Local Cave and Mine Leadership Awards scheme run by the British Caving Association. The company also has a direct relationship with the UK government's Mines Inspectorate.

Go Below offers underground adventures in two different mines and a small café which doubles as the HQ. In total, it employs 15 full-time staff and up to 60 freelance underground guides and is one of the largest employers in the valley.

With infectious passion, they have continued to innovate. The latest is 'Deep Sleep', which offers the chance to stay overnight in a giant cavern 419 metres below the surface. It's a steep and challenging descent through the old workings to reach your accommodation, which comprises four cabins and a more luxurious grotto. Breakfast is served in a larger, communal area festooned with fairy lights. Micro-hydro-powered batteries provide electricity, and the otherwise chilly 10°C air is warmed by the residual heat from the computers in each shed. These PCs are data crunching for the UK's National Health Service and medical science projects via a 1.5-kilometre-long cable running to the top of the mountain above.

Overleaf: A deep sleep is guaranteed in Go Below's Grotto Room – over 400 metres below Snowdonia's mountains.

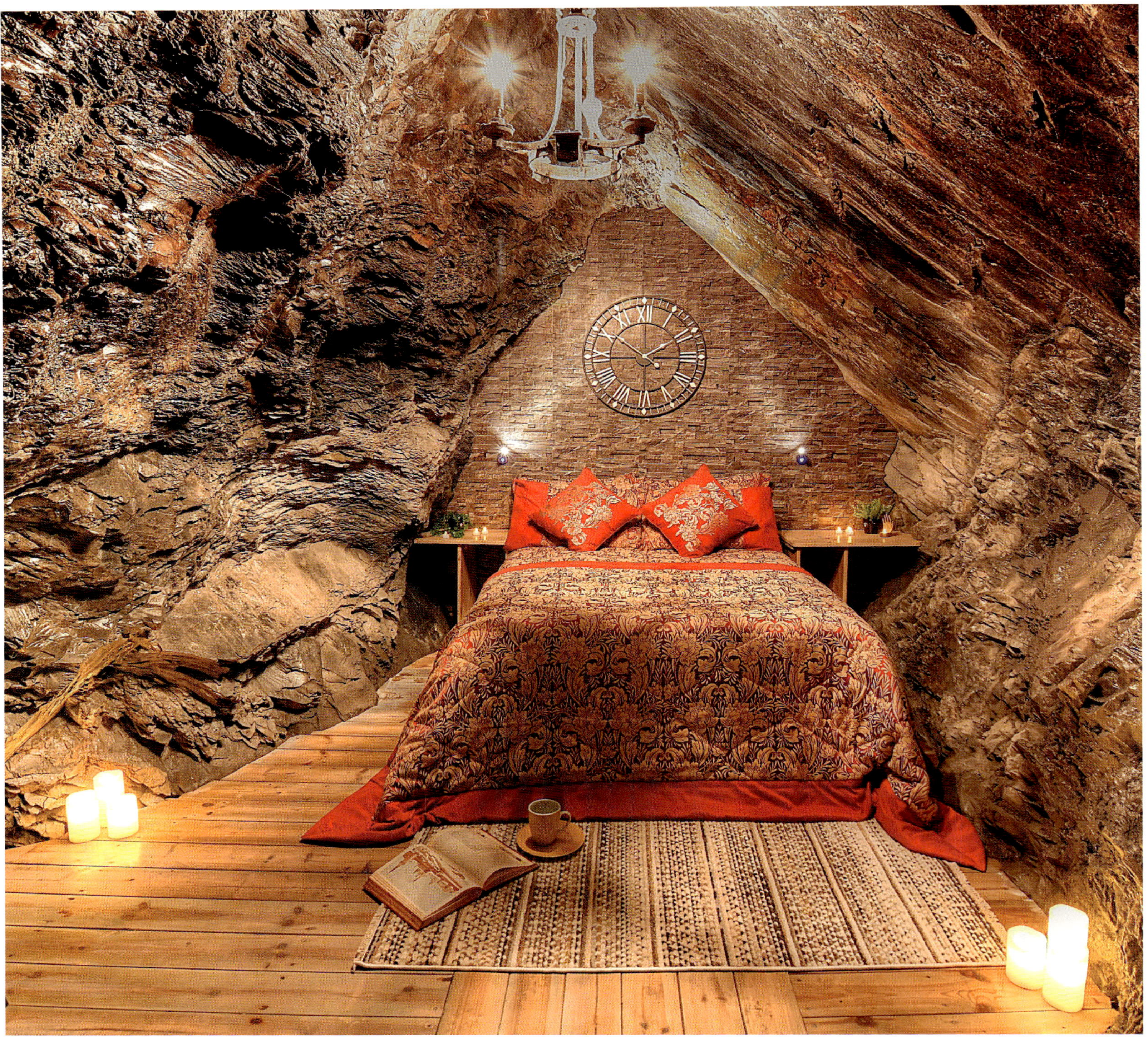

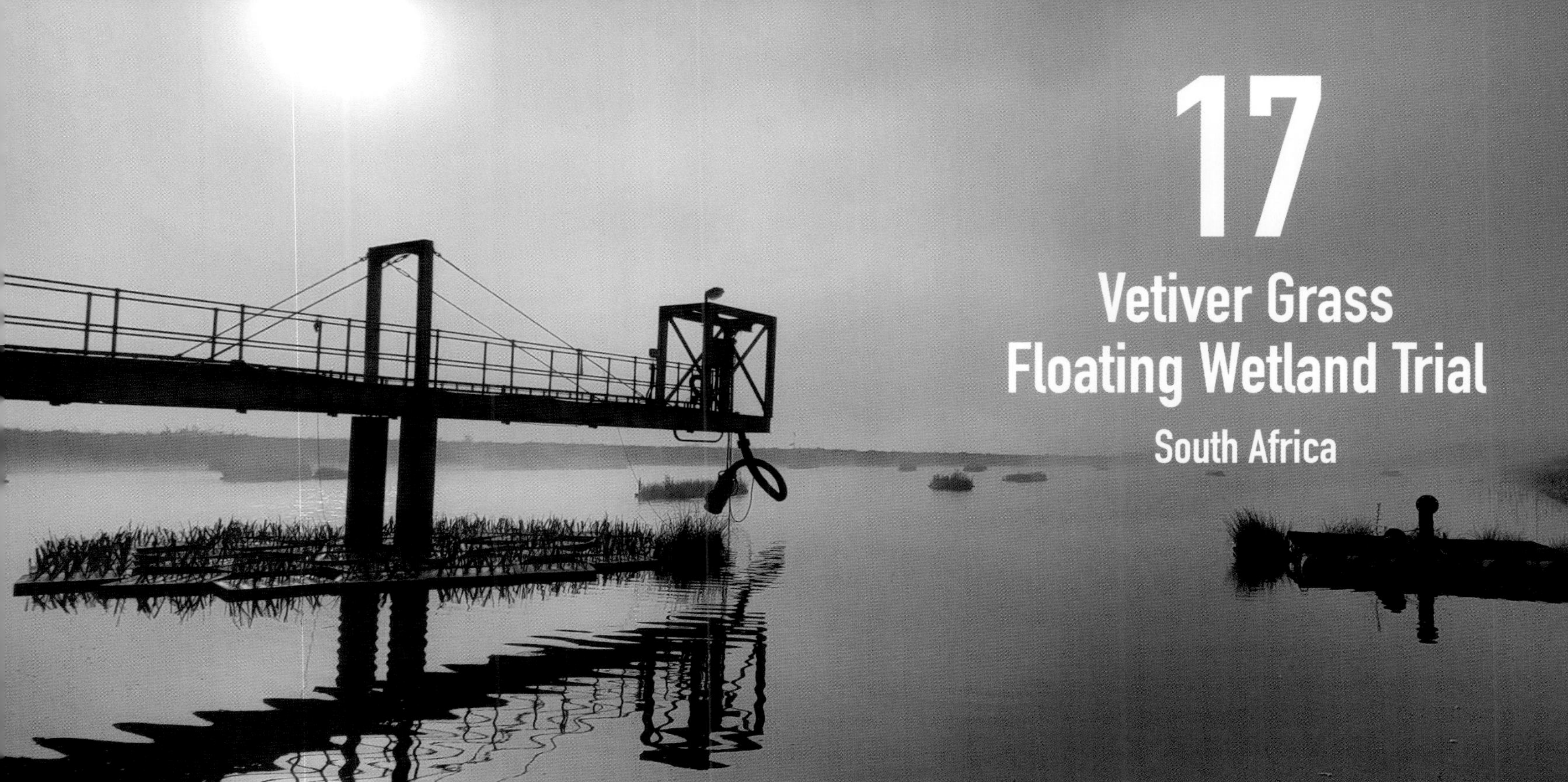

Vetiver grass is being trialled in Floating Treatment Wetlands (FTWs) at South Africa's former Voorspoed diamond mine in collaboration with the University of Pretoria. The mine closed in 2018. The site's stormwater ponds collect run-off, which is contaminated with nitrates from explosives residues, amongst other things. Originally the ponds were slated for rehabilitation, but Voorspoed saw potential in FTWs that could bring commercial benefits to the post-mining site. Although under certain circumstances it can be invasive, used in the right place, it is something of a 'super plant' that can protect and repair an environment by slowing soil erosion and decontaminating wastewater streams. It has many other potentially commercial applications too including animal fodder and even aromatherapy making vetiver a unique tool for mine site rehabilitation.

The Voorspoed vetiver floats in polystyrene rafts anchored against the wind. Noticeable ecological benefits include an increase in insect life attracting more bird life, while underwater, the plants' dense root niches have been colonised naturally by fish. Trials are underway to determine how effectively the roots purify the run-off. The FTWs will also decrease evaporation and reduce pit lake refill time, thus increasing water availability in this drought-affected region. It's early days, but the concept has been proven in principle.

The next step is to develop another FTW in Voorspoed's 250-metre-deep, flooded open pit. It is known as an 'ugly pit' because the crumbling geology and subsequently unstable slopes that have made it asymmetrical also make it difficult to safely access the water's surface. Innovative cable conveyors have been installed to place vetiver rafts on the surface and sample water. The faster the pit lake fills, the quicker the slopes will stabilise.

Establishing small, local vetiver production businesses offering supplies and services, such as raft manufacturing, plant harvesting and processing, is being considered.

18

Bounce Below & Underground Golf

UK

As well as still producing small quantities of high-quality slate, the subterranean caverns of Llechwedd quarry in Eryri, North Wales, offer two very different ways to enjoy the underworld.

In Bounce Below, opened by Zip World in 2014, six trampoline-style nets occupy a volume twice that of London's St Paul's Cathedral. You enter the giant spider's web with its three levels of trampolines 55, 20 and 6 metres above the cavern floor, connected by a huge slide, net tunnel and a spiral staircase, as users descend ever deeper into the bowels of the earth.

The site is also home to the world's first subterranean 18-hole adventure golf course. It is accessed by Europe's steepest cable railway, which was originally installed to take quarrymen underground to mine slate.

The bizarre course runs over four floors supported by an impressive steel structure. In a nod to the heritage of the place the 'crazy' golf course weaves its way through obstacles of quarrying machinery and artefacts, all theatrically lit in deep blues, reds and green. This underworld is an otherworldly experience! The Llechwedd site employs a diverse team, including Welsh speakers of all ages and experiences from this economically depressed area – some of whom are the fourth generation to work on this site (except now in rather different circumstances to their forebears).

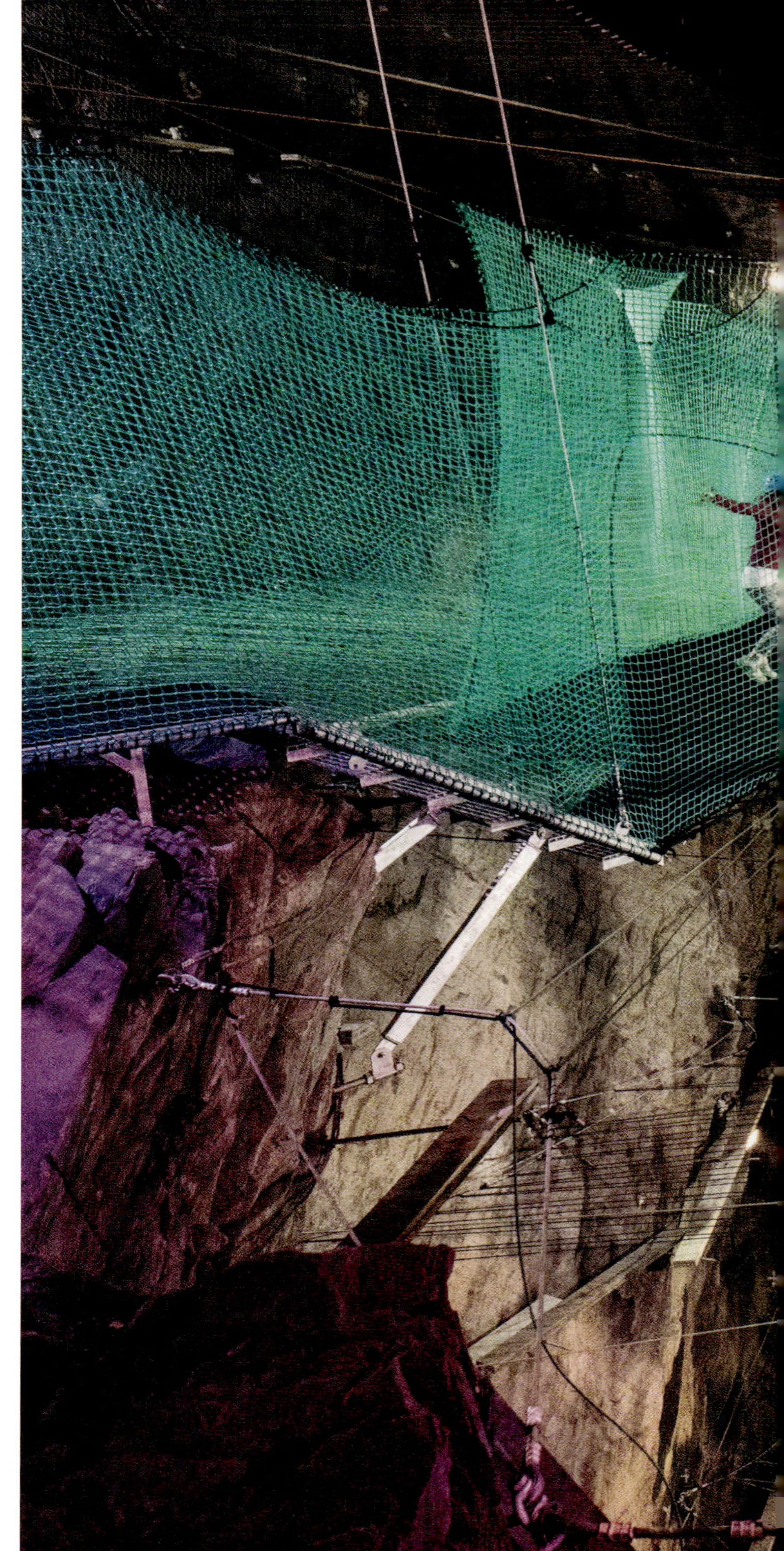

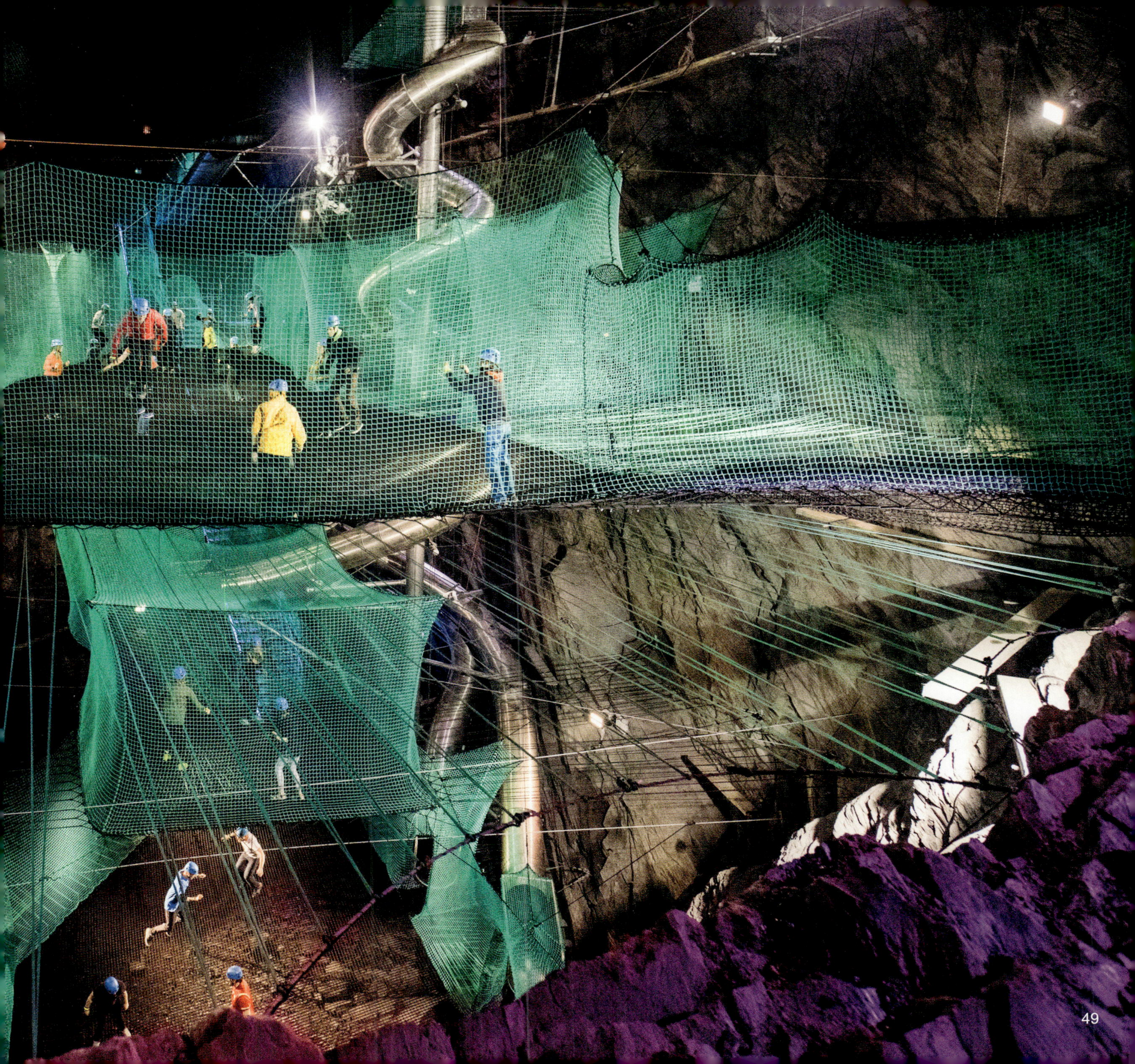

19
Tetraeder Sculpture
Germany

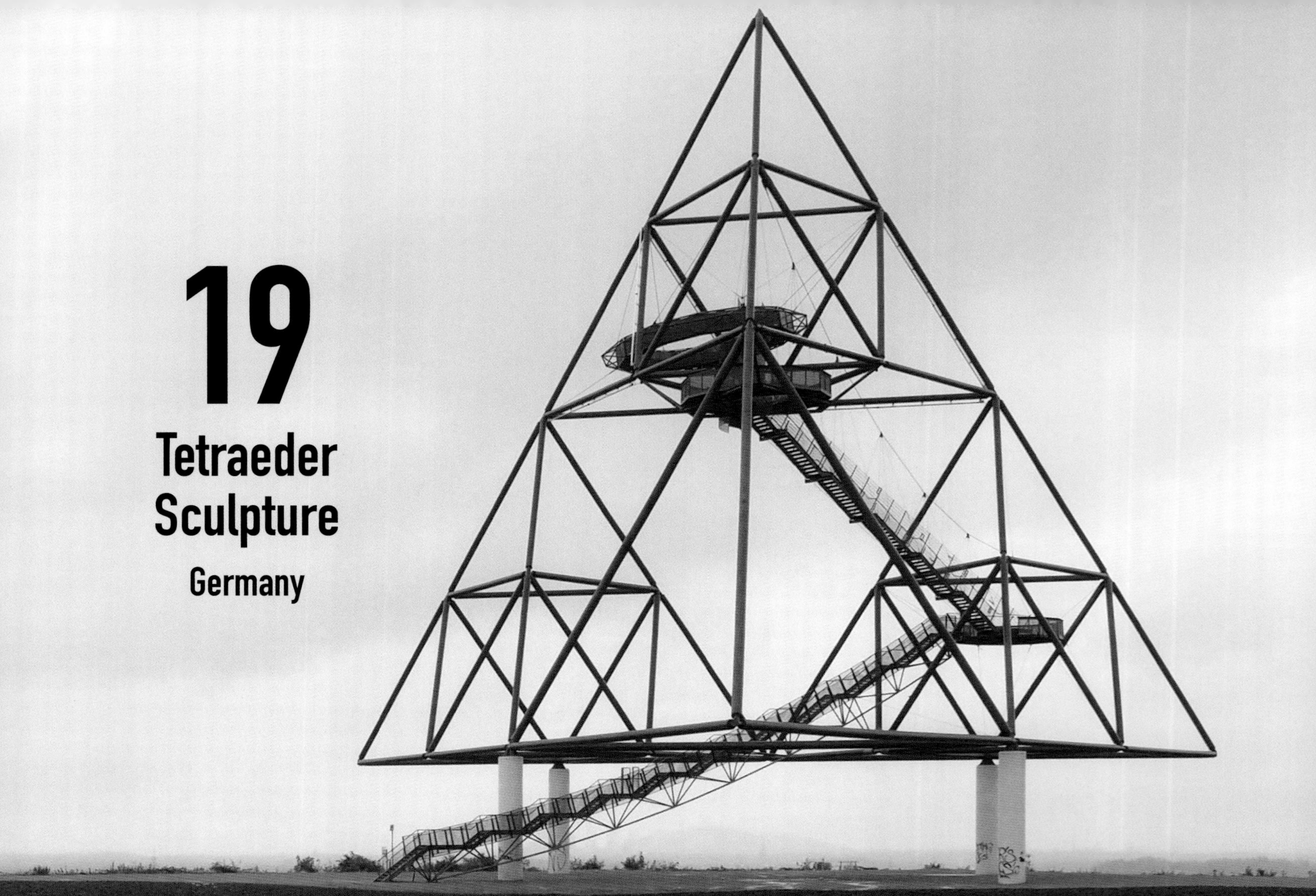

After a century and a half of operations in Germany's Ruhr region, hard coal mining at the Prosper II mine ceased in 1974 when the mine closed. It is honoured by the spoil heap, now a popular forested green space for the enjoyment of residents and former miners, and the otherworldly *Tetraeder* sculpture.

Standing 60 metres tall and weighing 210 tonnes, the three-sided *Tetraeder* (Tetrahedron), was designed by architect Wolfgang Christ, and resulted from collaboration between International Building Exhibition (IBA) Emscher Park, the Ruhr regional authority, the mining company and the city of Bottrop. It was opened on German Unity Day in 1995.

Tetraeder is a small piece in the jigsaw of an evolving new, post-industrial sense of place for the region and is one of many reclaimed waste heaps featuring interactive artworks. It offers unrestricted panoramic views across this part of the Ruhr, towards other industrial icons and city tower blocks rising above the surprisingly verdant conurbation.

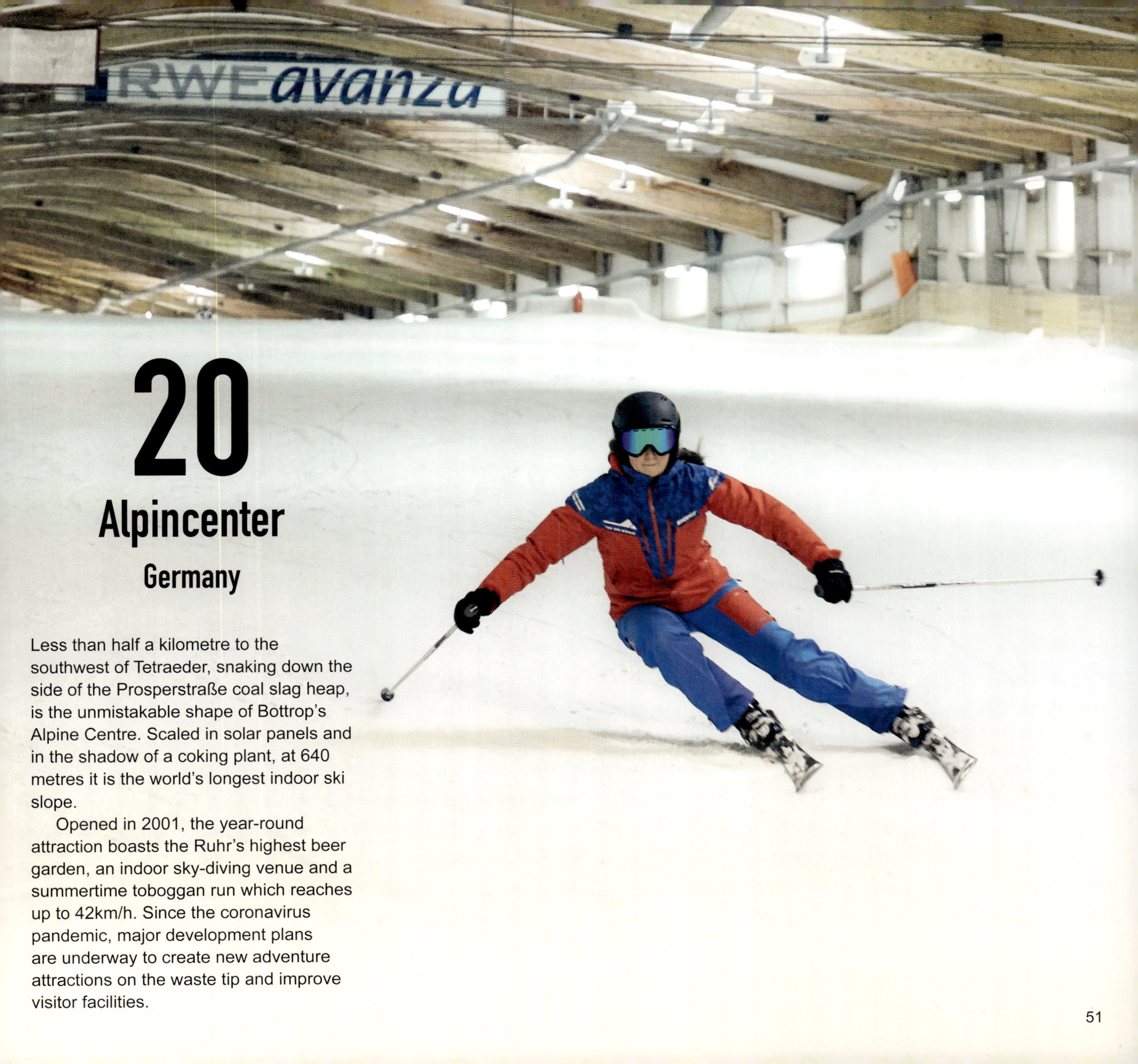

20

Alpincenter

Germany

Less than half a kilometre to the southwest of Tetraeder, snaking down the side of the Prosperstraße coal slag heap, is the unmistakable shape of Bottrop's Alpine Centre. Scaled in solar panels and in the shadow of a coking plant, at 640 metres it is the world's longest indoor ski slope.

Opened in 2001, the year-round attraction boasts the Ruhr's highest beer garden, an indoor sky-diving venue and a summertime toboggan run which reaches up to 42km/h. Since the coronavirus pandemic, major development plans are underway to create new adventure attractions on the waste tip and improve visitor facilities.

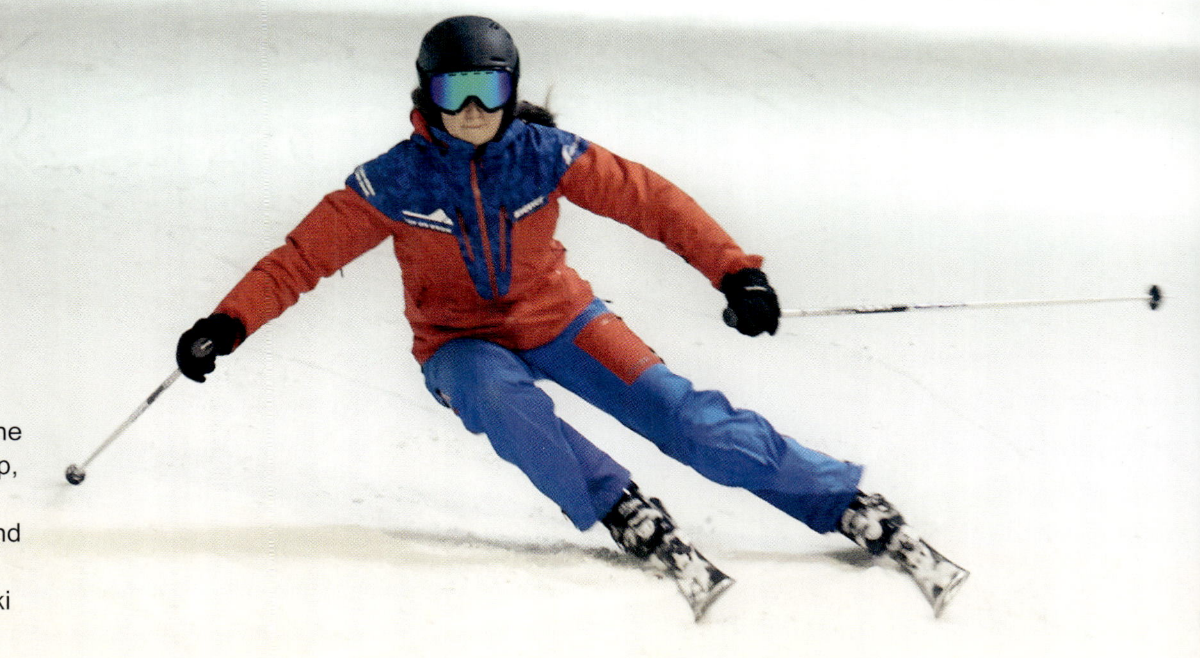

21

Gold Mine Tailings Reprocessing

South Africa

Gold was discovered in South Africa's Witwatersrand in 1886. The harsh, angular topography and dust of its waste heaps were produced throughout Johannesburg's gold mining heyday, at a time when environmental and public health controls were lax at best. Some were used for motorbike racing, shooting ranges and even a drive-in cinema, while thousands of homes were built close by. They continue to contaminate both air and water. Even after decades of remediation they are still visible from the air, often partly vegetated and increasingly integrated into the expanding conurbation.

There is still gold in those hills: old mine wastes can be reprocessed using new, more efficient techniques to yield more ore. One of the world's biggest reprocessing projects is DRDGOLD's work on hundreds of millions of tonnes of historic gold tailings that disfigure the landscape of Witwatersrand.

DRDGOLD's business model is based solely on reprocessing these dumps. Hydraulic mining is followed by collecting and processing the slurry to recover gold. The company's operating footprint covers 1,000 square kilometres; in 2022 alone it produced nearly six tonnes of gold from 28 million tonnes of waste. The residue is then stored in vast engineered impoundments capable of holding hundreds of millions of tonnes of reprocessed tailings. These new hills are designed, constructed and managed to the highest standards, and they need to be. Their sides are clad with waste rock and topsoil before restoring them with native vegetation to reduce dust, mitigate against rainfall mobilising metals in the wastes below and improve local biodiversity. Ranks of green windbreaks and irrigation from recycled water help plant establishment. Any groundwater seepage is collected in boreholes and reused in the processing operation.

The original sites are decontaminated by removing the polluted soils, then profiled prior to handover and – hopefully – repurposed for development. The operation has created over 900 full-time jobs for local people, who also benefit from a sustained programme of social investment to empower, educate and provide alternative incomes – all funded by mine waste.

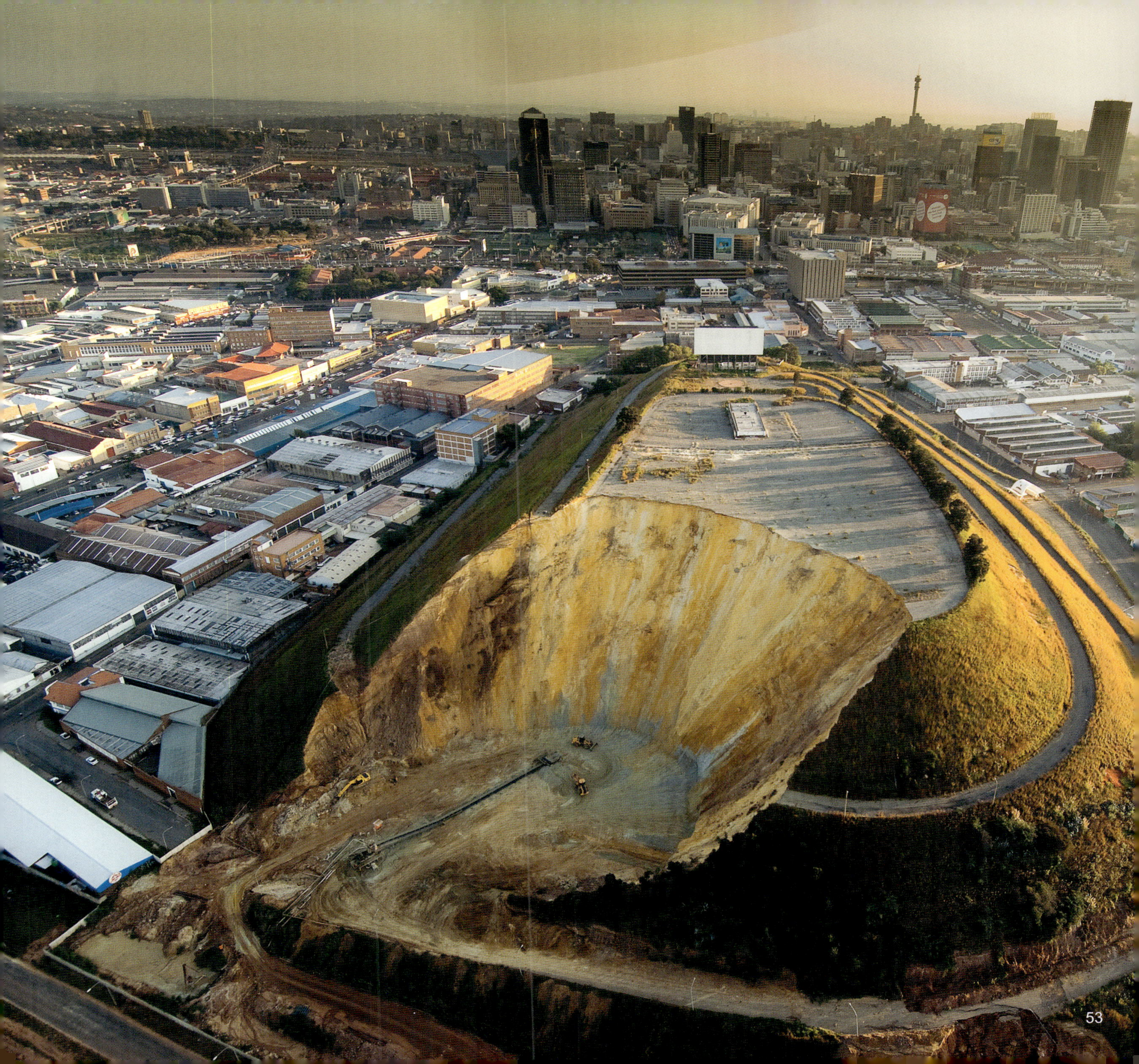

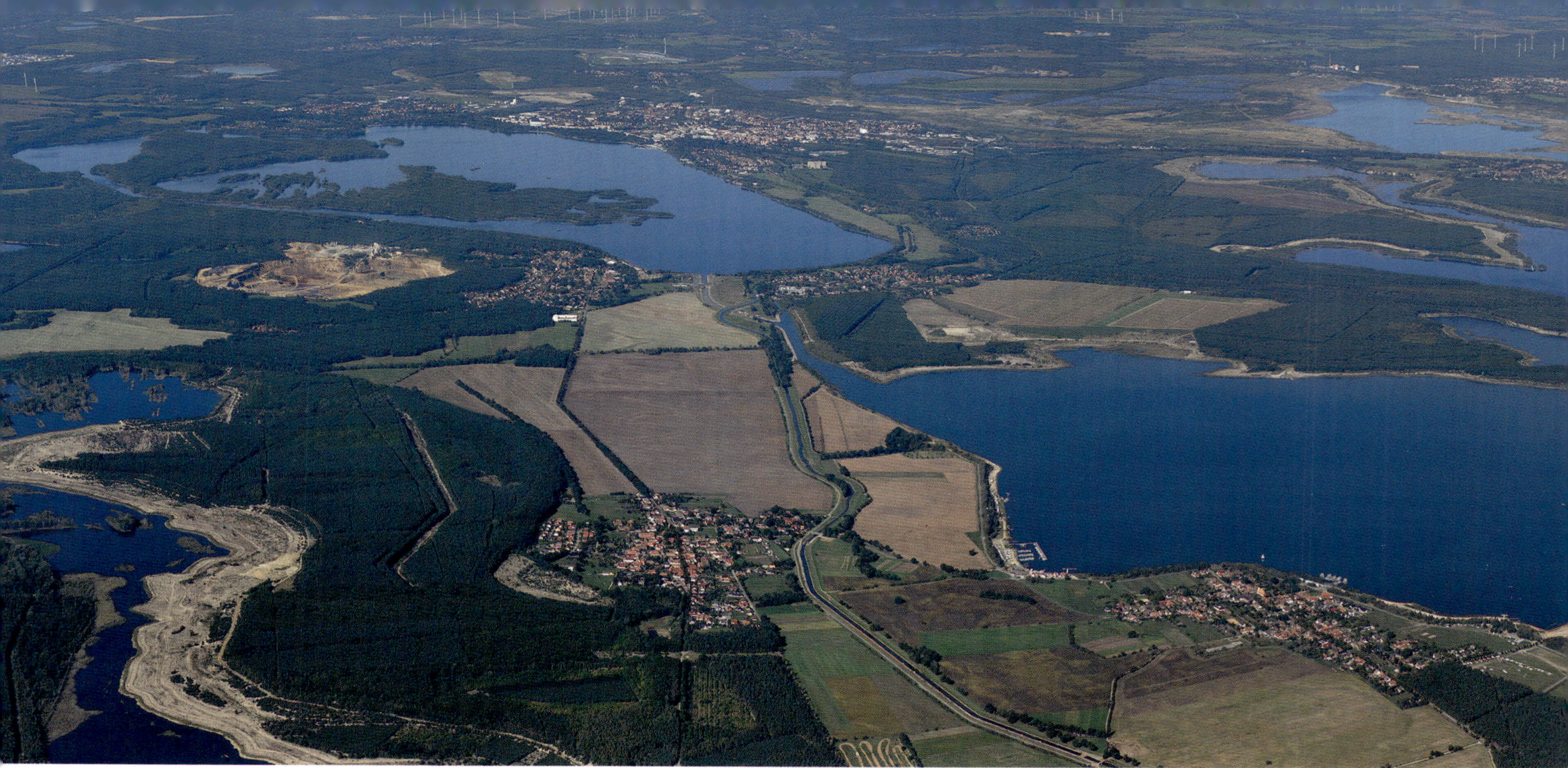

22
From Lignite to Lakelands

Germany

Tranquil forests, farmlands and lakes fill the windscreen for kilometre upon kilometre of empty east German country roads. Wind turbines rotate above the trees while, in the distance, coal power-station cooling towers generate clouds. The road passes through a well-kept village centre of stone buildings and small shops before a backdrop of looming apartment blocks. But only a generation ago, communist-era lignite mining dominated everything in this region, literally cutting swathes through the landscape – a landscape that is now slowly healing.

On German reunification in 1990, restructuring was urgently required to close uneconomic operations and upgrade and privatise the remaining pits in accordance with the 1982 Federal Mining Act. East Germany's state-owned, 300-million-tonnes-per-year lignite mining industry closed almost overnight, causing a major socio-economic shock.

The unified German government inherited the enormous environmental and socio-economic liabilities for those pits marked for closure in the Lusatian and Central German Lignite Districts across the four states of Brandenburg, Saxony-Anhalt, Saxony and Thuringia. The staggering challenge was to reconstruct industry, environment and society in two coal regions across a cumulative area of well over 1,000 km². In the ten years from 1989, 31 surface

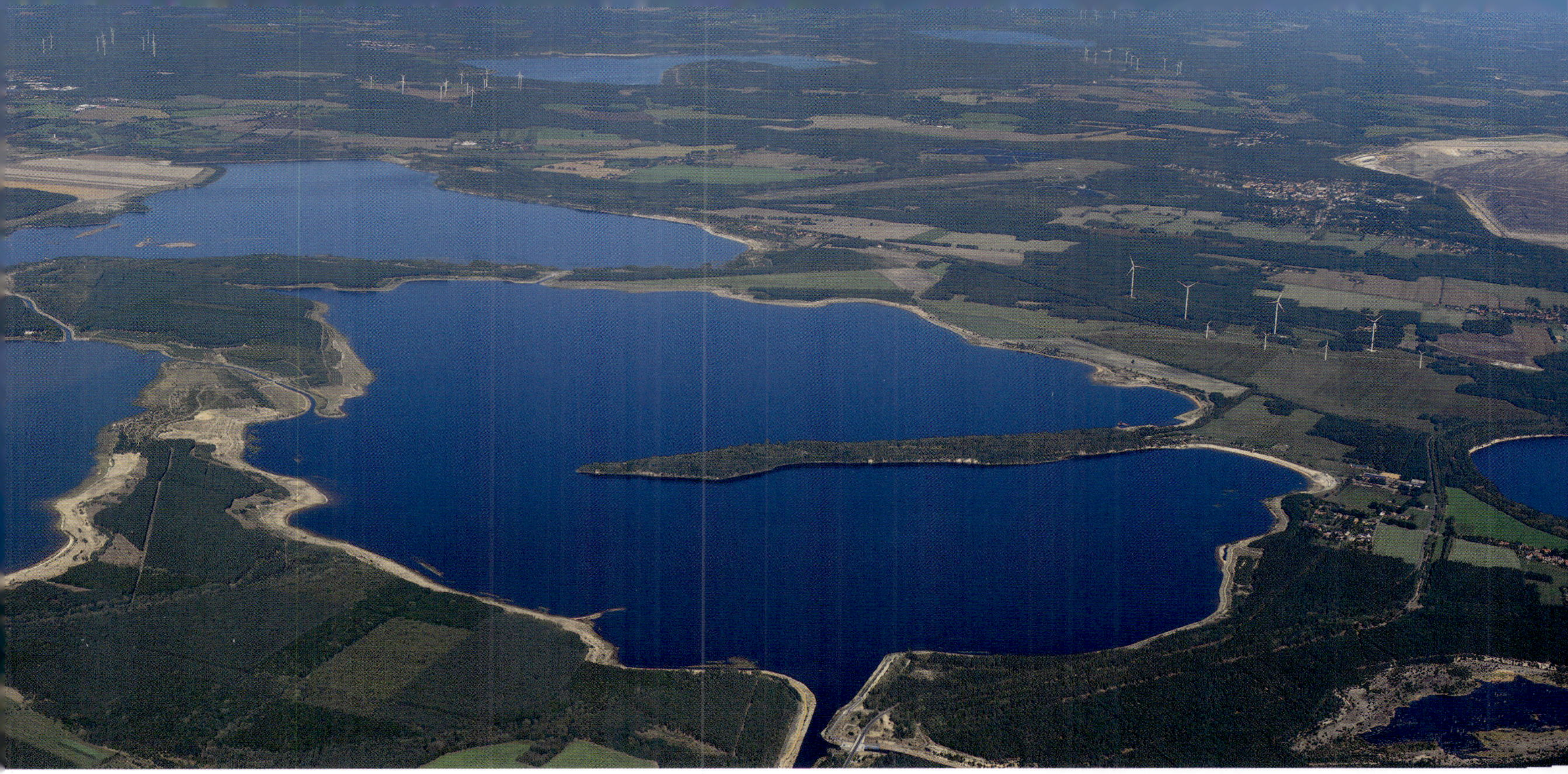

mining areas containing 207 surface voids and 43 lignite industrial complexes were decommissioned, excluding the enormous backlog of poorly rehabilitated areas under the communist East German regime, which had reserved no funds for their reclamation. The total area requiring rehabilitation covered 120,000 hectares (excluding more recent areas from the active mining that continued). The work remains ongoing and is complex, large-scale, long-term and costly.

Project planning and management are carried out by the Lusatian and Central Germany Mining Management Company (LMBV), which is wholly owned by the German government, overseen by the federal Steering and Budget Committee for Lignite Remediation (StuBA); 75% of the funding comes from federal sources, with the four states contributing the remainder. LMBV is responsible for creating a safe landscape for afteruse, including stable landforms and good water quality. It collaborates with four Regional Rehabilitation Advisory Councils that represent the interests of the affected regions. The afteruse projects themselves are implemented by private sector companies under typical commercial arrangements.

But why lakes? There are few pragmatic alternatives to lakes as a solution to the vast, unstable voids remaining after extraction. The voids are filled mainly with river water because relying on groundwater would take much longer and produce poorer water quality. The lakes are linked by canals and rivers to allow flow through the system, creating the world's largest, interconnected artificial lake landscape. This helps address water quality, but in some lakes, managing acidity and dissolved iron and sulphate levels is challenging. 'Barbara' and 'Klara' are specially built boats that apply lime to neutralise acid lakes.

Above: The flooded voids and reformed lands of the central part of the Lusatian former lignite district featuring new forests, farmlands and extensive solar and wind farms. Viewed north over Lake Geierswalder (middle foreground) towards Lake Grossrässchen (middle background).

>

About 1.7 billion cubic metres of earth have been moved to create new landforms surrounding the voids. As a lake grows, its lakeside land and slopes may become prone to liquefaction, causing subsidence and catastrophic slope failure. Before new land uses can begin, this ground needs to be stabilised, compacted and made safe using explosives, vibration or heavy rolling. To date, 1,200km of lakeside slopes have been stabilised, comprising 1.1 billion cubic metres of material. Rising groundwater can also destabilise overlying, manmade landforms. Access to these areas is restricted while this is being addressed, which can take years and allows such places to become wildlife havens.

Once stabilised, the new land is suitable for forestry, agriculture, commercial, industrial and/or residential development, wind and solar farms, nature conservation and a host of recreational and tourism facilities to maximise the use of the lands and their attendant lakes. Handing over the post-mining land and lakes is a difficult, time-consuming process. Some are given to the states, others to communities or to private owners; however, the federal government remains responsible for land stability and water quality. The stakeholders collaborate in planning to create different characters for each lake and community to avoid too much replication. Of the rehabilitated areas, 37,000 hectares of land have been reforested, 16,000 hectares have been turned to agriculture, and 21,000 hectares of water bodies and 50km of lakeside beaches have been created.

As tranquil new lake districts emerge from a dark, despoiled past, the cultural identities and perceptions of place among local people are also in flux. Creating the new physical landscape backdrop was one major challenge, but getting local people to see value and connect to that which they once shunned, or from which they were excluded, is a different challenge that cannot be solved by heavy engineering.

Based in the former Lusatian coal town of Grossrässchen, from 2000 to 2010, International Building Exhibition (IBA) Fürst-Pückler Land was a transformational, human-scale programme in the Lusatian Lakeland regeneration story. IBAs are a century-old German invention for exploring creativity and innovation in urban and regional development. Lasting ten years, they focus on a specific place and connect local regeneration planning and development with leading-practice community consultation and creative approaches, where solutions to dealing with the nitty-gritty details of enhancing people's lives are developed. With sensitivity and creativity, the IBA Fürst-Pückler Land explored new land and water use concepts while reconnecting the people of former mining communities with their healing, post-mining landscapes and broadening their horizons.

Increasing public access was a major issue for local people; for decades as the mines expanded, once-close communities were disconnected. The IBA created 30 demonstration projects focusing on changing perceptions of place and stimulating economic,

cultural and ecological revival. The IBA also provided a range of supporting services to affected communities, such as advocacy, marketing, business development and training, giving creative, ecological and economic impetus to the regeneration work.

Between 1993 and 2021, some EUR 12 billion was spent, with more billions due in the coming decades in the lead-up to and after the end of German coal mining in 2038. This commitment does not go unquestioned by other parts of Germany, particularly over how to justify spending such vast sums of public money in the east where there are relatively few people for such a small rate of return. However, the achievements of the programme are astonishing. Renewable energy and tourism industries have been created almost from scratch, exploiting the new landscape assets. Many former mining villages that turned their backs on their mines are now turning to face the water. A new cultural identity is slowly taking shape in a region once defined by coal mining and power generation, which will take a generation or more to fully form. Germany's experience presents a world-class paradigm of how to exit from large-scale coal mining and leave a positive – albeit as yet incomplete – legacy.

Opposite: Sailing on Lake Markkleeberger in the Lusatian Lake District. Watersports are a key part of the new economy.

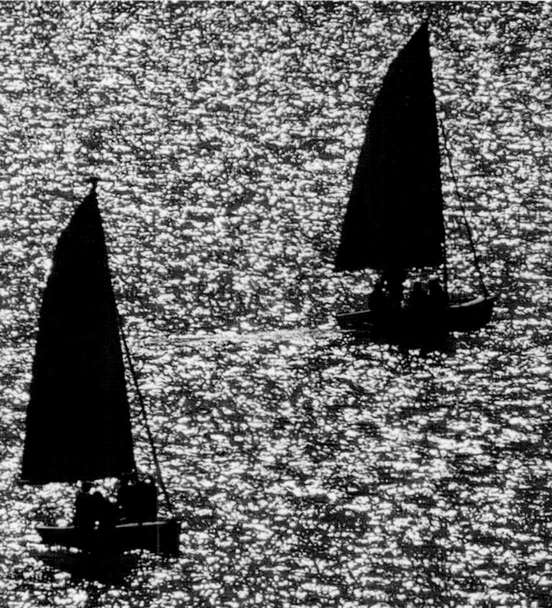

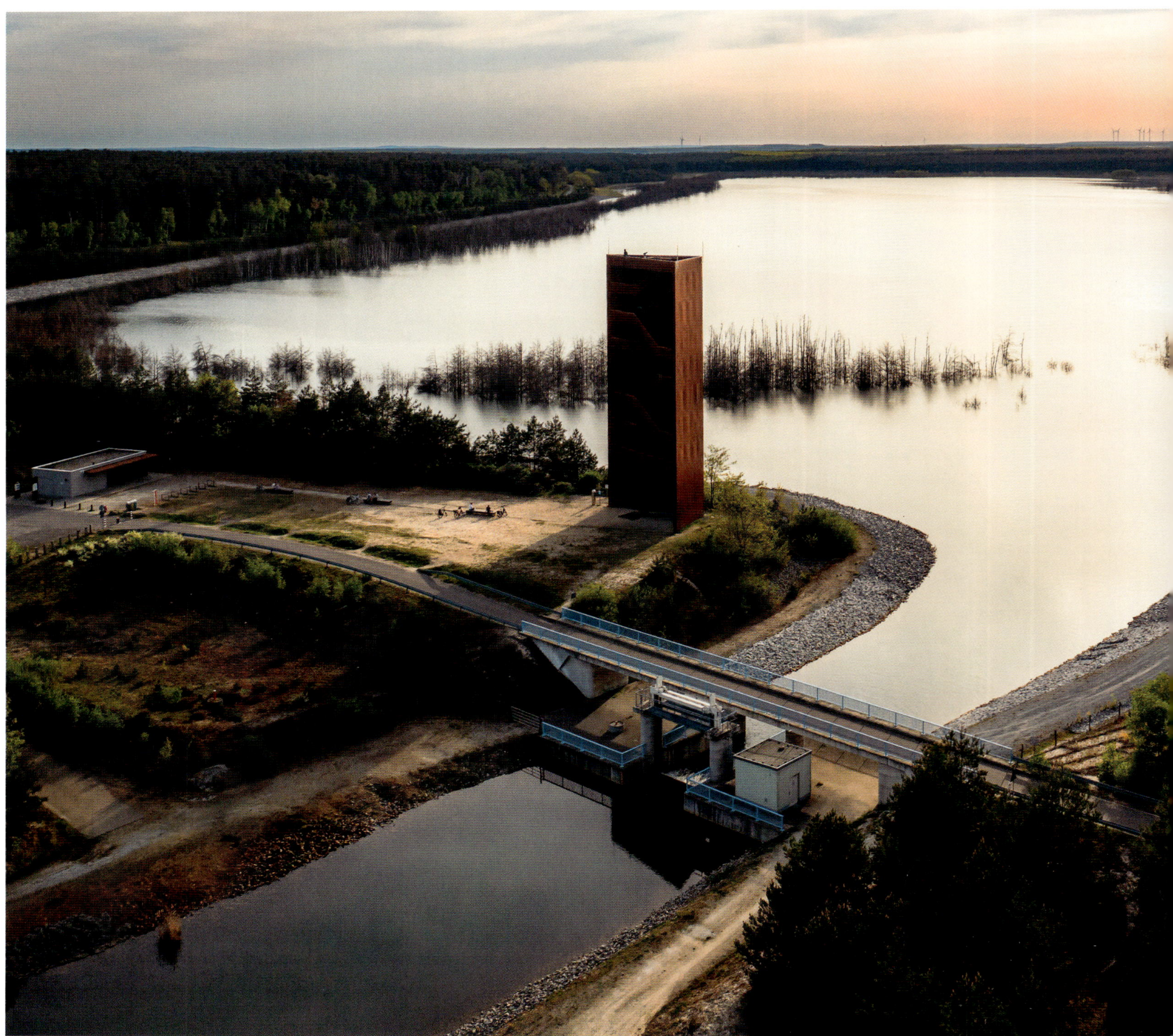

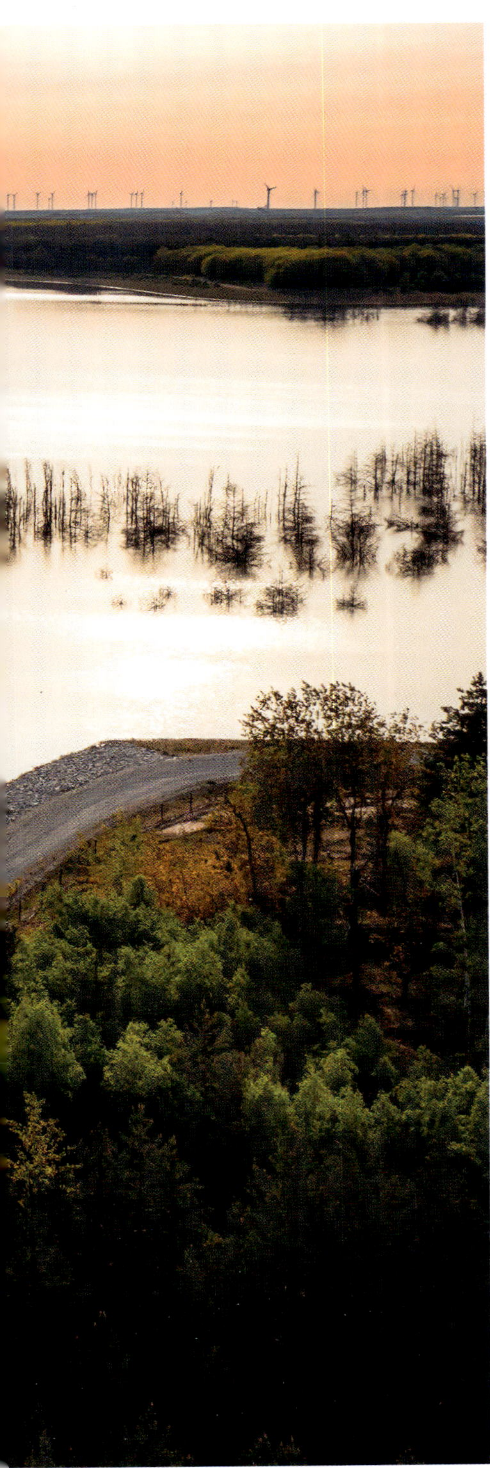

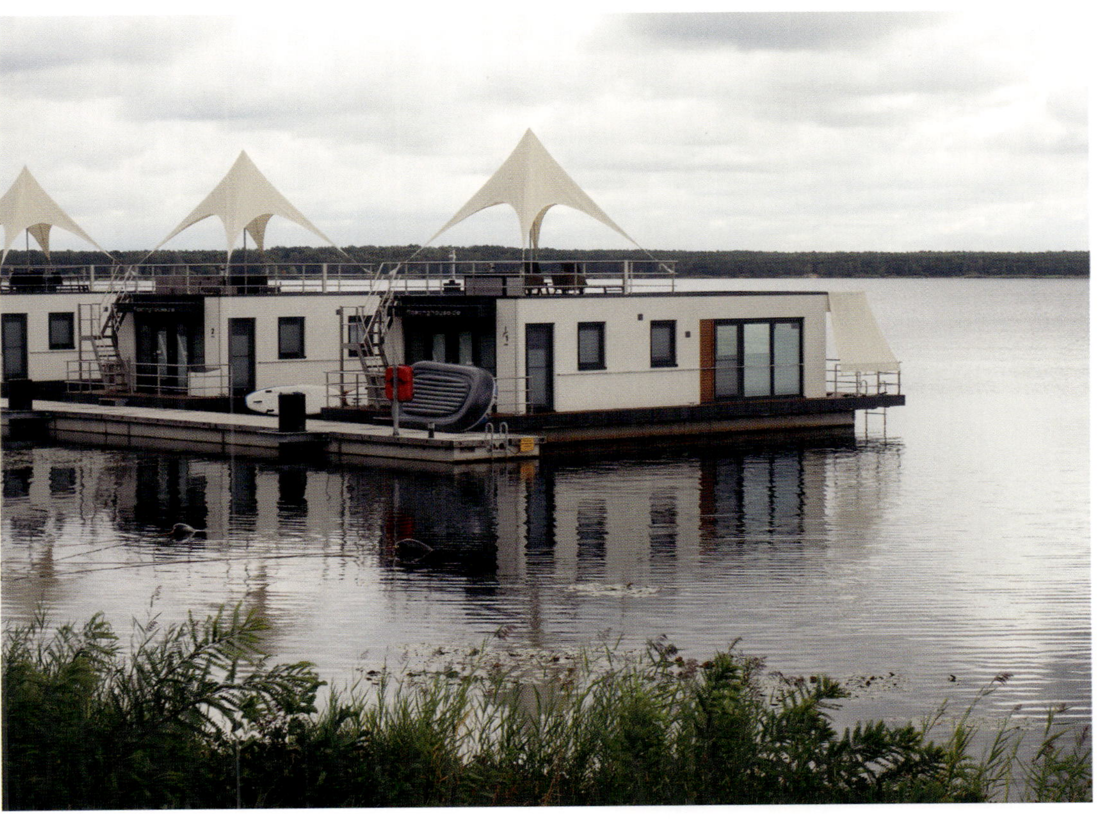

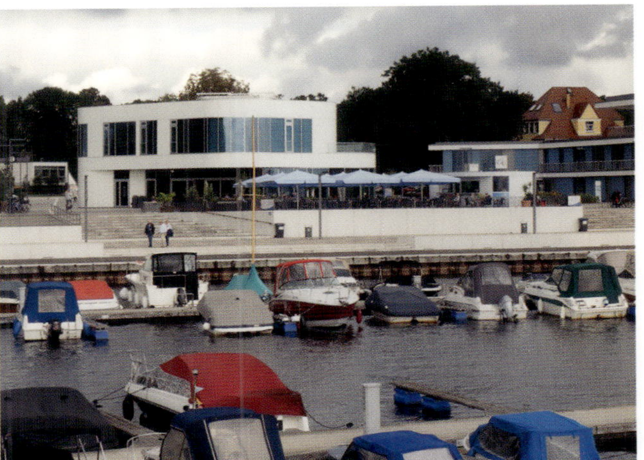

Above: Tourism is growing in both the central and east German lake districts. Floating holiday homes, such as these on Lake Gräbendorfer in the Lusatian Lake District, are proving popular.

Left: Senftenberg's marina and lakeside cafés and restaurants on Lake Senftenberg, in the Lusatian Lake District, are typical of the many waterside developments in the region.

Main image: The Rusty Nail landmark – a viewing platform at the intersection of several cycle routes where Lake Sedlitz, Lake Geierswalde and Lake Partwitz meet.

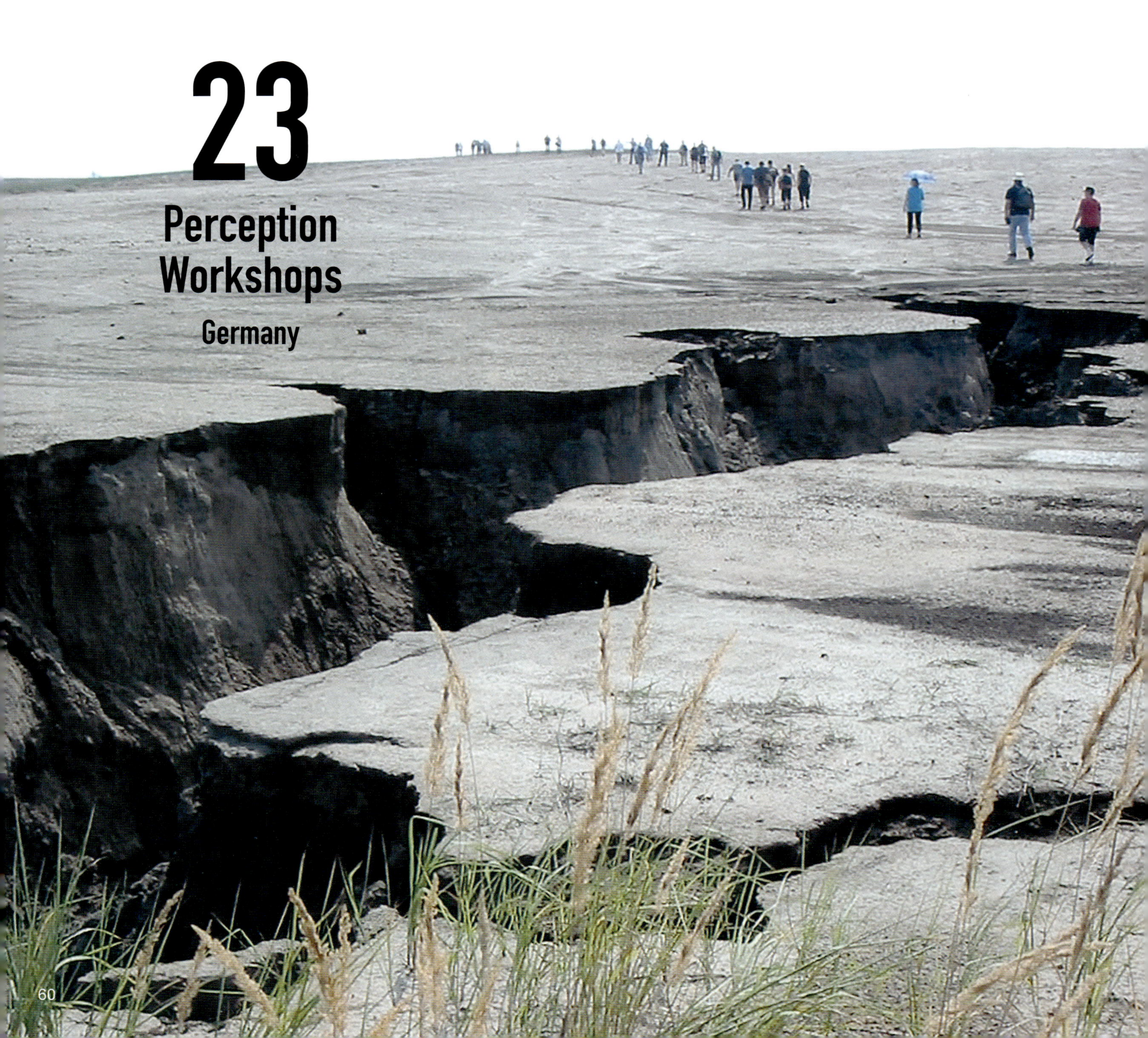
23
Perception
Workshops
Germany

Like many Lusatian lignite mining communities, the ancient town of Grossräschen has experienced trauma; half the population – over 4,500 people – were moved in the 1980s to allow expansion of the Meuro opencast lignite mine, devouring houses, shops, streets and public buildings. This, followed by the political uncertainties of German reunification and compounded by the socio-economic and cultural upheaval of large-scale mine closure, led to collective feelings of grief, distrust in institutions and loss of confidence. Even the devastated landscape on the edge of the remaining settlement had an end-of-the-world feel.

After closing in 1999, the post-mining landscape was stabilised, and since 2007 its expansive void has been filling with water to create Lake Grossräschen, due to be complete in 2025. But physical transformation does not cultivate a new society alone; there are many examples of ineffective social transition that were built on easy and lazy formulaic approaches, when what is required are more bespoke, creative and emotionally intelligent methods developed in response to the local context. Grossräschen provides a fascinating example of such approaches. For decades, local people were excluded from vast expanses of mine land – an ever-changing, lunar landscape interrupting long-used routes between neighbouring settlements. The people turned their collective backs on it. Their perceptions needed fundamentally resetting to connect to the new landscapes from which they will weave future employment and social and recreational opportunities while nurturing an emerging new cultural identity.

In addition to initiating 30 projects in an international exchange of ideas and experiences to shape the future of the Lusatian Lakeland, a major strand of the 10-year International Building Exhibition (IBA) Fürst-Pückler-Land's work programme, based in Grossräschen, was to address the disconnect between local people and their wounded landscape and explore potential new futures. Taking Marcel Proust's famous quote, 'The only true voyage of discovery… would be not to visit strange lands but to possess other eyes' as inspiration, the IBA organised many events to foster new perceptions and experiences and build comprehension. These included popular tours into once-forbidden mining lands, plays and participative theatre, musical events and much more.

Two members of the IBA team, architectural sociologist Karsten Feucht and sculptor/architect Rainer Düvell, developed the immersive Wahrnehmungswerkstatt® (Perception Workshop) methodology for discovering the character and potential of a place by exploring the senses and reinterpreting that place while bringing people together. Their theatrical promenades through the barren, post-mining landscape were very popular. *In situ* workshops took former miners, former residents of lost villages and people with no connection to the region around the site to explore the same reality. Coloured by their individual backgrounds, they all perceived this reality differently. They discussed their disparate perceptions with each other, encouraging mutual understanding of the relevance of different perspectives of the same place. In the same geographic place but separated by time, a former miner found himself in Opencast Pit Number 244, while resettled villagers remembered their homes forever lost. Through all these activities, the people contributed to shaping the future of their landscapes.

One memorable workshop, conceived by Swiss theatre director Jürg Montalta, painted a full-size map of the 'lost' village *in situ* in sand on the reformed land (before the lake). Former residents were invited to sit in the map at previously familiar locations and describe their thoughts, experiences, memories, hopes and fears to each other, rebuilding emotional connectivity between place and people and many fascinated visitors.

The creative finale to the IBA's Paradise 2 art project was an audio-visual spectacular directed by Montalta. IBA invited four thousand torch-bearing local people to form a 12-kilometre-long chain around the shoreline of the half-formed Lake Sedlitz. The lights flickered as their holders danced to the music of 500 drummers and trumpeters. The event celebrated the reconnecting of communities that had become separated by the mines and enabled people to see each other across the dark watery spaces while simultaneously ceremonially inaugurating the new lake district.

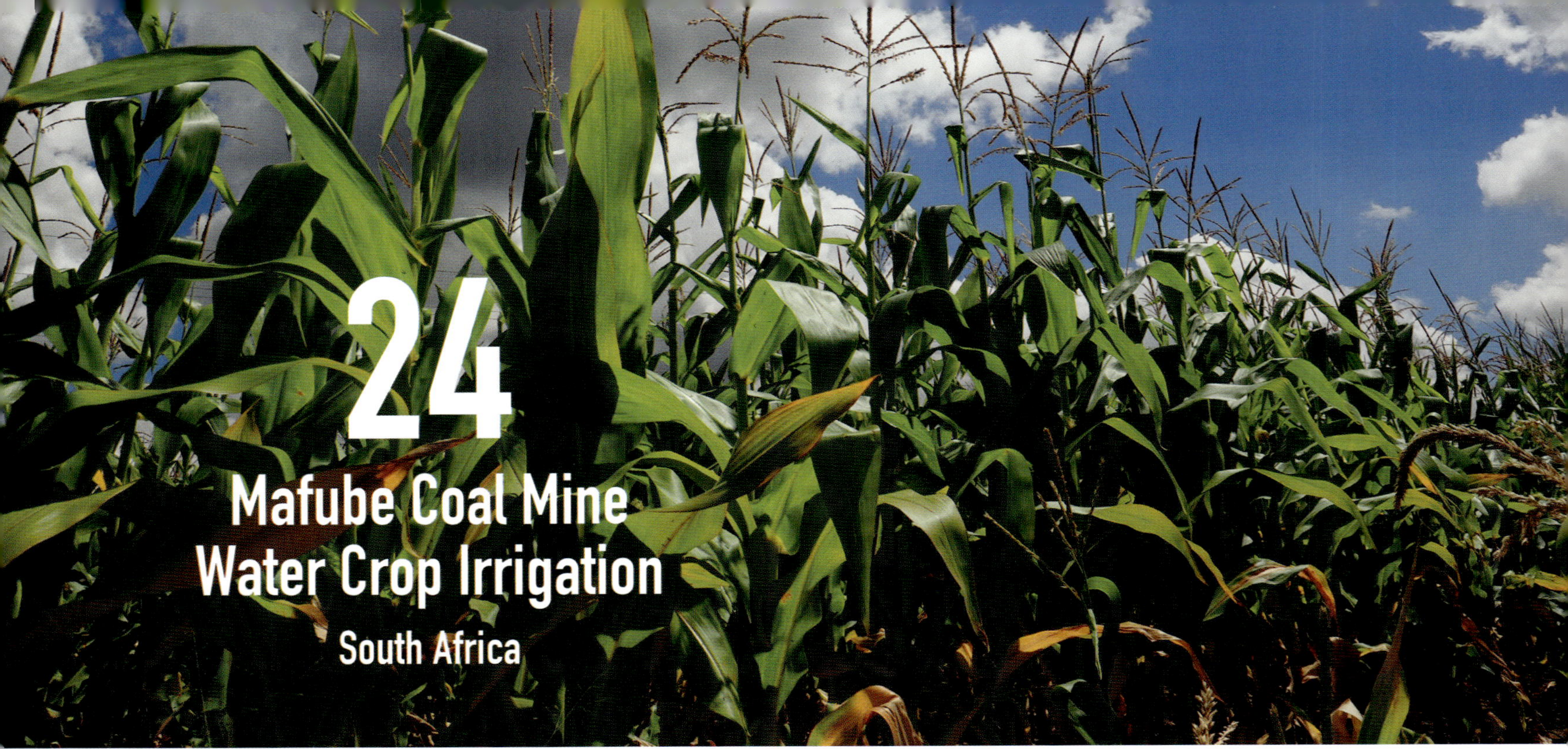

24

Mafube Coal Mine
Water Crop Irrigation
South Africa

Groundwater from coal, gold, and certain base metal mines is often too acidic or alkaline or contaminated to use untreated. If it reaches the surface after mining has ceased and the pumps have been turned off, it can damage water supplies and ecosystems for decades, even centuries. This is a significant problem in South Africa, where gold and coal mines have been closing on a large scale in recent decades, with more set to close in the future. Simultaneously, rising demands for water and severe droughts are affecting agricultural production.

A neat solution to both problems has been researched for years by the University of Pretoria, Thungela and others at the Mafube mine in the Mpumalanga coal region, where most of its mine water is alkaline and suitable for irrigating farmland without treatment. The main impurities, calcium and sulphate, together make gypsum. Many farmers already apply gypsum as a soil improver to their fields.

The government has taken an understandably cautious approach. In 2016, they agreed to a demonstration trial to test the opportunity. Tolerant crops, such as wheat, maize, soya and ryegrass, were considered, with wheat and soya being particularly suited to the conditions, but the farmers preferred to grow maize.

Two 30-hectare maize plots were set up at Mafube, one on rehabilitated mining land and the other on non-mining land as a control for comparison. Since then, the yields from mine-water irrigated plots have been consistently several tonnes per hectare more than for rain-fed plots, and the maize was proven to be safe for human consumption. If shown to be sustainable, this concept could become commonplace and not just for coal mines; closed South African gold mines also have similar water management challenges. Furthermore, crops can be grown even in the dry season leading to dramatic increases in production. On Mafube's rehabilitated mine lands, research on various grass crops, such as lucerne, continues.

Mafube's groundbreaking work is changing the national perception of mine water from a liability to an asset. Such thinking is applicable to other countries where water-scarce agriculture and suitable mine water quality intersect.

25
Jinyun Quarries
China

There are at least three thousand abandoned quarries in Jinyun County in Southeast China. 'Tuff' – porous rock formed from volcanic ash – had been mined in the area for building materials since the Tang Dynasty 1,300 years ago.

Most of the quarries are abandoned following decades of over-exploitation, but nine – all within walking distance of each other – have been selected by local government to be reimagined by a Beijing-based architecture firm as a tourist destination. Three of the nine have been developed so far, guided by a minimalist principle that aims to augment the quarries' existing features.

Quarry 8 is now an open-air library. Concrete stairs have been installed with enclosed bookshelves that rise to stone platforms where people can practise calligraphy or sit and read. Quarry 9 is a performance space consisting of a concrete stage and seating area, while its high porous walls provide excellent acoustics. Quarry 10 stays true to its roots as a demonstration quarry where visitors can watch stonemasons at work. Plans for the remaining quarries include a water garden featuring vast avenues of reflections, a teahouse and a further performance space in Moonlight Quarry.

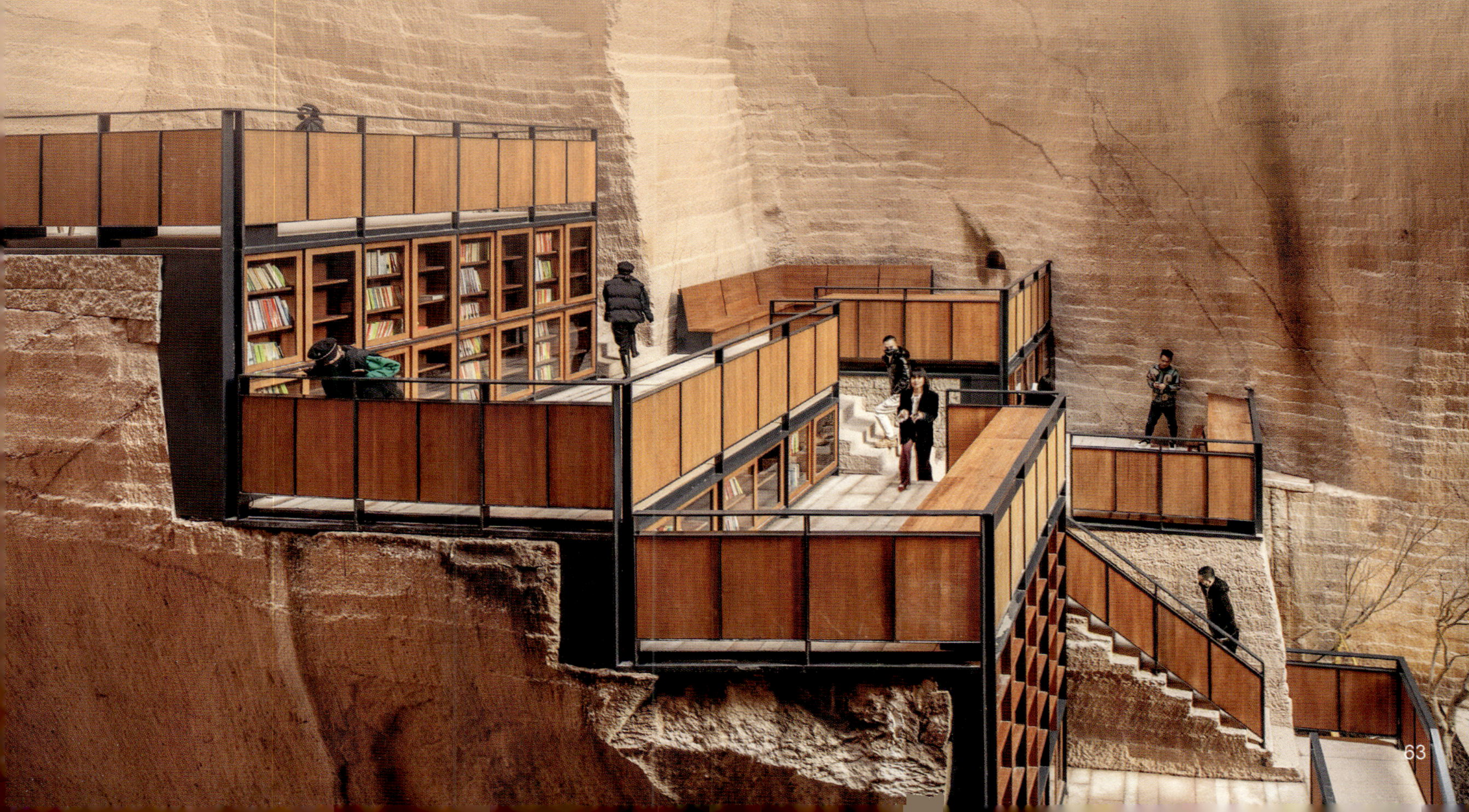

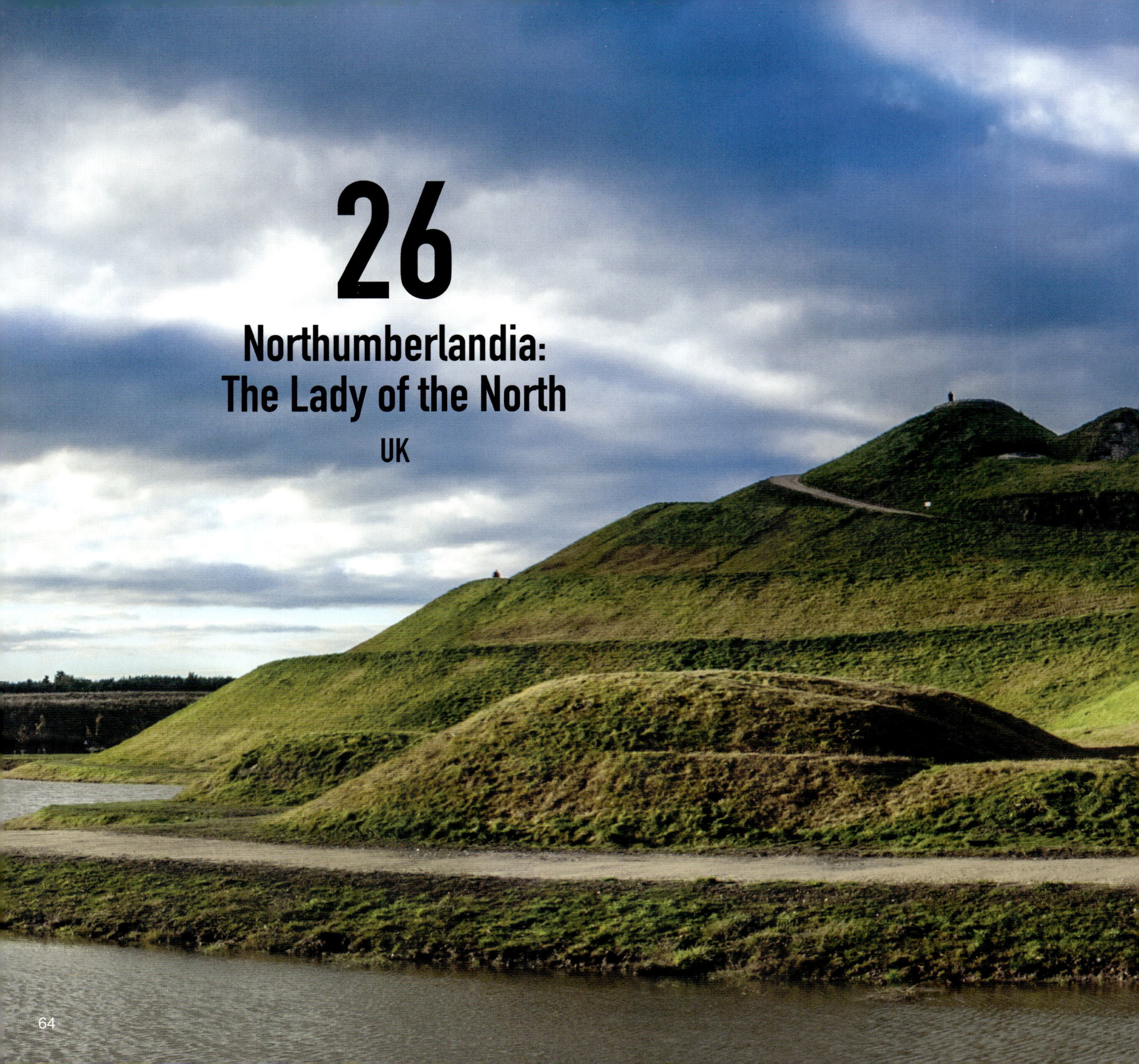
26
Northumberlandia:
The Lady of the North
UK

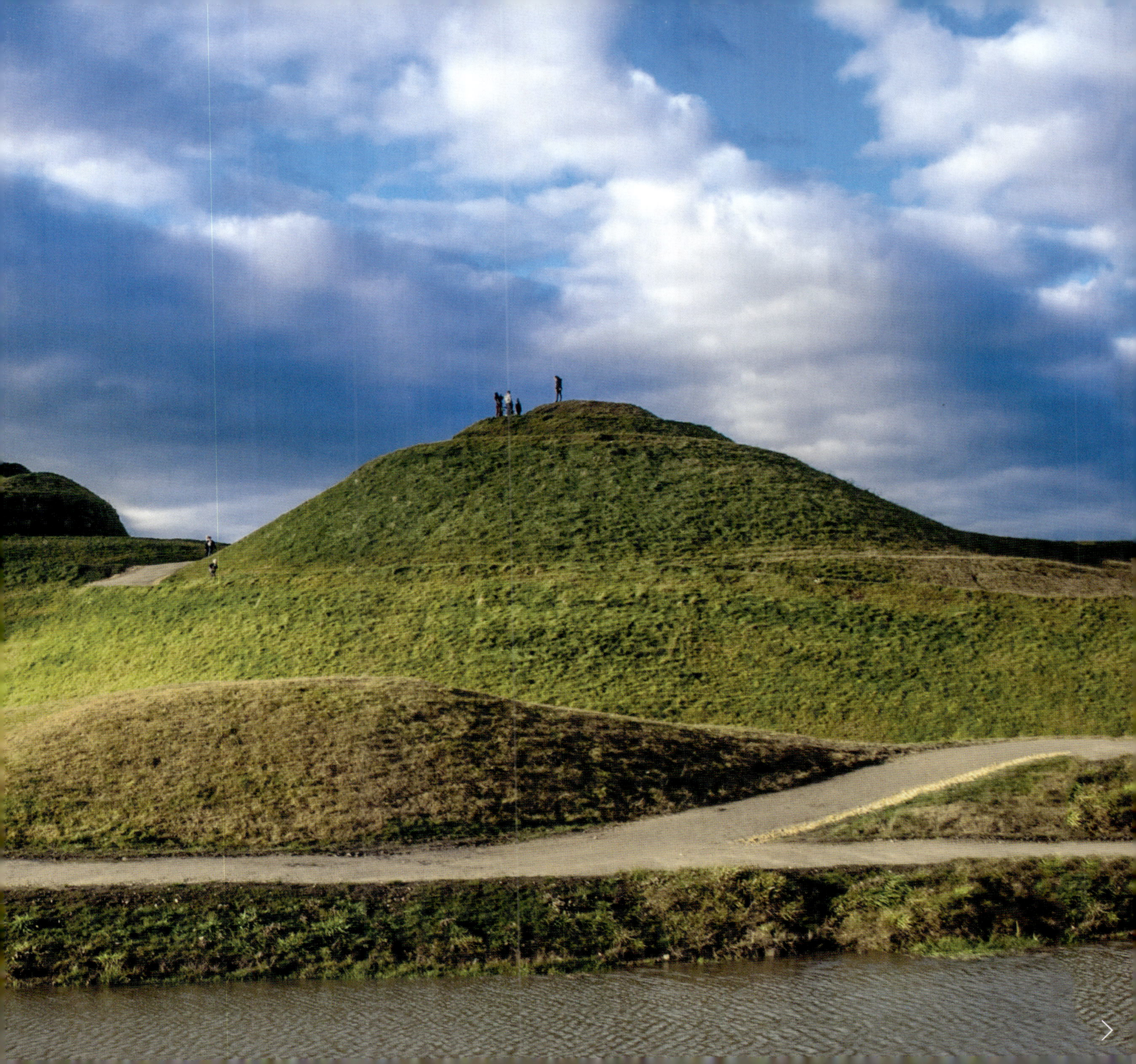

Her voluptuous curves are free for all to survey and, only upon wandering closer through the trees, does her full beauty become apparent, inviting you to explore further. Her name is Northumberlandia: the Lady of the North, and she is an enormous landform sculpted from a coal mine's overburden – a public asset created from unpromising beginnings.

She reclines within a 19-hectare community park in the English county of Northumberland. The park aims to be a tourism gateway, while boosting the local economy and providing a community amenity. Access is free and it is located close to an arterial road, so acting as a welcome rest-stop.

Northumberlandia's right hand points towards the former Shotton surface coal mine from where she was born. The initial idea for a land art response to the visual intrusion of a waste tip on low-lying topography arose from the mining company Banks Group. They aimed to leave a tangible benefit associated with a 'restoration first' approach to the development of the mine. The company worked closely with the landowner, Blagdon Estate, and the late Charles Jencks, the internationally renowned artist. Constructed from about 1.5 million tonnes of overburden,

Northumberlandia was formed while the mine was still operational. After two years of construction, it was opened in 2012 by HRH The Princess Royal.

Responsibility for managing a site's liability over the long-term is a common obstacle to many such projects. Banks insisted on the long-term management of the site being carried out by either the estate or a charity, rather than the local authority, and this became the agreed approach in the planning permission.

The Land Trust assumed the long-term ownership and management of Northumberlandia in 2012, funded by a planning condition payment by Banks. This was invested to provide long-term funding for the park's management and development in perpetuity; neither the company nor the council wanted this obligation without insurance.

The Land Trust is a UK national land management charity working with the public and private sectors to solve challenges for land in their ownership, leading to sustainably managed open spaces and green infrastructure for community benefits – usually from post-industrial, brownfield sites. It provides a secure exit for landowners from their land management responsibilities by taking on the maintenance, risks and

liabilities, such as planning conditions, public liability, health and safety and environmental risk. Such a model should be transferable to other jurisdictions.

Further funds sourced by the Land Trust have enabled the construction of a visitor centre and café, which generate additional income while supporting modest employment opportunities and space for social gathering and educational materials.

The Trust has developed a detailed model that calculates the Social Value (charitable outcomes) it delivers across its 85+ sites. It looks beyond the pure economics of a project by assessing the financial aspects of the social, cultural, health and educational values created while taking a long-term, pragmatic approach to understanding its real costs and benefits.

The park is well-used and much-loved by local people. Day-to-day site management is carried out by the Northumberland Wildlife Trust on behalf of the Land Trust, who have not only enhanced the site's biodiversity, but also deliver a wide range of community events and educational activities and run the popular café.

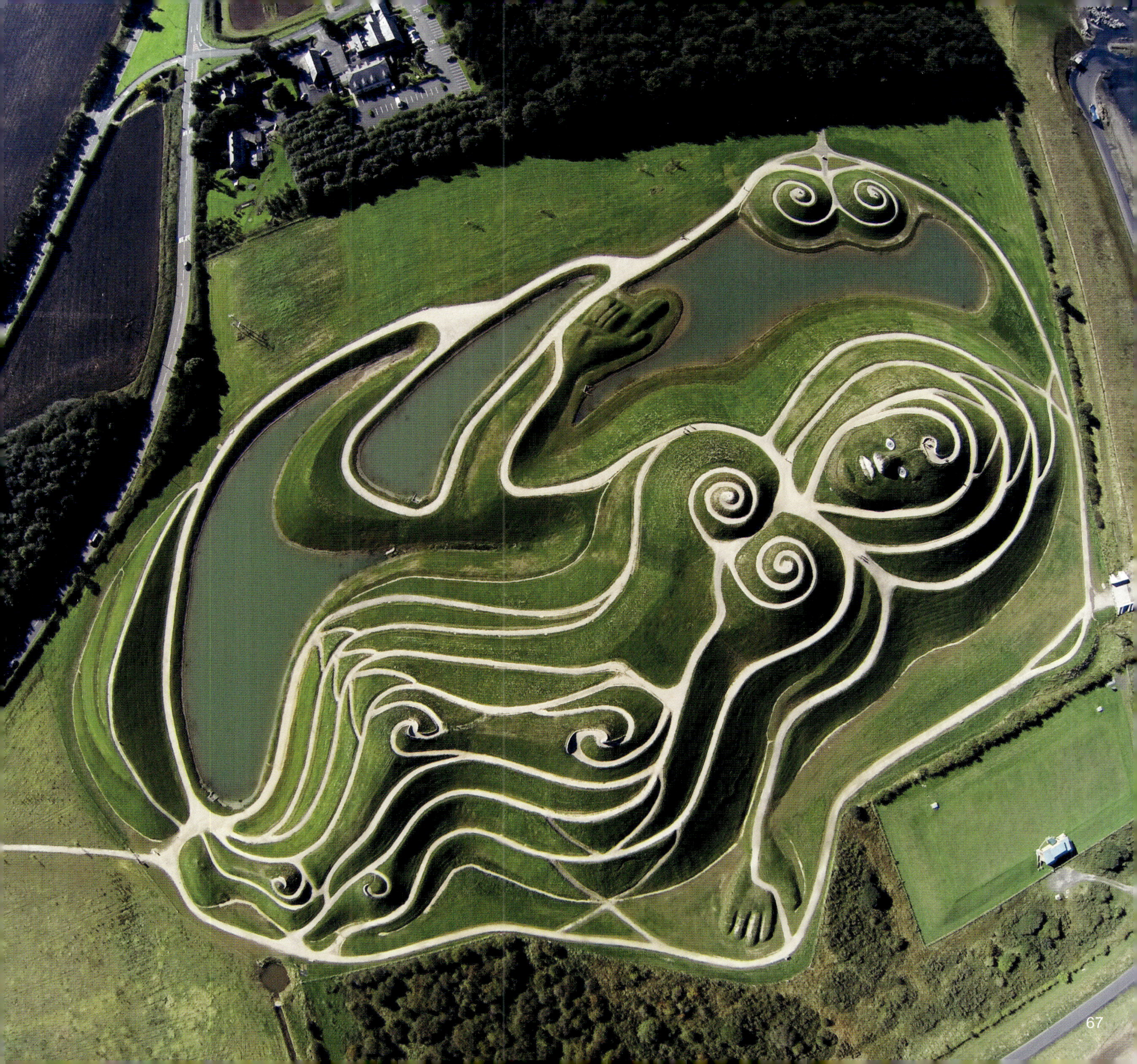

27
Kelian Lestari Sumatran Rhino Sanctuary

Indonesia

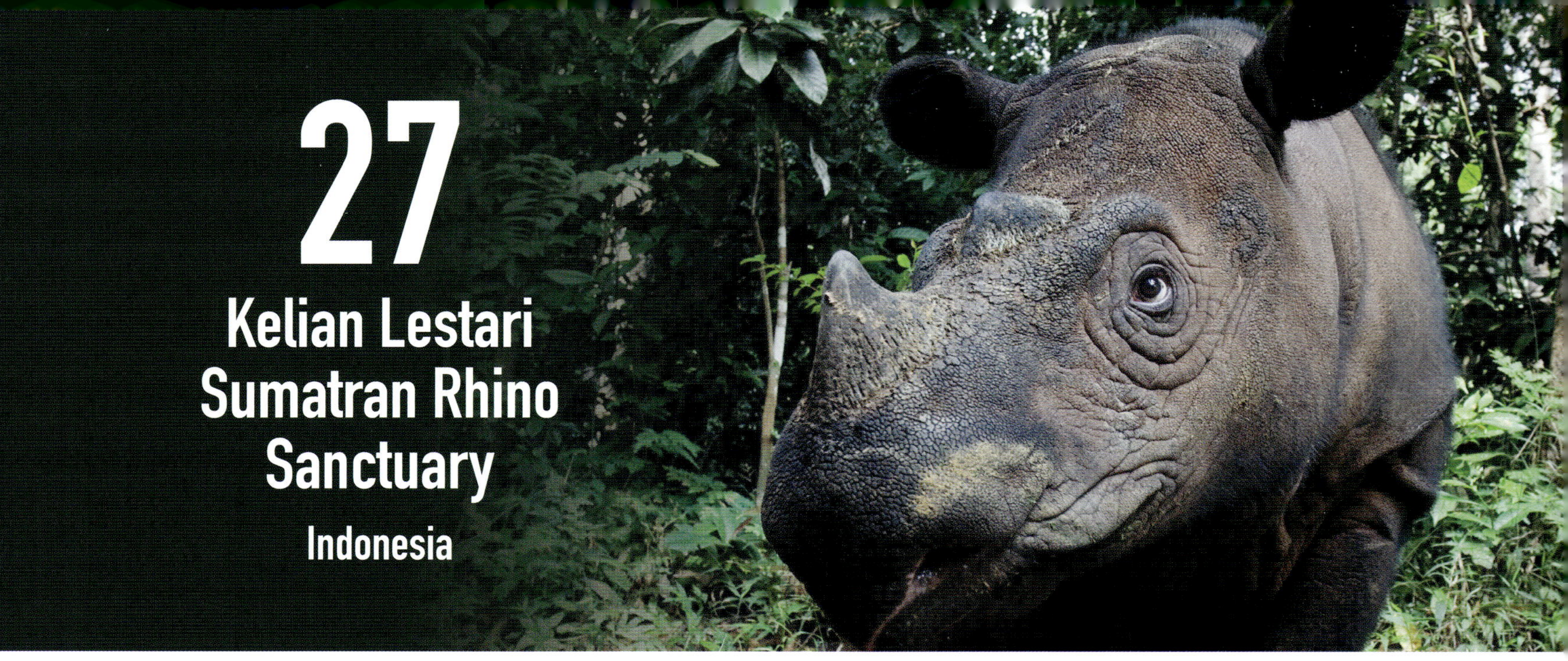

Kelian gold mine, in the dense equatorial rainforest of central Borneo, was one of the world's largest when it was operated by Rio Tinto subsidiary PT Kelian Equatorial Mining (KEM) between 1992 and 2005.

As a result of the closure planning process, stakeholders agreed to protect the forest on the mining lease area to discourage illegal logging and thus limit environmental degradation, potential damage to the remaining permanent structures and the degrading of water management systems due to sedimentation. In 2007, Rio Tinto set up a trust fund to support the management of the almost 6,000-hectare protected area and to aid community and education initiatives. The Kelian Lestari Protected Forest was established in 2013, and KEM is allowed access for ongoing monitoring and maintenance of permanent structures such as the dams, pit lake and wetlands.

According to WWF, the critically endangered Sumatran rhino population consists of around 40 wild-roaming individuals in non-viable and fragmented populations, mainly on Sumatra, with a very small number of individuals on Borneo. In 2013, orangutan researchers were surprised to find rhino footprints in the West Kutai district of Kalimantan, where the animal was thought to be extinct. At least 15 rhinos were subsequently identified inhabiting two forest pockets under threat from agriculture, hunting and mining. Following an approach from the Ministry of Environment and Forestry, KEM provided an area within its existing lease to set up a rhino sanctuary and captive breeding facility within the protected forest. The first 50 hectares have since been expanded to 500 hectares. The main stakeholders are national and regional government bodies and local NGOs; KEM facilitated the protection of a larger continuous area by allowing part of its mining lease area to be included.

In 2018, after several failed attempts to capture a rhino and relocate it to the sanctuary, a female, later named Pahu, was caught and transported to her new home. Pahu is now enjoying the protected forest's luxuriant vegetation under watch by vets and rhino specialists. The scientists are crossing their fingers for Pahu in the hope that IVF, using sperm from males at two other Indonesian sanctuaries, will result in offspring.

The mining company played only a minor role in setting up the rhino sanctuary, but the example raises a broader point: that non-operational lands may occupy a large part of a mining concession's footprint and can act to protect habitats and species that may otherwise succumb to other pressures.

28
VINETA
on Lake Störmthal
Germany

Hungry for cheap energy, Germany expanded its eastern brown coal mines in the 20th century, devouring communities and displacing thousands of people. The village of Magdeborn, thought to have been founded in the 7th century, was consumed by the Espenhain opencast mine in the 1970s, displacing around 3,200 people. When mining ceased in 1996, the void was flooded to form Lake Störmthal.

Today, the only sign that a village ever existed here is the VINETA sculpture, a church-like structure floating in the middle of the lake. Created by Krystallpalast Varieté, a Leipzig theatre company, VINETA hosts celebrations and cultural events.

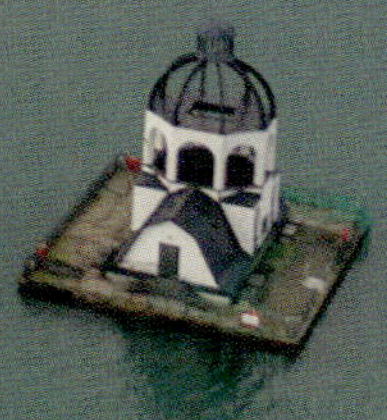

29
Shimao Wonderland Intercontinental Hotel
China

The 5-Star Shimao Wonderland Intercontinental Hotel was just an award-winning concept when it was featured in the original *101 Things to Do with a Hole in the Ground* in 2009.

Built in an 88-metre-deep flooded limestone quarry, 32 kilometres from the centre of Shanghai, the hotel is described by its creator, Martin Jochman, as a 'groundscaper'. It is powered by solar and geothermal energy and takes advantage of the quarry's thermal properties to help regulate its temperature. It took 12 years to complete the project, which the chief engineer described as a fight against gravity. The quarry hadn't been worked since the 1950s and had to be drained in order to attach the building's steel frame to the side of the cliffs. Over 40 patents were filed – derived from the engineering methods developed to build it. Once construction was complete, the quarry was flooded again. Sixteen of the Shimao Wonderland's eighteen stories are below the rim of the quarry, and the lowest two are actually underwater.

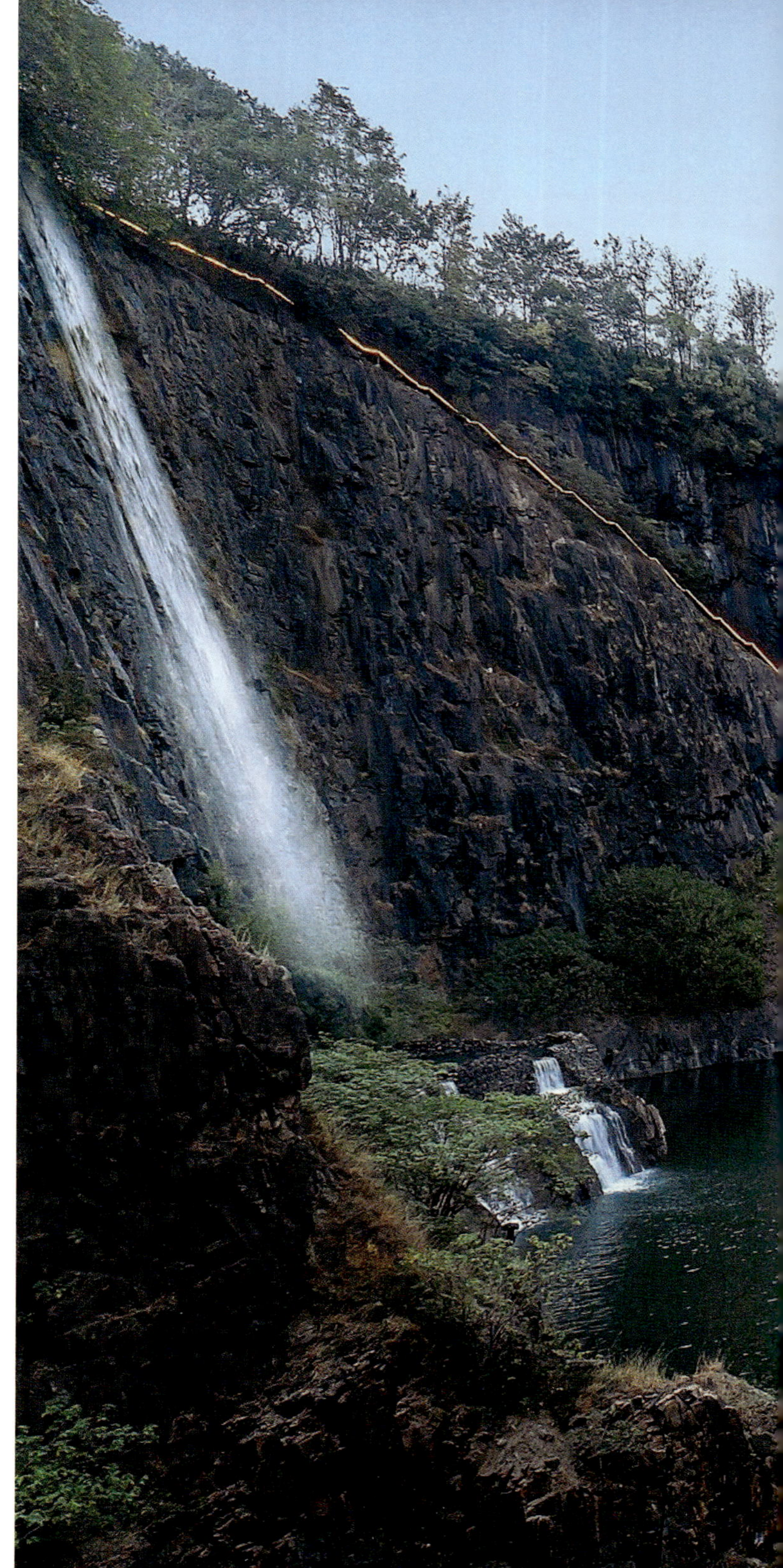

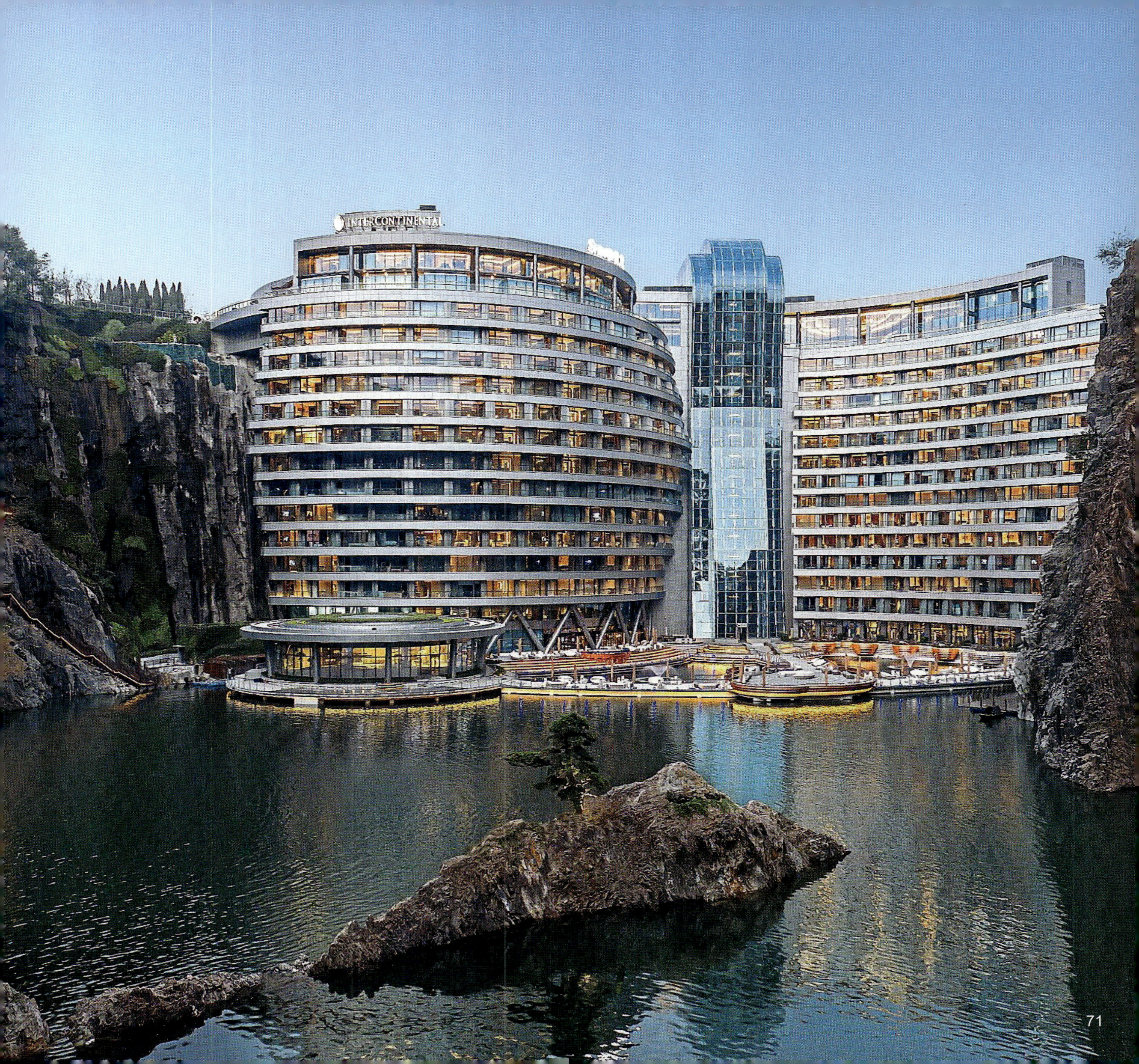

30
Nordgen Seedbank
Norway

Halfway between the North Pole and Norway, the Svalbard archipelago is probably most famous for its 'doomsday archive', the Global Seed Vault – a backup collection of many of the world's vital seed crop varieties. But there is another lesser-known seed bank just under a mile away inside the same mountain.

Unlike the Global Seed Vault, a purpose-built, temperature-controlled facility drilled into virgin rock, the NordGen seed bank was jointly established by the Nordic countries in a disused coal mine, Mine #3. The project was set up in the mid-eighties to provide a backup of the Nordic countries' crop seeds and investigate the degree to which low moisture and low temperature improved seed longevity. NordGen's backup is now held in the Global Seed Vault, but the experimental side of the project continues. The seeds in Mine #3 are stored in a steel container in the depths of the mine at an ambient temperature of -4°C compared to the Global Seed Vault's artificially maintained -18°C. The results so far suggest that it's an effective approach.

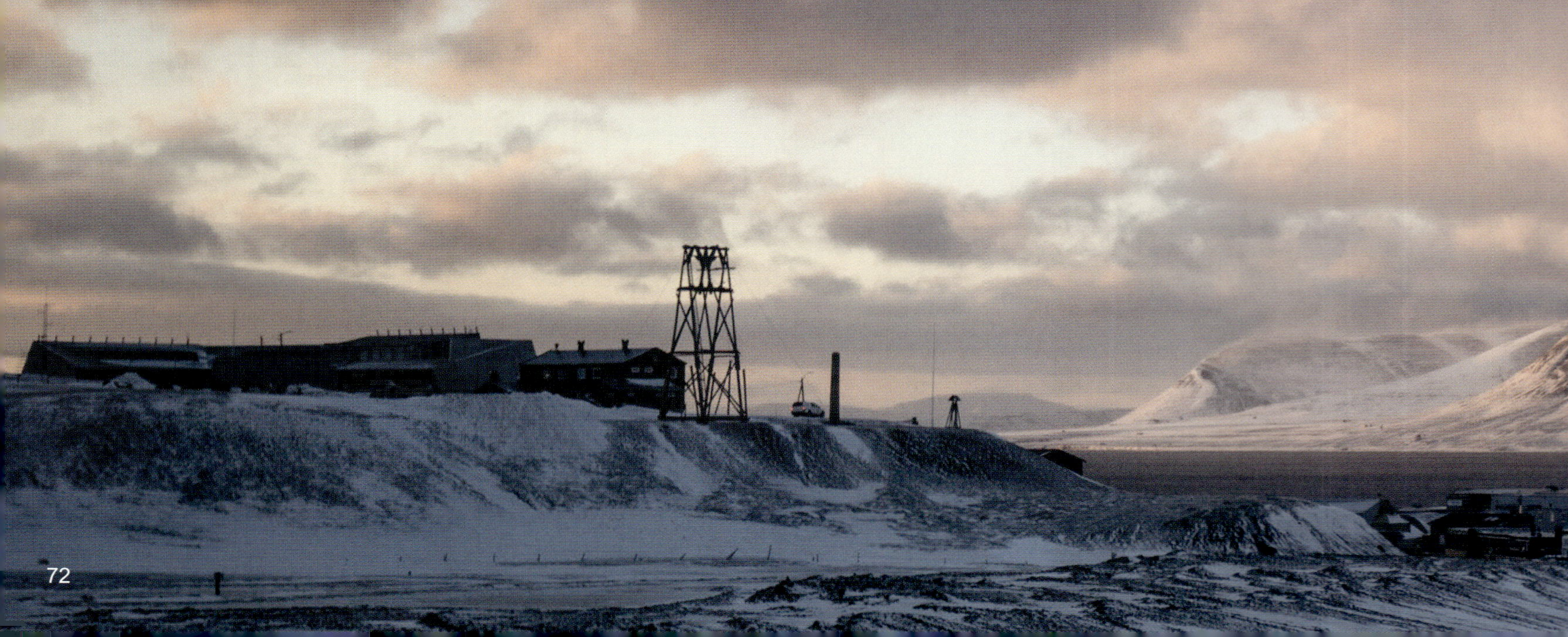

31
Arctic World Archive

Norway

Inspired by the nearby Global Seed Vault, the Arctic World Archive (AWA) aims to preserve a digital back-up of the world's cultural and historical artefacts – for a fee.

Operated by a privately owned company, the AWA occupies part of Mine #3 on Svalbard, which already contains the NordGen Seedbank described opposite. Like its neighbour, the AWA is 300m inside the mine, 300m below the mountaintop, safe from electromagnetic storms or attacks. The archive is held in shipping containers behind a concrete wall with a steel door. Surrounded by permafrost, the vault would remain safely below freezing point even if the facility experienced a loss of power.

The AWA uses a patented film technique intended to resurrect art and literature such as Munch's *The Scream* and Dante's *Divine Comedy*, historical documents like the Brazilian Constitution or data from the European Space Agency, should the need arise. In an age already swamped with deepfakes, physical copies of documents like these could prove essential in the fight against misinformation.

While depositors can access copies via the cloud, the physical film remains untouched, making the archive unhackable, offline and off-grid.

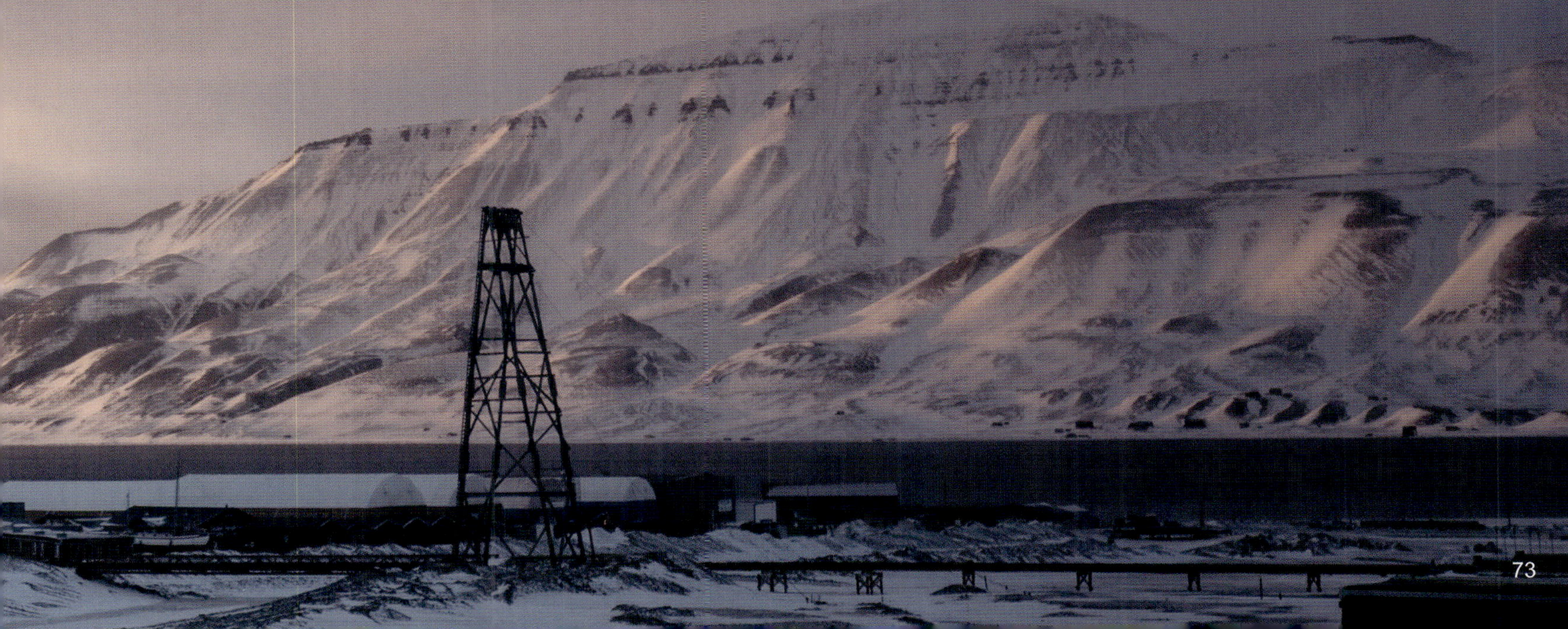

32
Pyramiden
Norway

Named after the mountain that overshadows it, Pyramiden on Svalbard is a former Soviet mining settlement that was acquired from the Swedish in 1936 along with Barentsburg. The Swedish had struggled to make a profit, and it seems the Soviet Union found it equally hard to make it prosper. Mining here was hard. The coal seam was 400 metres up a mountain, and extraction was a long and unrewarding process.

However, they persisted. Some say it was because Pyramiden offered a model Soviet society; others suggest that Svalbard's status as a free economic zone and its accessibility to the West gave it strategic importance. Whatever the Soviet Union's motives, the town was among the best equipped in the USSR, boasting a hotel, restaurant, concert hall and cinema, alongside recreational facilities such as a swimming pool, a gymnasium and a basketball court. It also has the world's northernmost statue of Lenin.

In the 1950s, Pyramiden, not Svalbard's administrative centre, Longyearbyen, had the larger population. Two and a half thousand people lived and worked in the town. When the USSR collapsed, so did the population of Pyramiden, and within a matter of years, it was a ghost town that only reindeer and polar bears visited.

In 2013, the company that owned the mine reopened the Hotel Tulpan, and tourists began to arrive in the still chilly 7°C summer months. In 2018, 36,000 people visited Pyramiden to see one of the world's best-preserved examples of a Soviet settlement. The polar bears still visit, too, which is why the guards carry rifles.

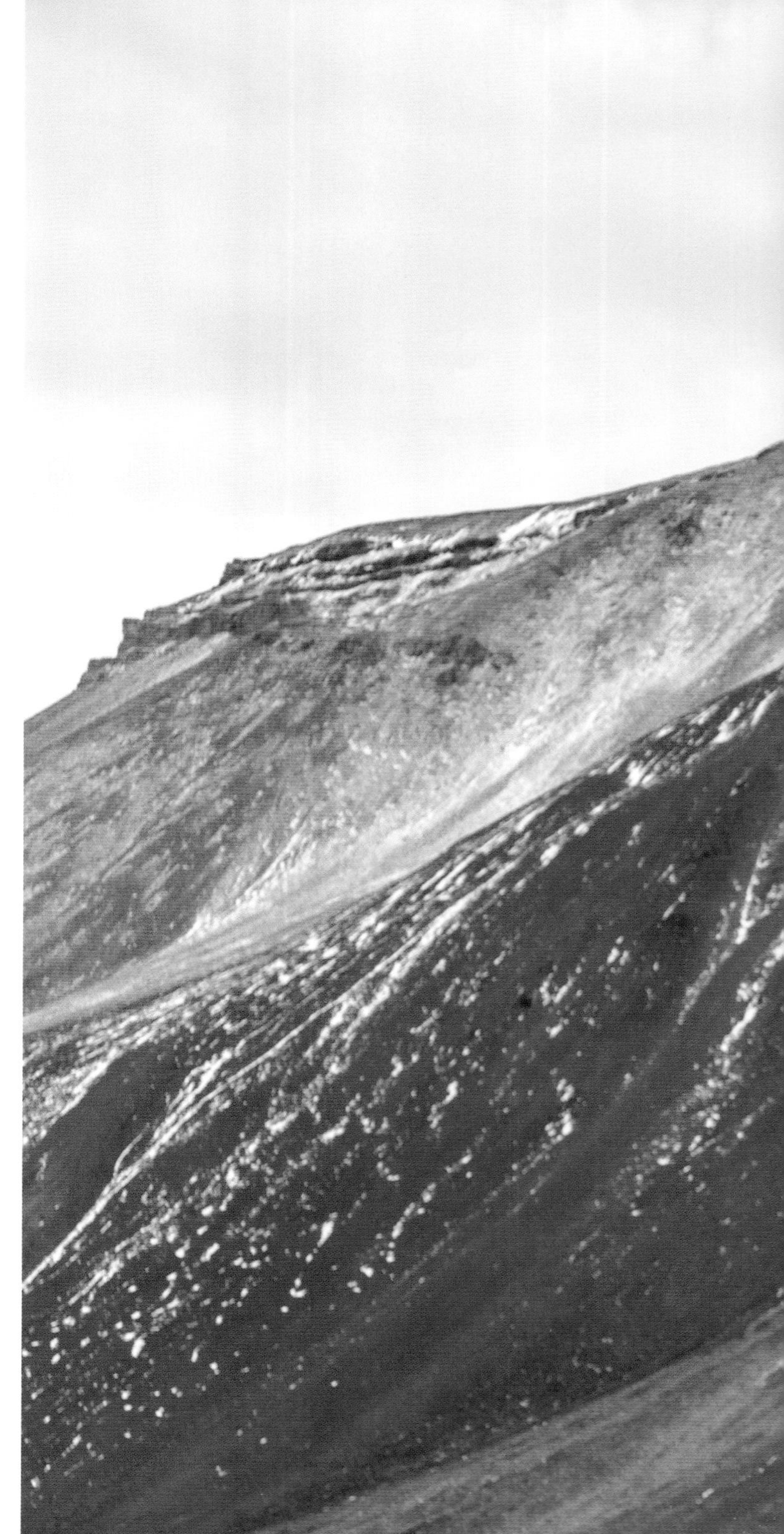

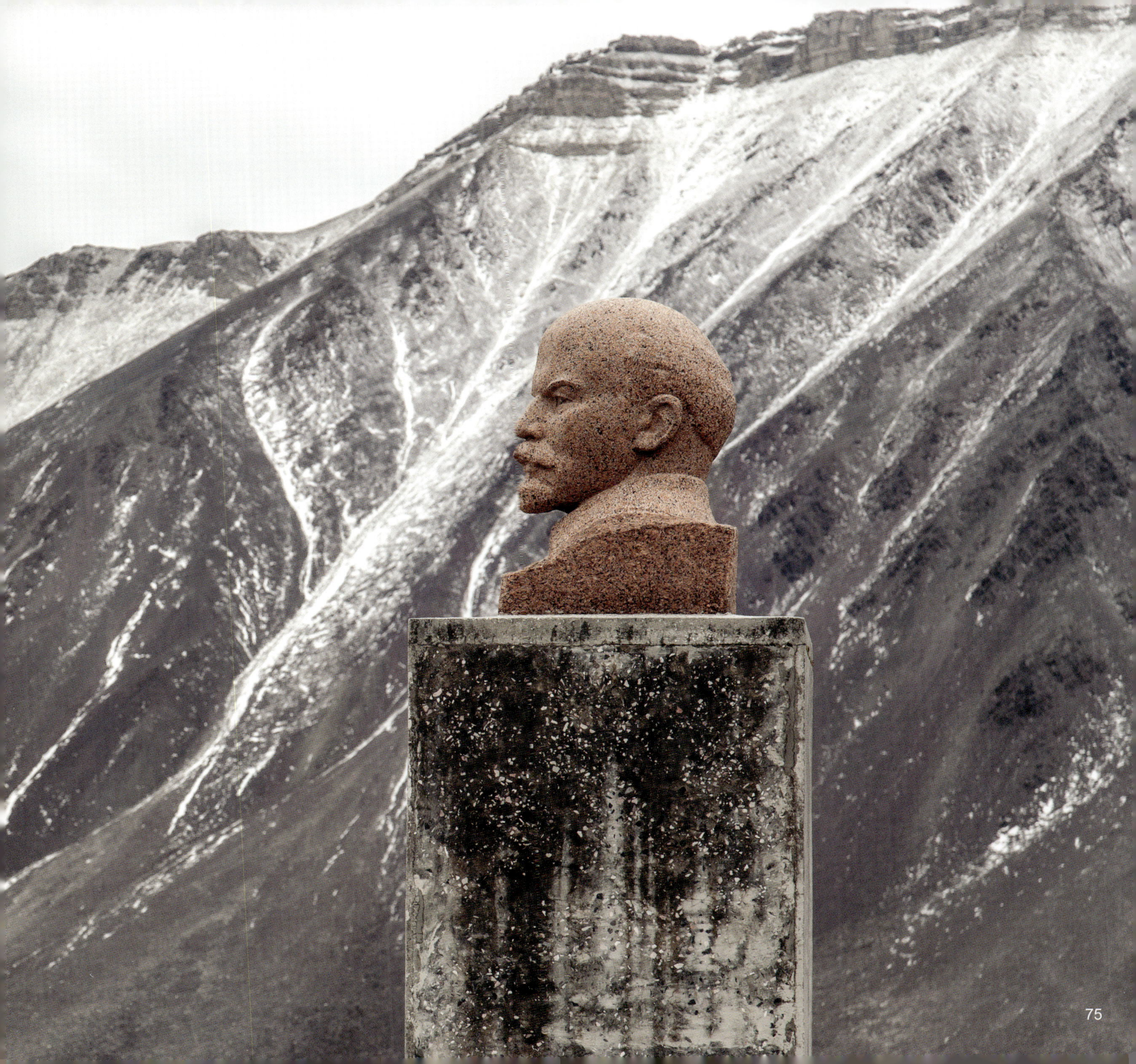

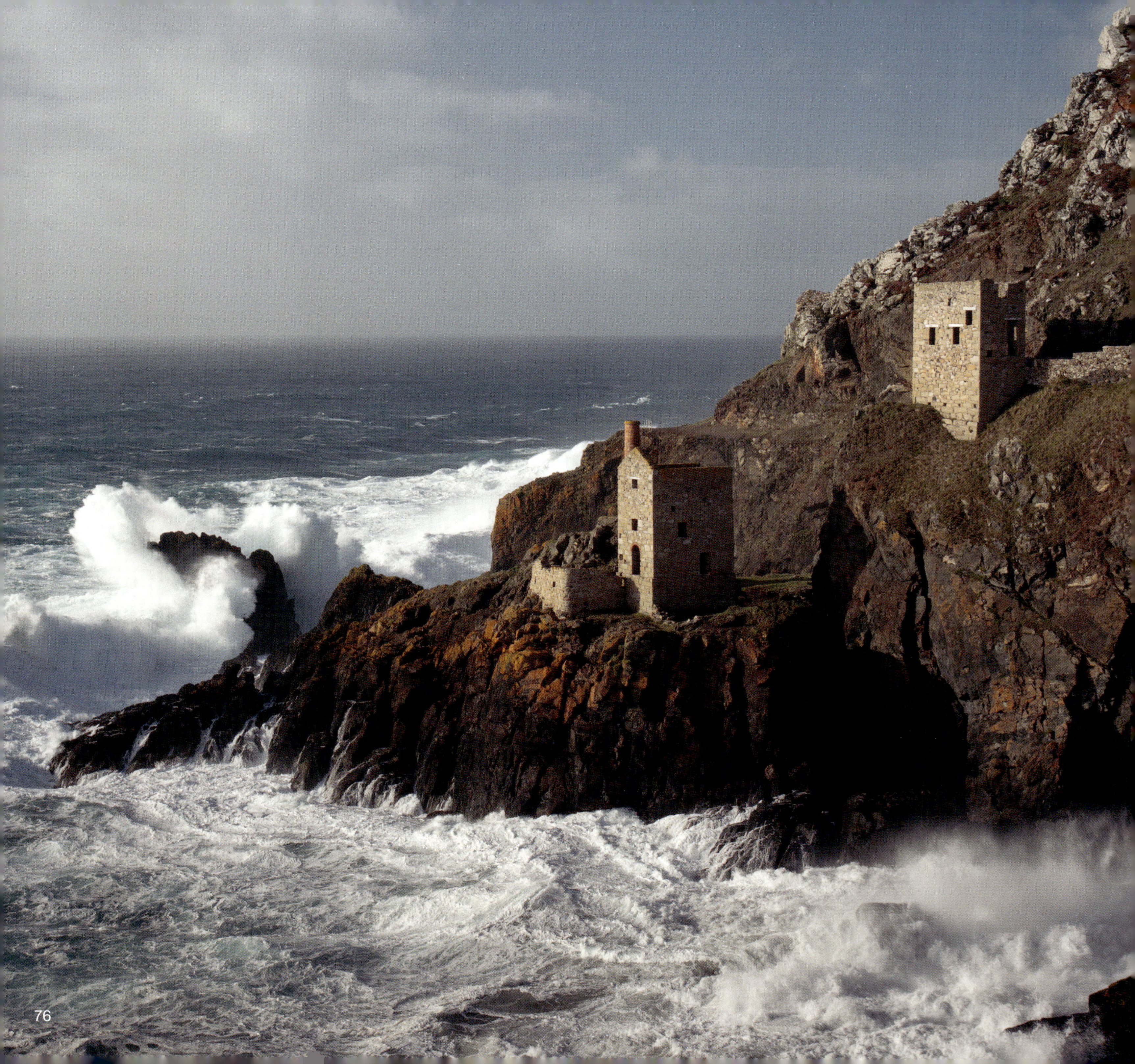

33
The Tin Coast Partnership
UK

Where England ends in unyielding granite cliffs, Cornwall meets the wild Atlantic in an endless battle between sea and stone; behind lies a tranquil landscape crafted by man and geology over millennia. Abandoned granite engine houses offer a tangible connection to the romantic but harsh copper and tin mining legacy and are a source of pride and cultural identity for local people.

Metal mining on England's extreme western coast dates to the Bronze Age. Its isolation has meant that much of the historic landscape remains unchanged. Since 2006, it has been designated as a Cornwall and West Devon Mining Landscape World Heritage Site, and it is also protected at a national level. Iconic mine sites, isolated farmsteads and small communities sit within a landscape of ancient field systems, Cornish hedges, dry stone walls and well-trodden paths.

Tourism is often touted as a way of replacing lost jobs, incomes and social and cultural capital after mining, but it often underperforms unless there is already tourism infrastructure in place. Although Cornwall attracts some five million – mainly summer – visitors per year, most most don't visit this part of the county.

The Tin Coast Partnership was initiated by The National Trust and draws on the expertise of the small local businesses, parish councils, NGOs and community groups involved in managing the landscape and tourism and with a deep cultural connection to the place. The partnership aims to extend the season, broaden the visitor base, and make it an easy place to visit and explore. The partners are keen to ensure that tourism in the area is sustainable and doesn't overburden the local community.

The Tin Coast's three 'gateway' mine sites are Botallack Mine, Levant Mine and Geevor Tin Mine Museum. Each relates a different chapter of the mining landscape narrative. Botallack Mine's iconic Crowns Engine Houses of BBC TV's Poldark fame cling defiantly to rugged cliffs. Its restored Count House is now used for functions and as a café, which – thanks to the Tin Coast Partnership – is now run by the Geevor Tin Mine Museum (previously run by the National Trust). Since it took over, the café has increased its local suppliers as well as its income.

The cliff-top Levant Mine features its original 1840 steam engine that ran until the mine closed in 1930. In that time, miners dug down over 500 metres and tunnelled 1.6km out under the seabed. The engine sat rusting for 50 years before it was restored by dedicated volunteers, going 'back into steam' in 1993. It now attracts over 10,000 people a year. Especially poignant is the memorial to the man-engine tragedy of 1919, in which 31 miners were killed underground. Anguish still resonates through their descendants who, with many other local people, attended the 2019 centennial commemoration in a powerful evocation of a sense of place.

Geevor Tin Mine Museum offers a more recent connection to the mining industry. To the despair of the local community, Geevor closed in 1990, but it was saved from demolition by Cornwall County Council and then lovingly turned into a museum by the former mining community. Many of its features are as they were on the day the miners left for the last time.

Innovations by the Tin Coast Partnership include green transport, free public Wi-Fi, smartphone apps, and the application of geosense technology to manage visitor movement. The group is also investigating the potential for heat recovery from underground mine water at Botallack, Levant and Geevor to warm local houses and offices.

Opposite: The Crowns Engine Houses at Botallack Mine. **Overleaf:** Levant Mine.

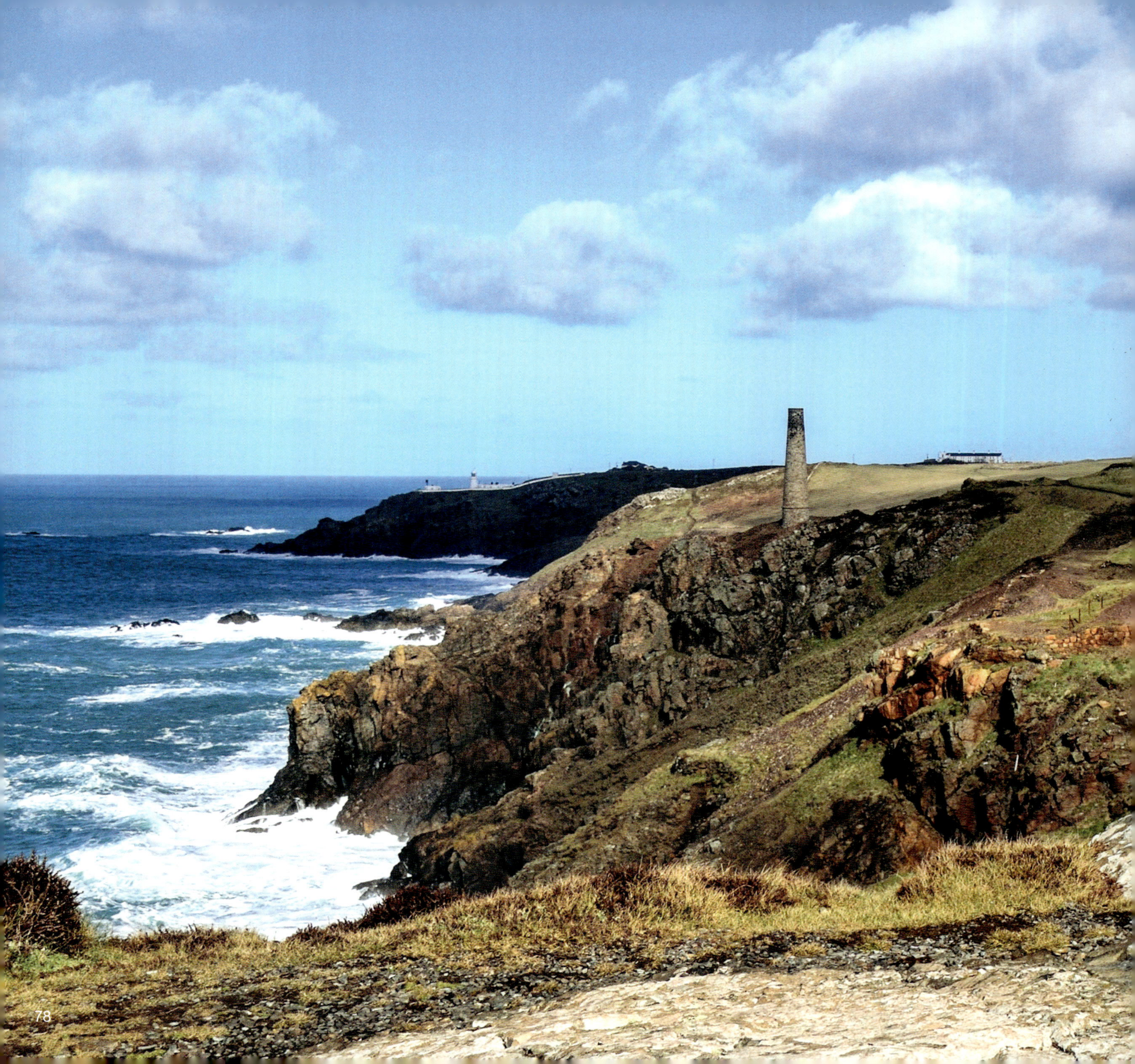

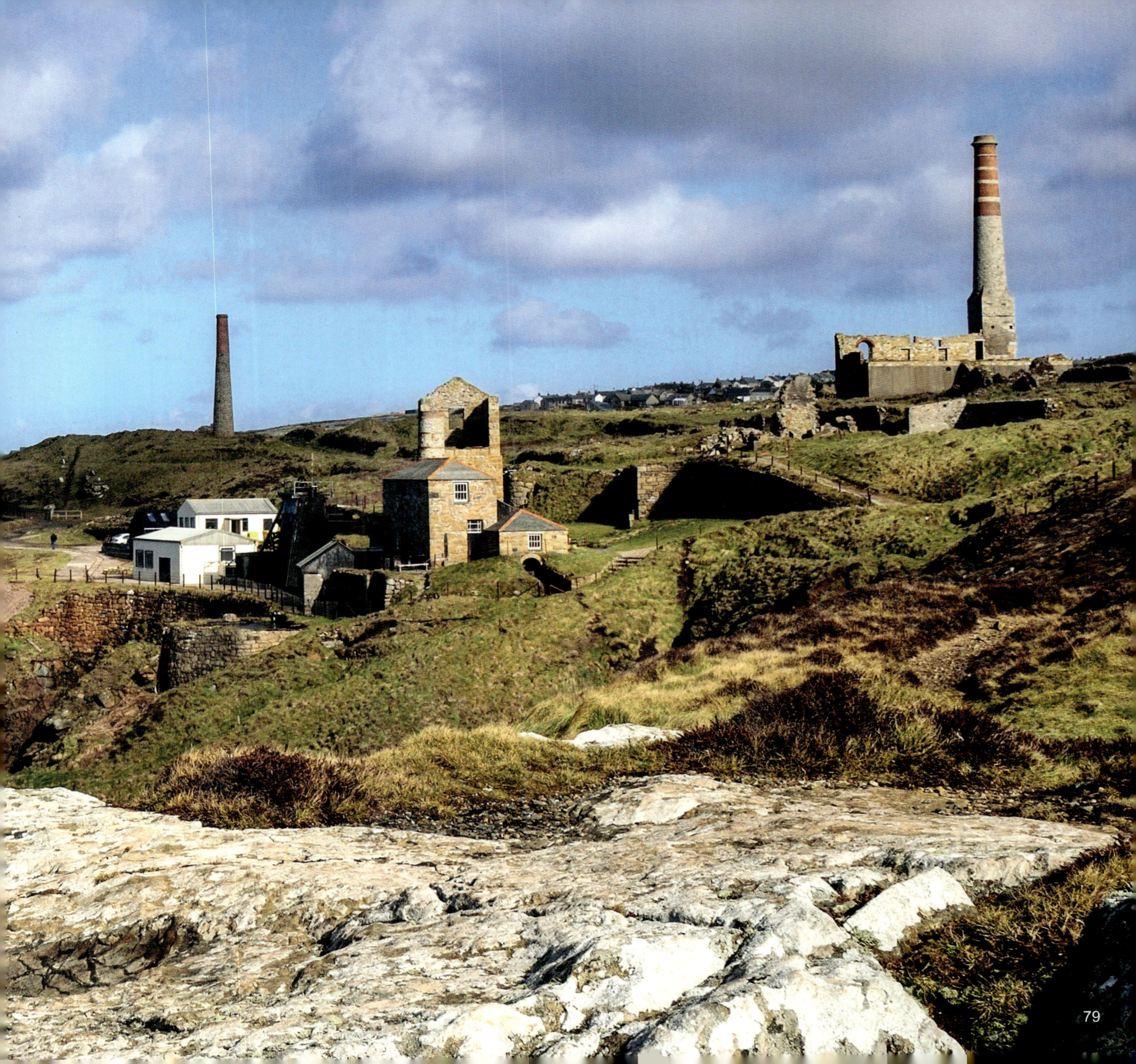

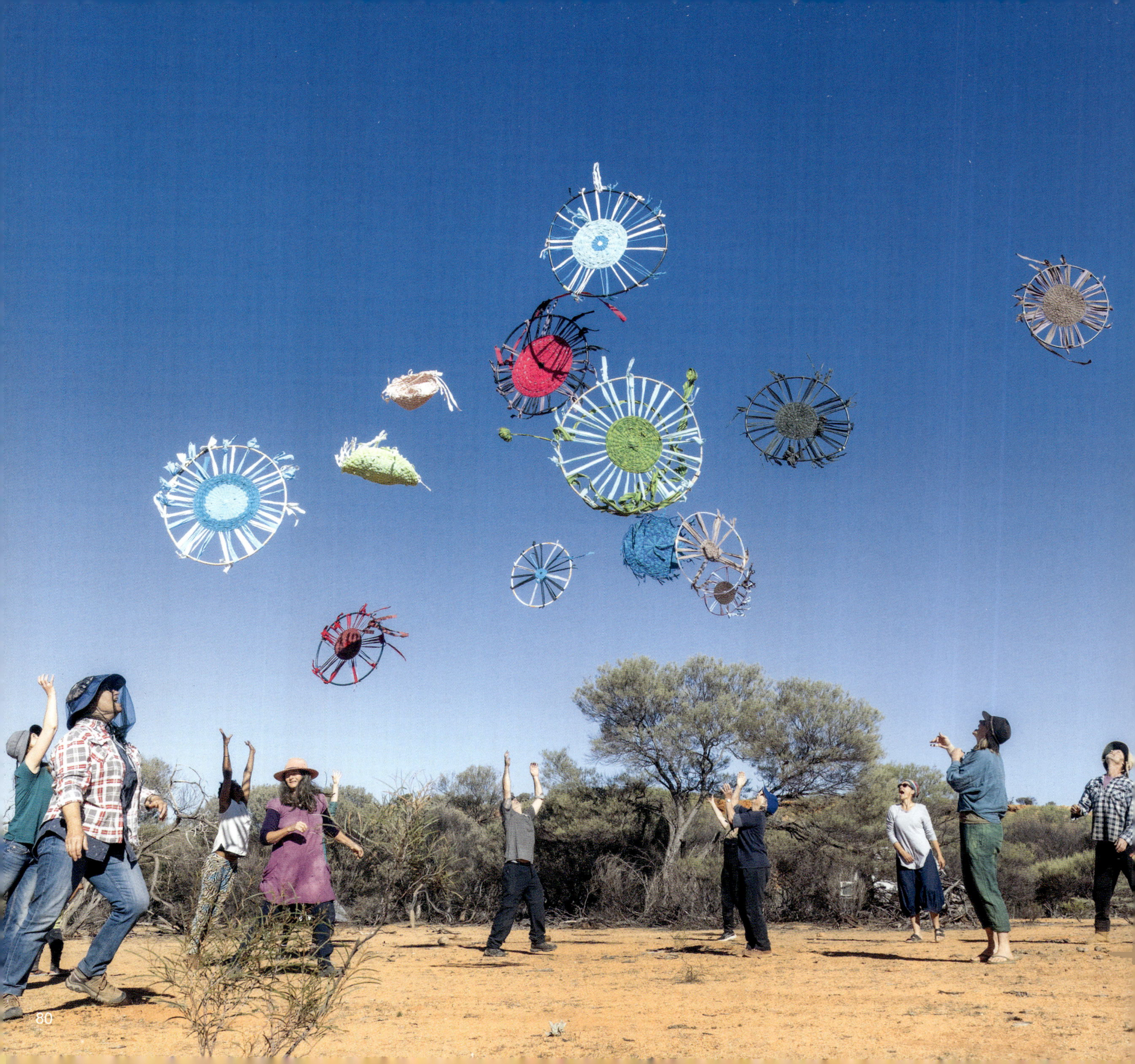

34
Reclaim the Void

Australia

Reclaim the Void is a contemporary art project that will be assembled on land disrupted by mining in the northern Goldfields region of Western Australia. It arose as a response to the pain felt by First Nations elders in the small town of Leonora, at 'those gaping mining holes left all over our Country'.

'Country' in the Aboriginal sense is a holistic way of speaking about the landscape, sky, earth and water of a place that includes animals, rocks, plants, song, story, ancestors, songlines and dreaming.

The huge 'dot artwork' will be made from thousands of small, circular rag rugs made from discarded fabric by people all over the world, and based on an original painting by Ngalia artist Dolly Walker. She spent her life fighting to protect Country and painted Country to preserve and tell the stories.

Walker's son, Kado Muir, cultural spokesperson for *Reclaim the Void*, explains that prospectors and miners have been digging up the land in this area for over 100 years. Everywhere there's some level of disturbance: earth turned over, small shafts, massive open pits and very deep underground mines. This creates an imbalance and unsettling in the earth that disrupts the vibration of the land and the spirit. It disrupts the stories and songlines, the dreaming tracks where the ancestral beings travelled and left their essences in the land.

While sometimes miners are gone within a few months, the hole – and the injury to Country – stays forever. Significant places that are the physical embodiment of ancestral beings are now disappearing as a result of mining activity. The only way those places continue is in stories, songs and art.

Reclaim the Void is not anti-mining – indeed, First Nations peoples have used resources taken from the land for a very long time. However, these activities were always done with respect, care and moderation, based on a practice of reciprocity and exchange. When something was taken from the earth, ochre, for example, special objects were made and kept at that location. In Australia industrial-scale mining, by contrast, removes millions of tons of ore, and uses – and sometimes contaminates – huge quantities of water. It scars the landscape by leaving many enormous voids.

Reclaim the Void is about bringing art back in – art that's about Country. It is about connecting the *Tjukurrpa* or dreaming tracks. The artwork will draw attention to the impact of mining, acknowledging the over-extraction of resources from sacred lands and the wounding that results. The project is also about learning that all Country has a story and offering people everywhere a chance to give back to Country in a spiritually grounded, culturally guided way. The project seeks to give back in a way that speaks to the heart and the spirit.

As each rug is created, *Reclaim the Void* is bringing together a community of people who are weaving their stories, giving their time and creating healing through art. After the artwork is laid on Country, there will be a digital version where viewers and contributors can click on rugs and hear or read the story of the maker(s). One rug may be made by a few people – passing it on from one to another in a gesture of collaborative making that underlies the project. Without the contribution of thousands of hearts and hands, the final artwork, and what it says about reclaiming the spirit of earth, the spirit of community, cannot be made.

35
MountainRose Vineyard
USA

MountainRose Vineyard is an idyllic, family-owned estate nestled amongst the mountains of southwest Virginia, covering 11 hectares in total, including the farm and a small orchard. Reflected in the winery's small blue lake, lines of vines flow over the undulating terrain offset by considered plantings of roses and American dogwood and a background of broadleaf trees. This serenity belies the terrain's tortured coal mining origins, although clues remain such as the lake, which was the mine's former settling pond.

David Lawson manages the vineyard. This parcel of Appalachia has been in his family since at least the mid-19th century when it was used for subsistence farming. Since then, it has undergone three different periods of coal mining – both underground and surface mining. Once the mining had ended, David was driven to create a positive new use for this damaged place and – relishing a challenge – he set about creating the first-ever vineyard on a Virginian mine site.

Before the vines could be planted, the heavily compacted and waterlogged, infertile, post-mining substrate required improvement. Cover cropping for up to four years added organic matter and nutrients before planting the vines, which take three years to bear fruit. It takes another couple of years to mature the wine before bottling and selling. The first commercial area of vines was planted in 1996, and the first wine was harvested in 2004.

Although the geography and climate of the site are well suited for vines, the soils lack key micronutrients in some areas, so a trial-and-error approach is used to treat these and select the most suitable variety. This combination of unusual soil and climate adds richness and intensity to the wines' colours, aromas and flavours. Late spring frosts in this region can be challenging, though, which restricts the varieties of vines available to late bud-breaking varieties, including some French-American hybrids.

In some areas, compaction still causes waterlogging, which is managed by digging channels. David trials and employs sustainable viticultural practices wherever possible; for example, Caliente Mustard was planted amongst the vines to suppress an infestation of soil-borne nematodes instead of conventional fumigation with poisonous gases.

The development work is ongoing and is proudly self-funded. David is preparing three more hectares to plant over a five-year period, depending on how long it takes to improve the soil. If something is not working well, he takes an experimental approach to determine whether the mine soil, climate or variety is the cause before it can be fixed. To this end, new grape varieties are being tested to select those most suited to the new conditions.

MountainRose Vineyard produces around 30 tonnes of grapes annually, which are made into about 1,200 to 1,500 12-bottle cases per year. The wines have won numerous state and regional awards. Some in their 'Mines to Wines' series speak to a sense of place, being named after some of the site's old coal seams, such as Jawbone and Pardee. These are available in local restaurants and local grocery stores through their Partners in Wine initiative and are also available online.

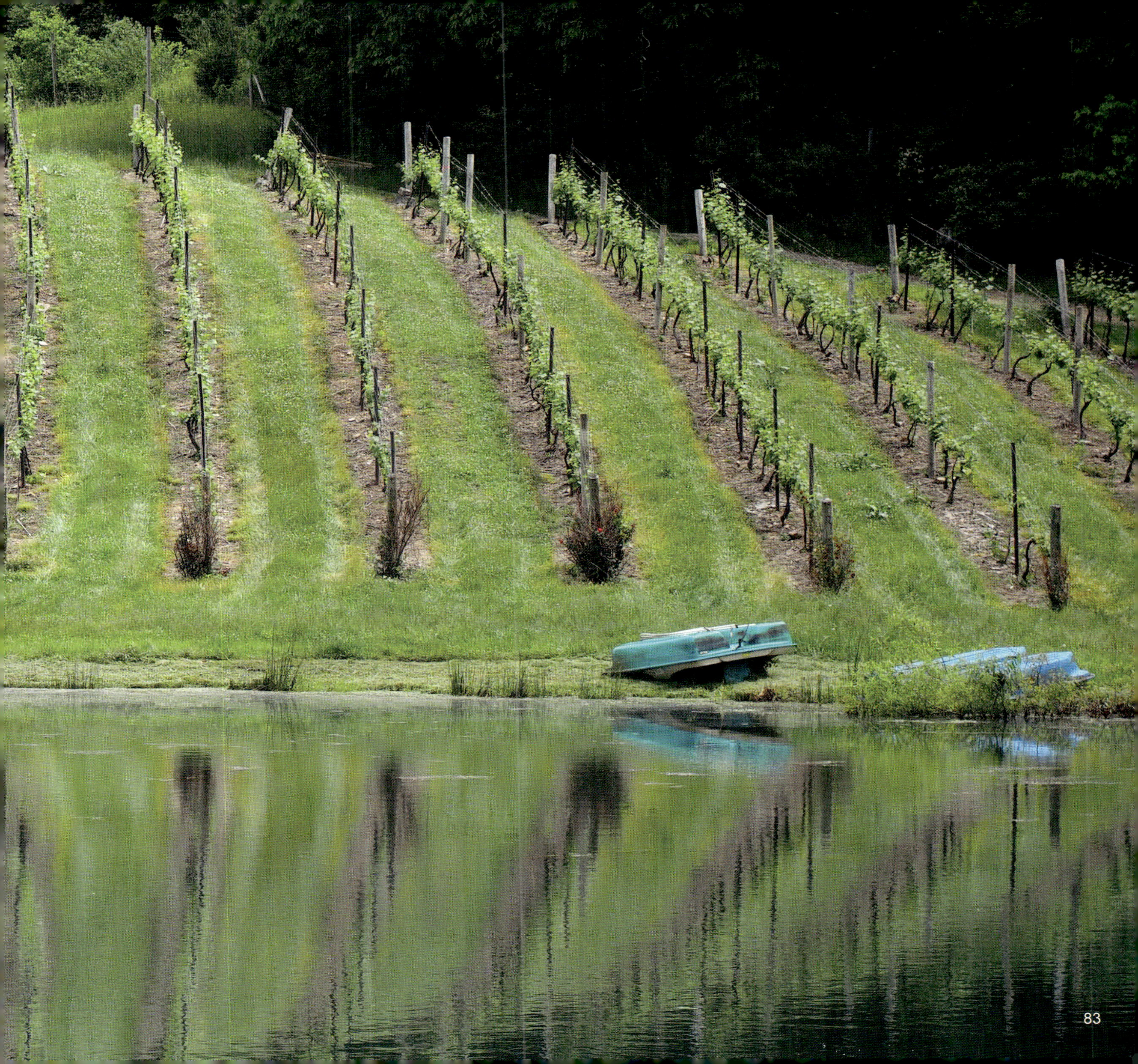

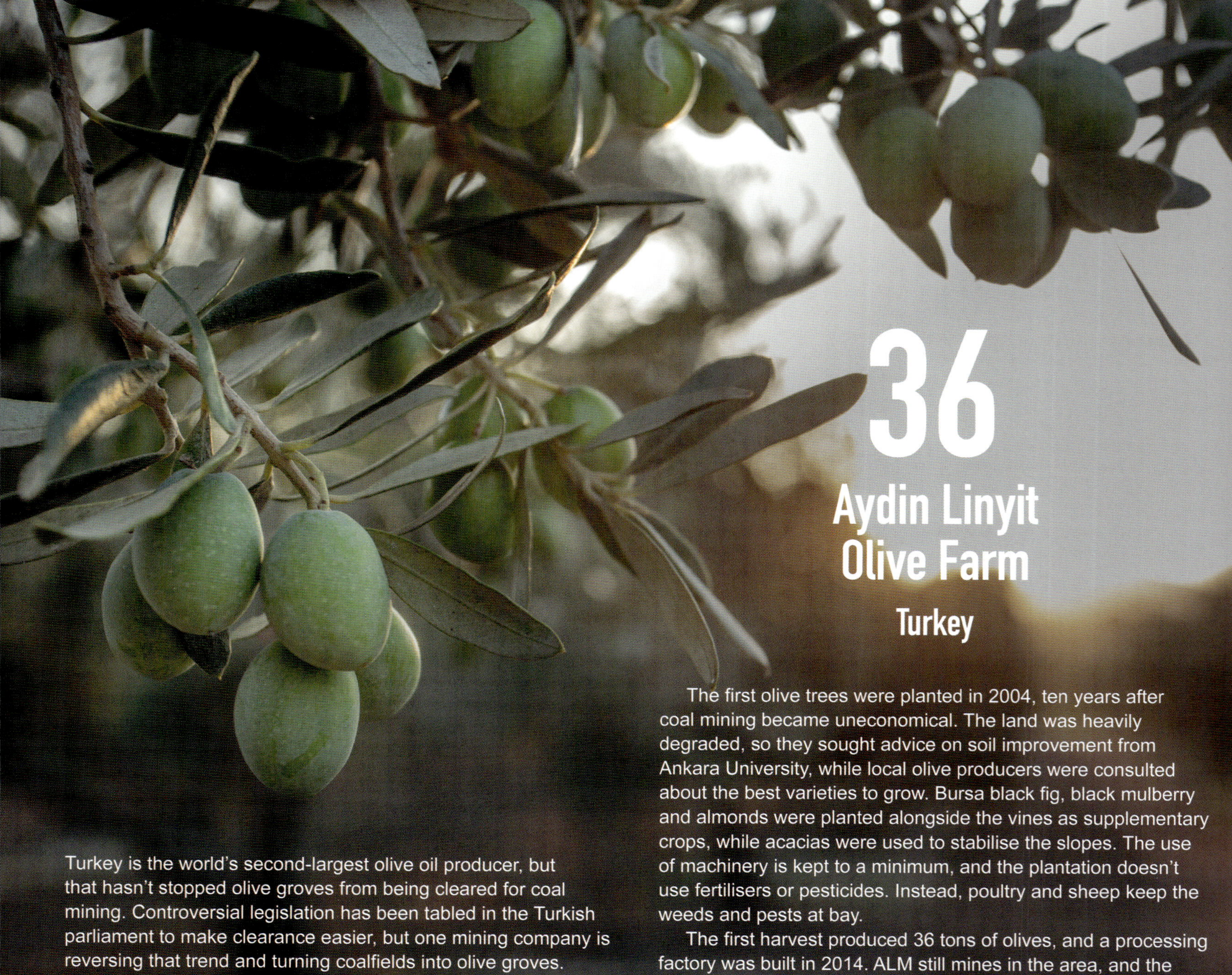

36
Aydin Linyit Olive Farm
Turkey

Turkey is the world's second-largest olive oil producer, but that hasn't stopped olive groves from being cleared for coal mining. Controversial legislation has been tabled in the Turkish parliament to make clearance easier, but one mining company is reversing that trend and turning coalfields into olive groves.

The Ankara-based Aydın Linyit Mining company and Zetay Tarım, an olive oil producer, invested TRY 258 million in converting its former opencast lignite mine in Aydın province into a lucrative 330-hectare olive farm.

The first olive trees were planted in 2004, ten years after coal mining became uneconomical. The land was heavily degraded, so they sought advice on soil improvement from Ankara University, while local olive producers were consulted about the best varieties to grow. Bursa black fig, black mulberry and almonds were planted alongside the vines as supplementary crops, while acacias were used to stabilise the slopes. The use of machinery is kept to a minimum, and the plantation doesn't use fertilisers or pesticides. Instead, poultry and sheep keep the weeds and pests at bay.

The first harvest produced 36 tons of olives, and a processing factory was built in 2014. ALM still mines in the area, and the trees they've planted help offset the coal mine's carbon footprint, but the success of the olive farm means the company is already planning to convert the remaining 60 hectares of opencast mine in the next few years.

The founding role of re-settled British convicts in the early colonisation and development of Australia is infamous. Incarcerated labour quarried rock to construct Australian towns and cities. Common in the state of Victoria was bluestone, a type of basalt, which was used to build much of old Melbourne.

There were 41 bluestone quarries in 19th-century Pentridge (later Coburg, now a suburb of Melbourne), in one of which toiled inmate quarrymen who extracted the durable stone to build the dark, foreboding walls of their own prison – the notorious HM Prison Pentridge, aka the 'Bluestone College', the final resting place of the outlaw Ned Kelly. Pentridge village was renamed Coburg in 1875 due to local embarrassment over the prison's notoriety.

The old quarry and surrounding land were acquired by Coburg Council in the early 1900s, when a weir was constructed across Merri Creek, flooding the quarry to create a recreational lake for swimming, lifesaving, diving and boating. Under the supervision of City Curator John Gray, the landscape was carefully shaped into an urban park for the rapidly expanding metropolis.

Today it is a much-loved and popular urban oasis and nature reserve and an integral part of the Merri Creek environmental, heritage and recreational corridor. Much of it is dedicated to threatened flora and fauna, including indigenous plantings and new wetlands. Its significant cultural features include mature English elms, Monterey pine and Canary Island date palms, the weir and early bluestone retaining walls and old quarry outcrops. The erstwhile prison's once dominant dark grey walls are now softened by the pale bark of the park's gum trees, while new residential units built within the prison grounds overlook both walls and park.

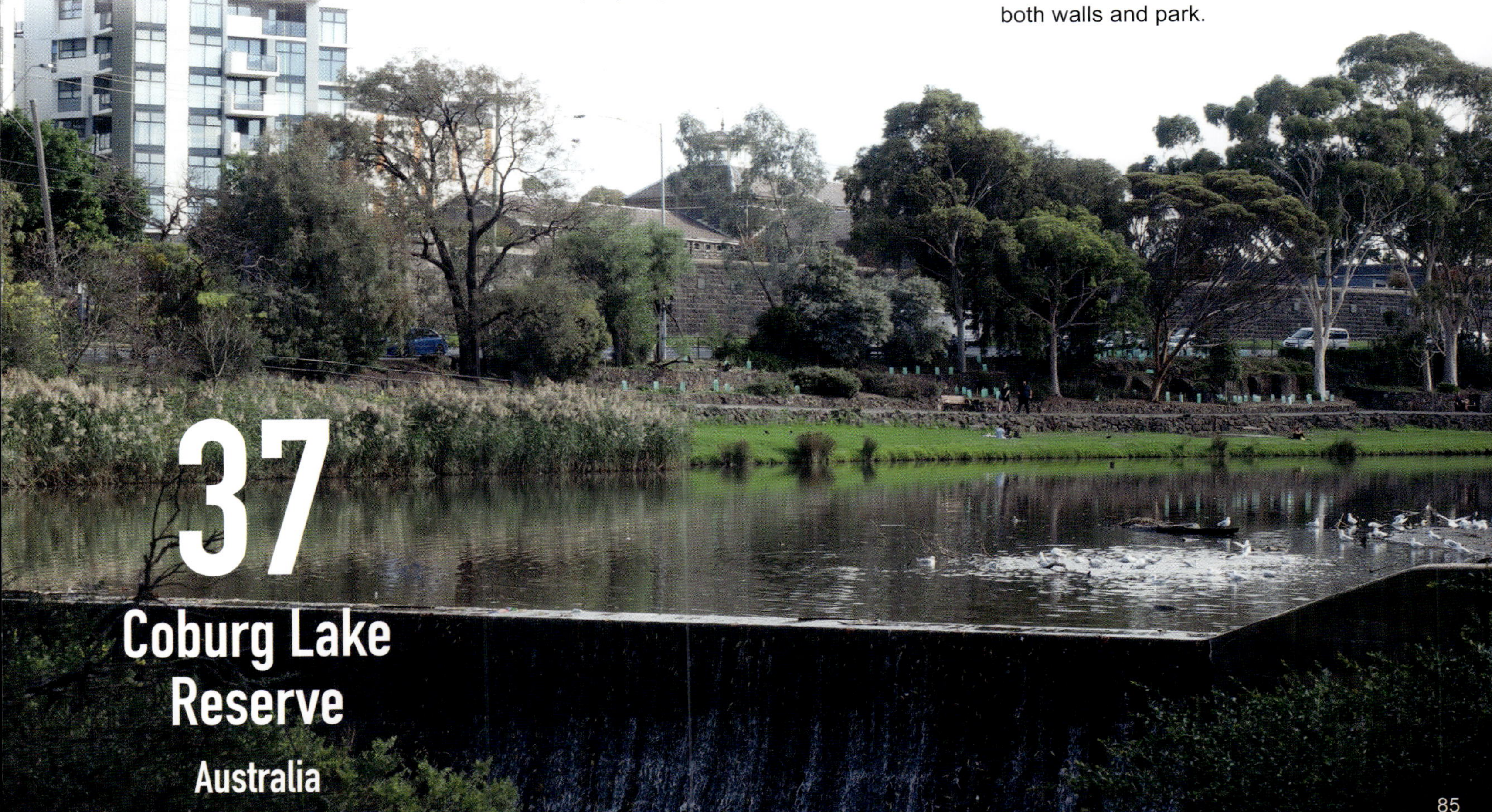

37

Coburg Lake Reserve

Australia

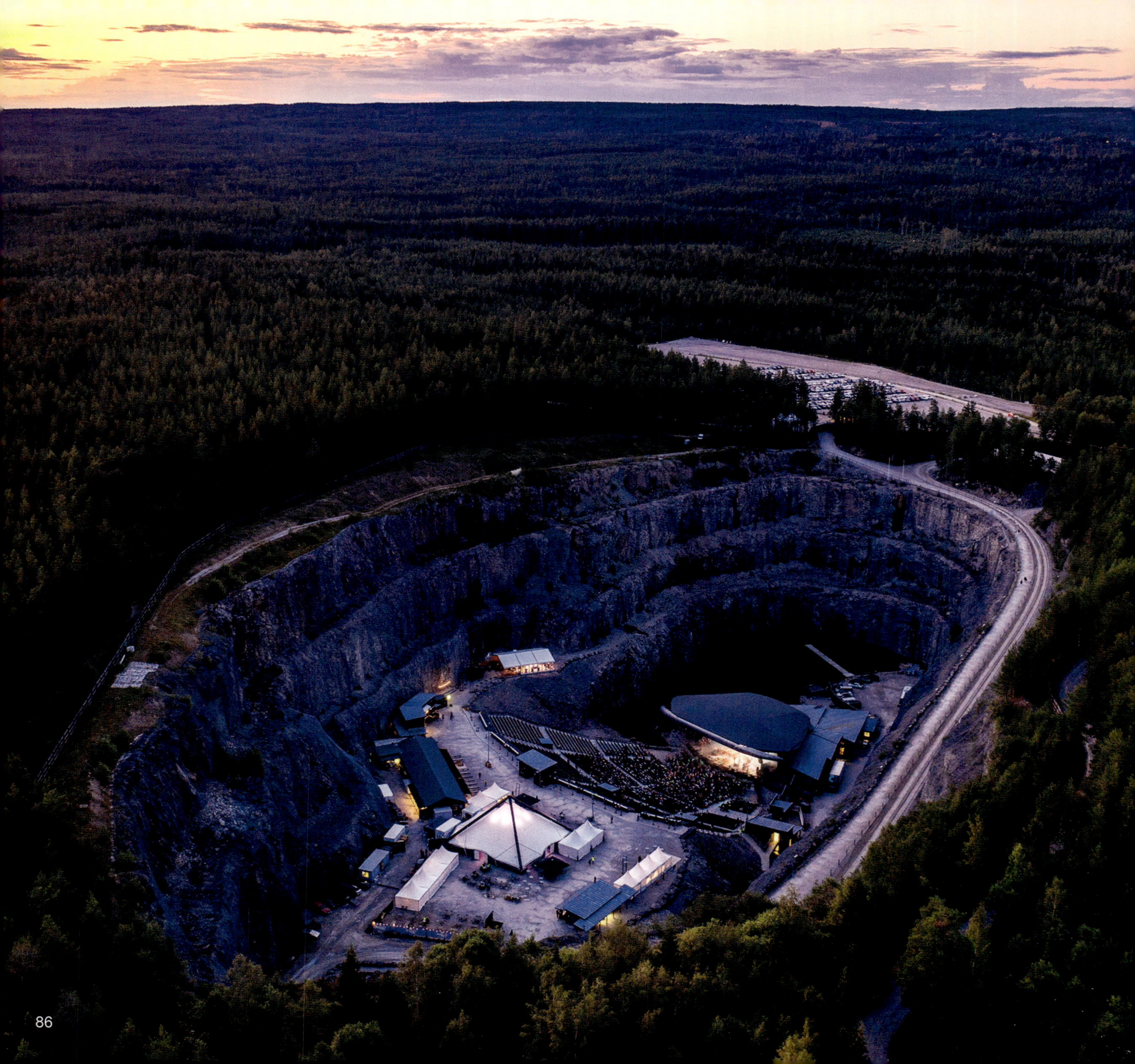

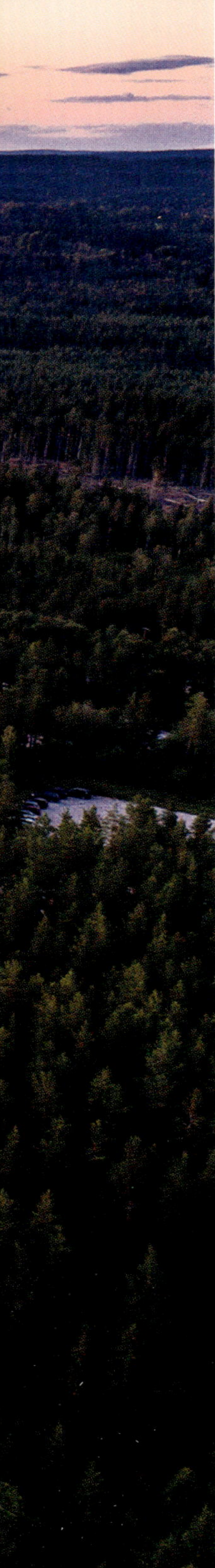

38
Dalhalla
Concert Venue

Sweden

Very few post-mining projects become world-famous; Sweden's Dalhalla concert venue, secreted among the forests and lakes of Dalarna County, is one of these. 'Dalhalla' is a hybrid of 'Dalarna' and Wagner's vision of the 'Valhalla' of Norse mythology.

The vision for converting the closed Draggängarna limestone quarry into a world-class music venue was the brainchild of the celebrated Swedish soprano Margareta Dellefors. She was searching for the perfect outdoor amphitheatre, and with Åsa Nyman, then Head of Culture in Rättvik Municipality, she found the disused quarry in 1991 and set about creating an icon. Dellefors projected her powerful voice to prove the quarry's exceptional acoustics derived from its dimensions and roughly hewn rock walls – as a prelude to the construction of a stage and seating. The first 'test' concert with an invited audience, press and broadcast media successfully proved the concept in 1993, and Dalhalla has grown from strength to strength ever since.

Architect Erik Ahnborg's simple lines and curves contrast with the rugged grey quarry walls and teal-hued pit lake. Since opening, its capacity has expanded from around 2,000 to 6,000. Employing almost 400 people during the summer season, it is one of Sweden's most popular attractions. Dalhalla's summer performances draw over 100,000 people a year, featuring opera, jazz, rock, pop and more, including, Robert Plant and Alison Krauss, Sting, and the Beach Boys. Dalhalla celebrated its 30th year in 2023 with a record 129,000 visitors, including its biggest-ever audience of 6,146 to see the rock band KISS on their final tour.

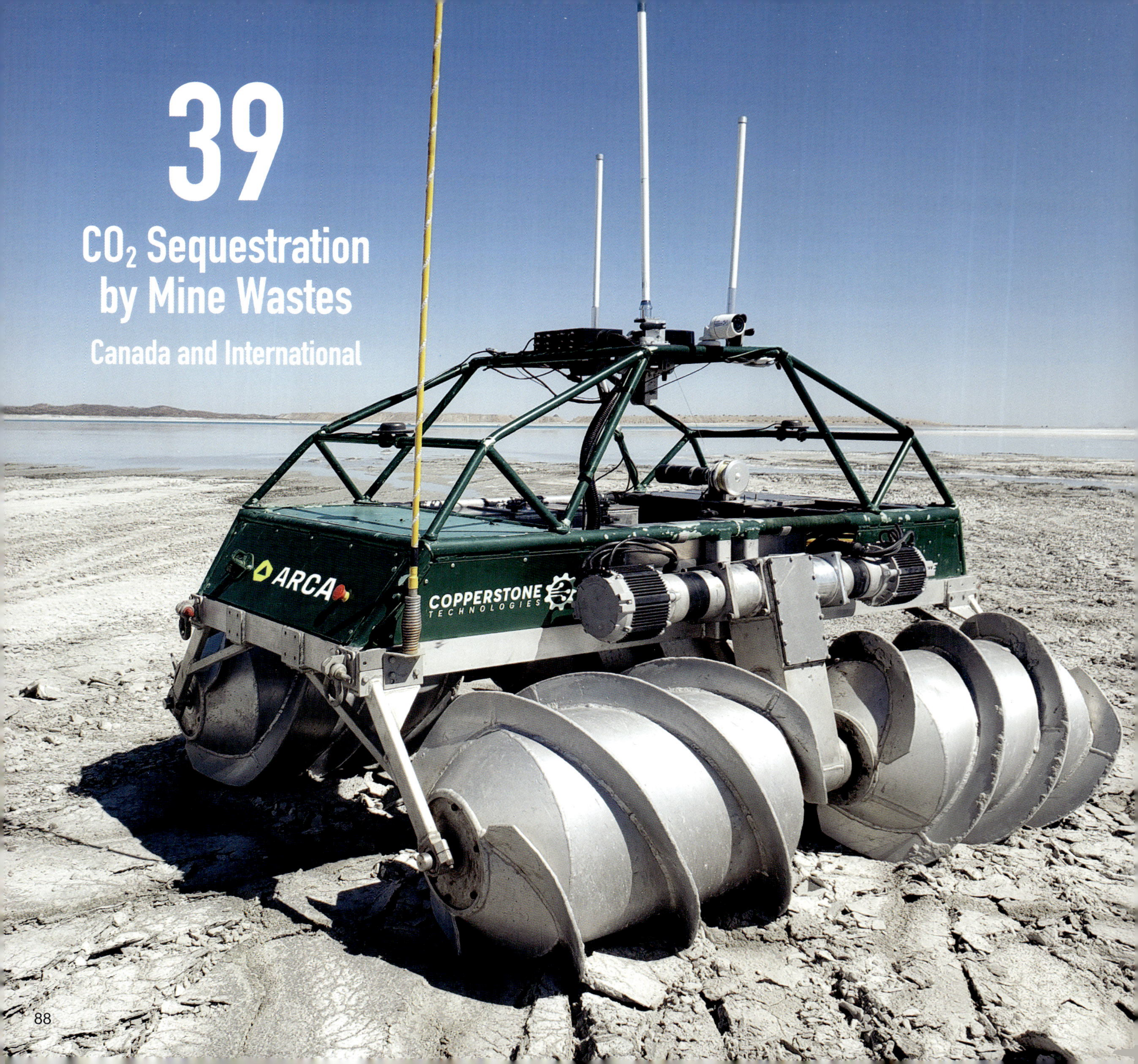
39
CO$_2$ Sequestration
by Mine Wastes
Canada and International
ARCA
COPPERSTONE
TECHNOLOGIES

Decarbonising global society means the large-scale closure of coal mines while massively ramping up the mining of minerals and metals critical to green technologies. Yet, mining is a very carbon-intensive industry – therein lies the paradox. But some mine wastes consist of igneous (and some metamorphic) rocks that are extremely low in silica but enriched in magnesium (aka ultramafic rocks), and such rocks have the potential to remove billions of tonnes of CO_2 from the atmosphere – forever. Incidentally, over 99% of all the carbon on Earth is estimated to be stored permanently in rocks.

Greg Dipple, the University of British Columbia's Professor of Geological Sciences and Arca Head of Science, has been studying this natural phenomenon for years. Two key events stimulated his interest: a visit to the USA's Los Alamos National Laboratory, where scientists were researching the absorption of CO_2 in vats of crushed ultramafic rock, and the presence of efflorescences (a sugary, white crust) on the surfaces of some Canadian mine wastes produced when magnesium-containing minerals react with atmospheric CO_2 in a process known as carbon mineralisation. The reaction is stable and effectively permanent.

There are over three billion tonnes of waste rock and tailings in Canada alone, mainly in asbestos mines in Quebec (most of which are ultramafic in nature). Over the past decade, the potential for mine wastes to act as a carbon sink has been a popular area of research. Dipple's team received backing from BHP and Rio Tinto to investigate natural carbon mineralisation in ultramafic tailings at BHP's Mount Keith nickel mine in Australia and Rio's Diavik diamond mine in Canada's Northwest Territories. The research at Mount Keith showed about 40,000 tonnes of CO_2 sequestration per year – equivalent to 11% of the mine's emissions – were being absorbed by the crushed rock on-site. The main limiting factor was burying the tailings before it had a chance to react.

Commercial interests in the concept led to Dipple spinning off a new company, Arca Climate Technologies (aka Arca). Carbon mineralisation by tailings is a slow process, so work began on developing methods to speed it up, including injecting carbon-rich waste gases into tailings and creating 'smart churning' technology. Like the lovechild of a Mars rover and an autonomous lawnmower, the robotic machine churns its way across the flat, grey tailings, propelled by a system of Archimedes screws. Developed by Canada's Copperstone Technologies and in collaboration with Arca, the rover's roamings disturb the tailings surface, increasing its surface area for carbon mineralisation. It also has the potential to remotely measure various factors that affect the reaction.

In practice, most ultramafic mine sites don't possess the required mineralogy, so the team has trialled microwaving the wastes to produce a 'microwave popcorn' of ultramafic minerals to enhance their surface area and break their chemical bonds. Several other technologies to help speed up carbon mineralisation are also in development.

Many mining companies are interested and its first commercial deployment was at Mount Keith mine in late 2023. If it becomes mainstream, it could influence mineral exploration activities to consider not only ore minerals but also to locate the most promising carbon-sequestering rocks and modify the subsequent mining plan to accommodate and make the best use of these rocks.

The technology makes the most sense when allied to other decarbonising systems, such as renewable energy. It is aimed at hard-to-remove-CO_2 in mining operations and should be regarded as 'additional' rather than as an alternative; it will always be preferable to avoid producing the CO_2 in the first place. There is also the prospect of generating carbon removal credits to offset parts of an operation, or to offset external emissions by others, or packaging with the product and selling it at a premium. Potentially, pre-purchase agreements could also be entered into with other organisations looking to offset their carbon emissions, known as carbon removal credits.

Arca is now a growing company of 20 people. Its work has already received high-profile accolades, such as a million-dollar award from the Musk Foundation's X-Prize Carbon Removal initiative.

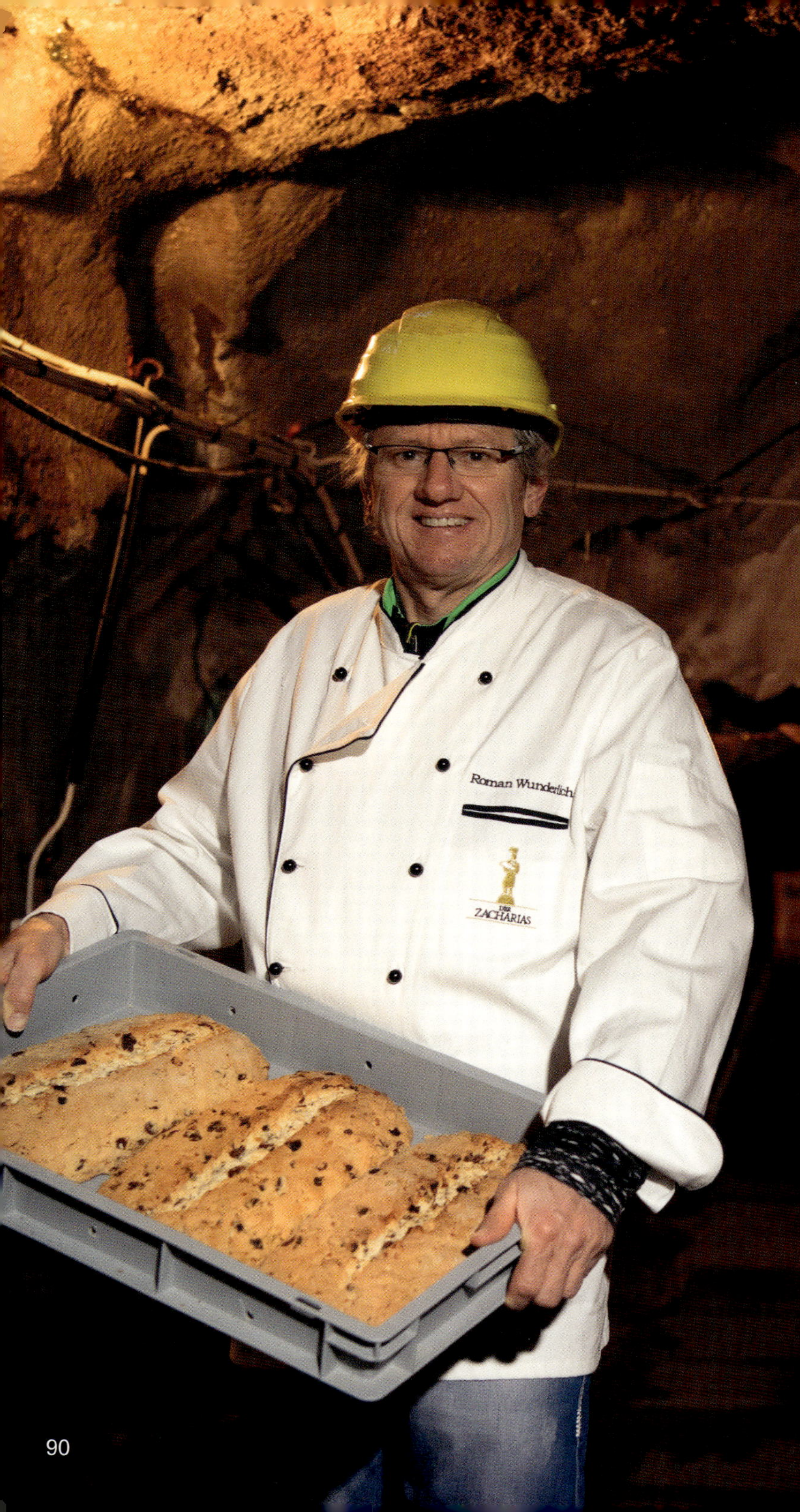

40
Cake Mine
Germany

The ancient orefields of the Erzgebirge, or Ore Mountains, sit along the German-Czech Republic frontier and have been mined for centuries for all manner of metals and minerals. Today, one old tin mine also produces cake – or, more precisely, stollen!

Stollen is a delicious traditional German festive cake-like bread containing – amongst other things – dried fruit and nuts, marzipan, almonds and even a nip of rum or brandy. To enhance the flavour further, as with a fine wine or cheese, after baking the stollen is left to sit or 'ripen' for days to weeks in a cool, humid place with good air circulation. Where better to find such conditions than down an old tin mine in Saxony?

Saxony in eastern Germany is famous for both its mining heritage and its stollen. The Tannenberg mine began in the 15th century and eventually closed in 1964. It reopened as a visitor mine in 1996 to try and stimulate tourism after German reunification. Roman Wunderlich of Wunderlich Bakeries was visiting the mine with his family and its manager, Steffen Gerisch. Together, they hatched the idea of combining mining and other local cultural traditions, notably stollen.

The collaboration began in 2006 with a year of experimentation, testing different storage methods and locations in the mine to find the optimal conditions and timings.

Today, the baked stollen is driven by electric buggy to a restricted area a kilometre inside the mine. Here, at a depth of 80 metres, it is left to mature for around four weeks at a constant 3 to 5°C and 92 to 94% humidity. Approximately 3,000 stollen are prepared this way annually. It is marketed with the catchy German slogan 'Der Stollen aus dem Stolln', which roughly translates into English as the not-so-catchy 'Stollen from the Mine'.

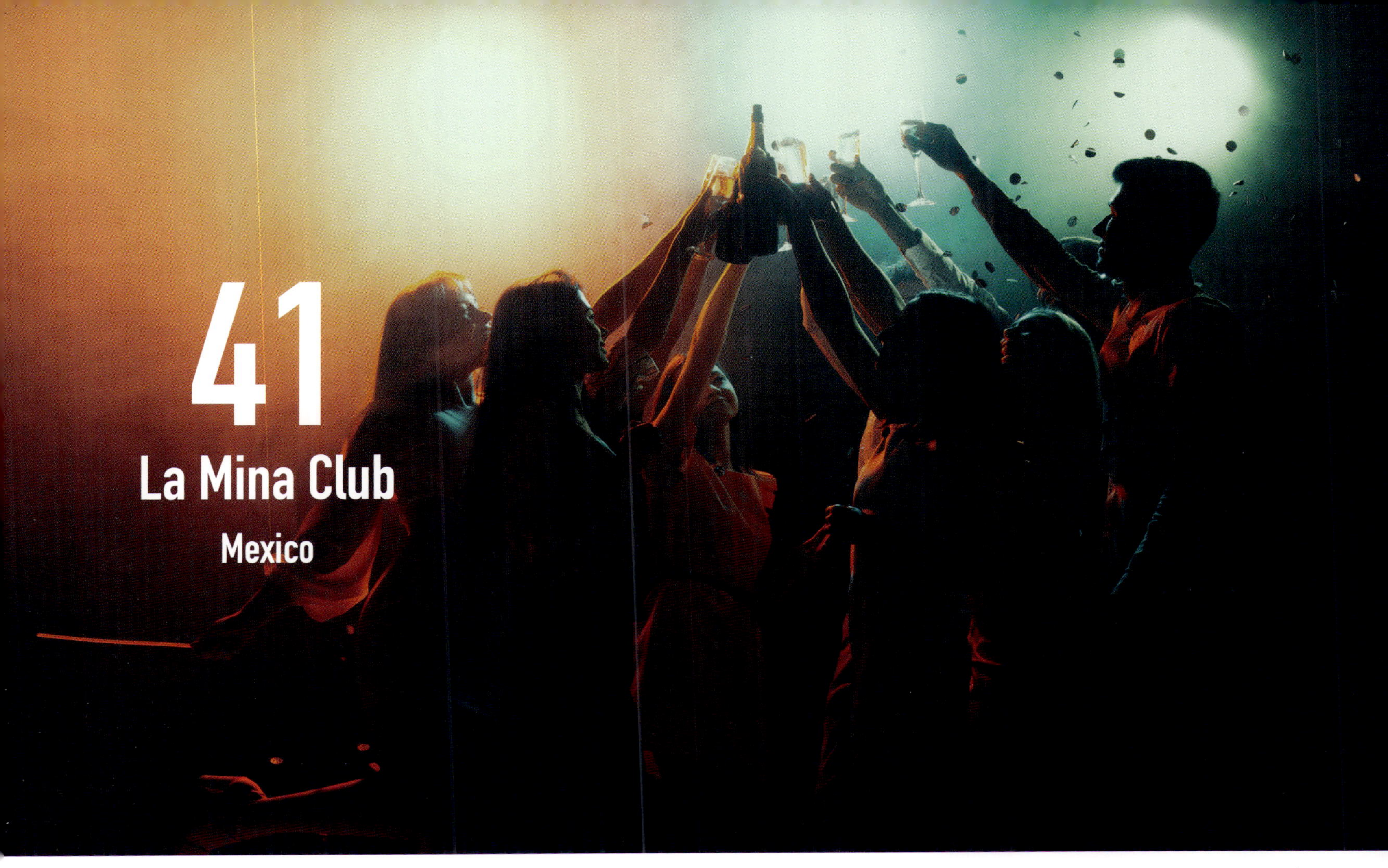

41

La Mina Club

Mexico

Many clubs aspire to be part of the underground scene, but Bar El Eden Mina really is – it's 184 metres below the surface in what was the El Eden mine in Zacatecas, central Mexico.

The mine is said to have been established in the 16th century and produced gold, silver, zinc, copper, iron and lead for the next 400 years. El Eden was first mined by slaves from the indigenous community, who endured hellish conditions and a high death rate.

Even in the 20th century, life expectancy for miners was only 30–40 years. The mine closed in 1965 due to the increased risk of flooding – the lower two levels are underwater – and its proximity to the city, which had encroached on it over the centuries. It reopened in 1975 as a mining museum – the bar and club were later additions.

The club is only open a few nights a week, but it's popular. To reach it, you board a land train that ferries you half a kilometre through the darkness to the club entrance – it only takes a few minutes, but it sets the scene. There's a small cavern containing a bar, and beyond it is a larger one containing the dancefloor. The club has a capacity of 400. Its rough stone walls are strobe-lit, and its unique acoustics are well-suited to DJing. This is one 'rock club' that doesn't have to worry about annoying the neighbours.

42

The Gold Coast

Australia

From afar, the gleaming high-rises of Surfers Paradise appear to rise from the sea into an azure sky, towering over the white sand beaches and sparkling sea of one of the world's classic surfing and beach life destinations. Visited by millions of sun-seekers annually, Queensland's Gold Coast and its iconic Surfers Paradise are possibly one of the most successful post-mining regeneration tales anywhere — yet few know the back story.

The coastal territory of the Aboriginal Yugambeh people that became southern Queensland/northern New South Wales was first industrially exploited when alluvial gold was discovered in the 1860s, triggering a gold rush. While the beaches were being mined for gold, metallurgical innovations elsewhere initiated a market for titanium for use in alloys. Titanium is found in ilmenite, which is found in the coastal sands, and later, ilmenite's use as a brilliant white pigment was developed, further raising demand. Other important heavy minerals present include zircon, used in metal castings, and rutile, which is used in electric arc welding. The heavy minerals had been concentrated by the wind and sea into easily distinguishable dark, thin beds laying horizontally in the otherwise pale beach sand. The beds were abundant and easily accessible in the beaches and dune systems where Queensland meets New South Wales.

Heavy mineral sand mining began over 100 years ago, with large-scale operations in the middle decades of the 20th century, including Burleigh-Broadbeach, Coolangatta-Currumbin, South Kingscliff and Tweed Coast. By 1955, there were six processing plants on the beaches between The Spit and Coolangatta, with three further companies on the Tweed Coast, along 60 kilometres of sub-tropical beach coastline from South Stradbroke Island to Kingscliff, over the border into New South

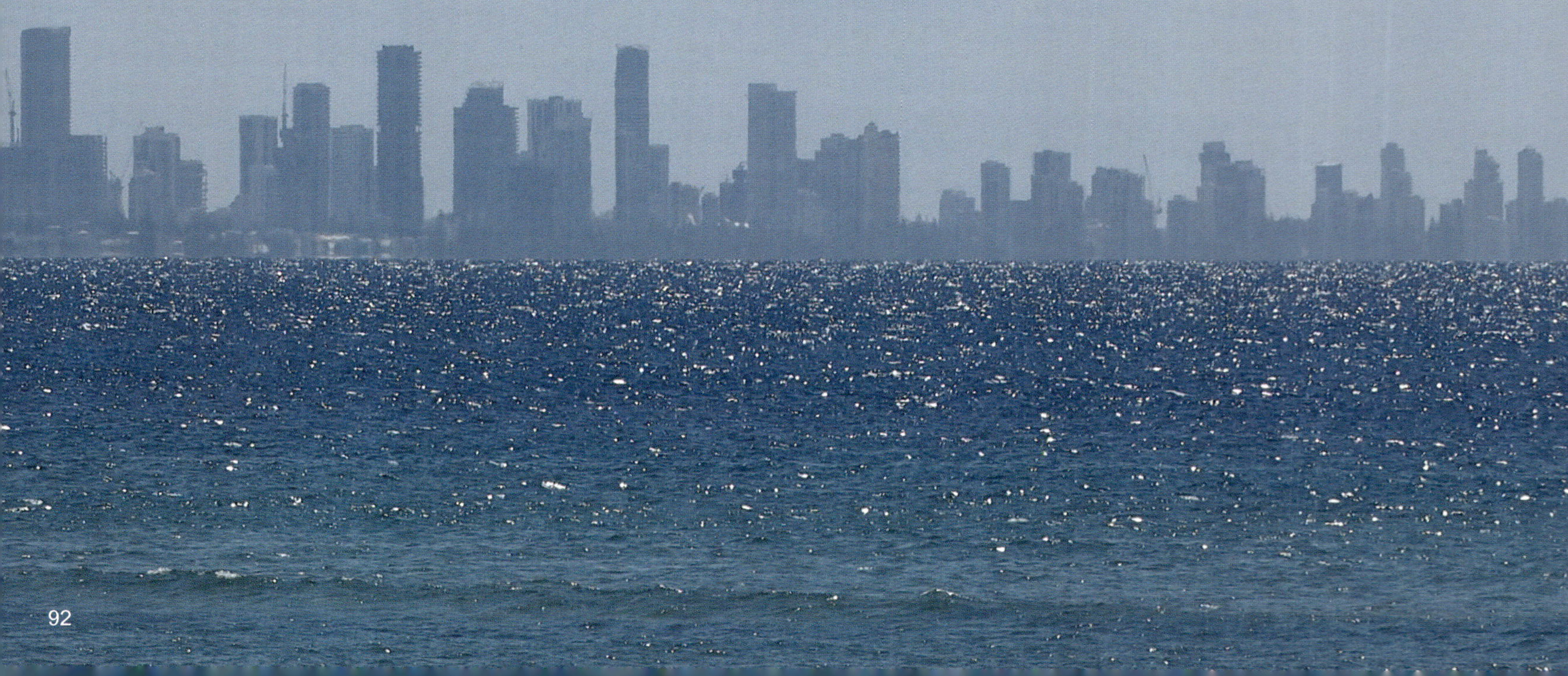

Wales. Mid-20th century black and white photographs show desirable beachfront housing behind a no-go sand mining zone – so much for sea views!

After extraction, the mining companies rehabilitated the surface, and the lease was relinquished by the government, which then prepared road infrastructure and sold allotments of development land, recognising its potential value and thereby becoming a beneficiary of the land sale.

Limited coastal tourism had begun to grow from the 1920s and really took off in the 1950s and 1960s, overlapping with the winding down of the sand mining boom. The burgeoning population of south-east Queensland, growing wealth and an ever-popular beach and surf lifestyle fuelled demand for new seaside accommodation and tourism facilities – especially around world-class surf breaks. Entrepreneurs from outside of the mining industry saw an opportunity and moved in – the flat, coastal landscape being ideal for holiday homes and tourism developments. The modern Gold Coast was born.

There are still over 1,000 people involved in mining here, and sometimes, this dovetails with construction; new high-rise developments require deep foundations dug into the sand, which often brings new mineral sands to the surface. This is collected by a local minerals company for processing at their Currumbin plant to extract the valuable minerals for sale.

In just two generations, an extensive coastal zone heavily disturbed by mineral sand mining, with limited infrastructure, sparse roads and a sparse population (it was 29,000 in 1950), has become Australia's sixth biggest city with a growing population of around 700,000 and a beach culture and reputation to die for. In 2019, over 14 million visitors pumped nearly AUD 6 billion into the local economy supporting 44,000 jobs in what is the third biggest industry after construction and manufacturing.

The Gold Coast World Surfing Reserve curves along 16 kilometres of coastline, much of which consisted of mine sites 70 years ago. In 2032, it will host the Brisbane Olympics surfing competition –not bad for a bunch of old mine sites!

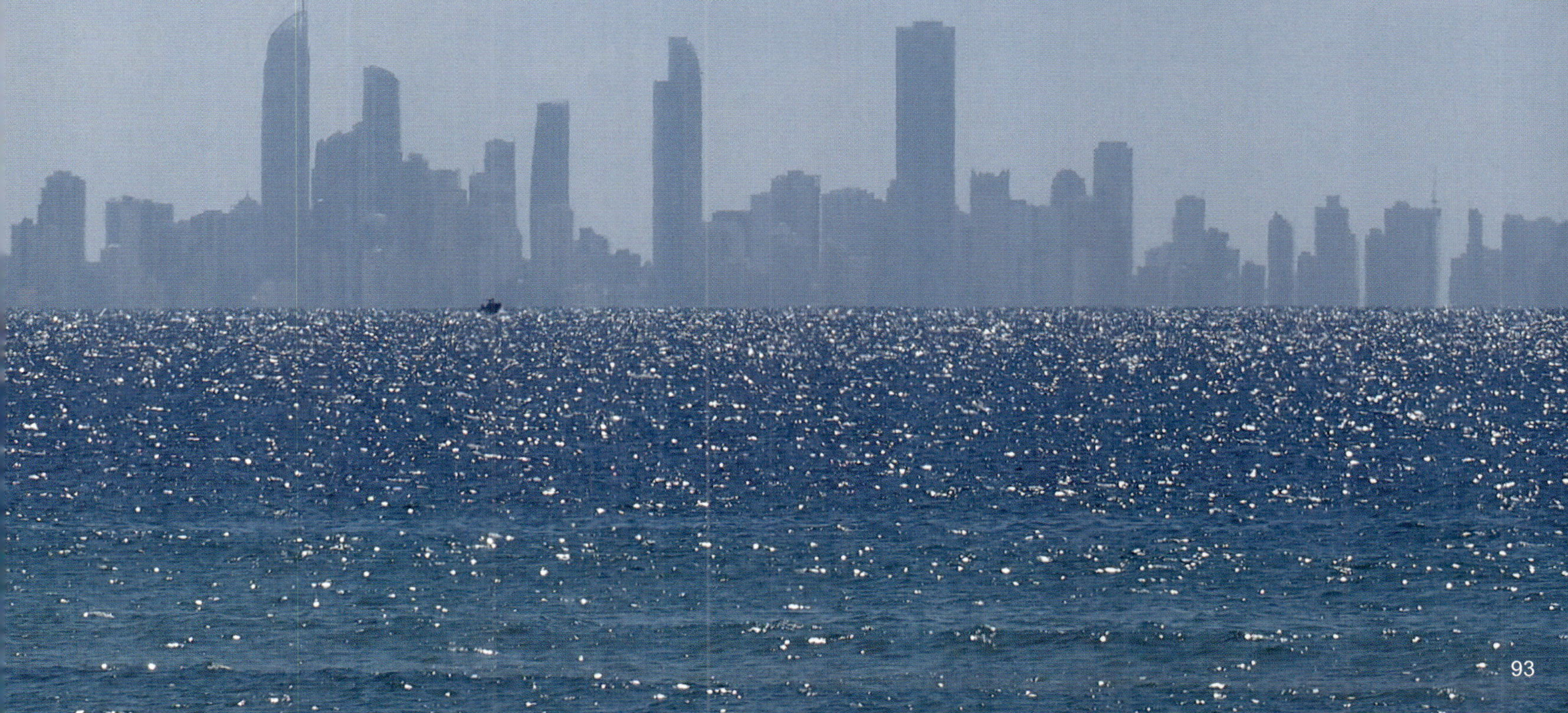

43

Paris Catacombs

France

Paris is known as the 'City of Light', but below its streets is a city of darkness mirroring the one above. There are at least 300km of tunnels and chambers, many of which share their names with the streets above.

The limestone, clay, sand and gypsum beneath Paris were first mined by the Romans. Over the centuries, as the City of Light rose and grew, so did the darkness below. And then, towards the end of the 18th century, the two abruptly met. On 17 December 1774, a sinkhole a quarter of a mile long opened on the eastern side of the Rue de l'Enfer (Street of Hell), dragging houses and people into the void.

Over six million corpses would follow, albeit by a more circuitous route, as the labyrinth below provided a solution to another equally urgent problem facing the city: it was running out of space to bury its dead.

Few Parisians were unaware of this problem; over the centuries, great burial mounds had been created in the churchyards, and the smell of putrefaction outcompeted any perfume, but it wasn't until 1780 when the retaining wall of the *Cimitière des Saints Innocents* collapsed and sent a tidal wave of corpses into the cellars of the adjacent houses that the idea of using the mines below the city to bury the dead occurred. Its architect was Charles-Axel Guillaumot, appointed by Louis XVI as France's first inspector of mines.

Over the next twelve years, bones were moved from the city's graveyards to the ossuary in a macabre nightly procession of carts, gravediggers and priests. The ossuary then became the de facto burial place for Parisians, fed in part by the guillotine thanks to Robespierre's brutal *Terror* – and then when the revolution began to devour its own, Robespierre himself in 1794.

The last bones were interred in 1860, by which time the ossuary had been open to the public for over fifty years, and it remains a popular tourist attraction today.

Although the ossuary is the most famous part of the mines, less than 0.5% of its tunnels are given over to the dead. In the 1820s, parts of the mine were used to farm mushrooms. By 1880, over 300 mushroom farmers were producing 1,000 tonnes of mushrooms a year. The arrival of the Paris Metro above the mines in the 1890s put paid to most of the farms, but a handful of micro-producers still supply the kitchens of upmarket restaurants.

Aside from the ossuary, the rest of the catacombs are off limits to the public and are policed by the 'cataflics' or 'catacops', albeit with limited success. In recent years, a pop-up cinema was run in the depths for several weeks, and the mines remain a popular destination for dedicated 'cataphiles' despite their often-unstable nature.

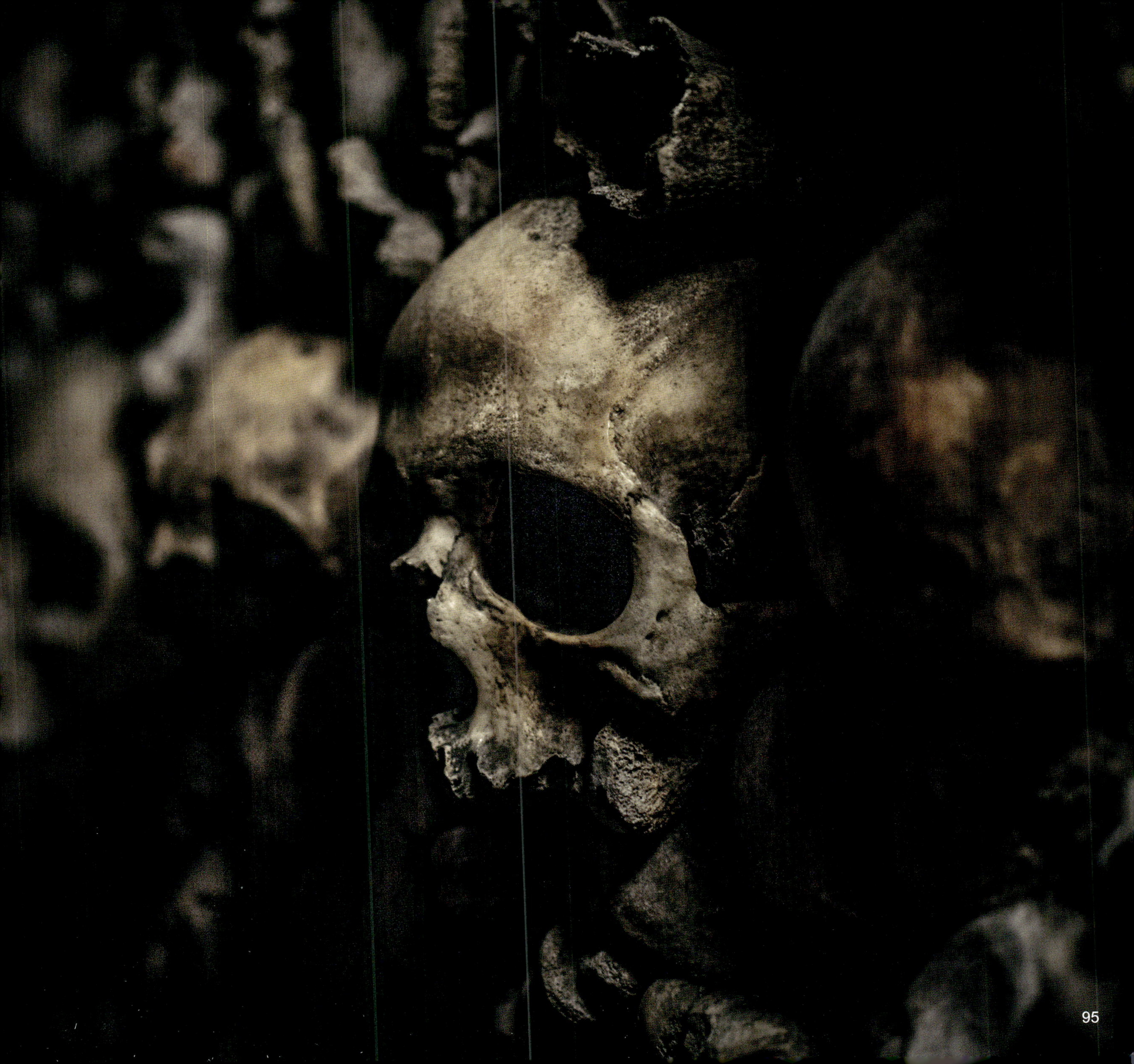

44

Shelton Abbey Pheasant Shoot

Ireland

Amongst Ireland's historic mining centres is Avoca, in the eastern foothills of the iconic Wicklow Mountains 55km south of Dublin. Until 1982, Avoca produced copper intermittently for over 230 years. During that time, it generated 7.5 million m^3 of mining residue, which was disposed of in an offsite tailings storage facility (TSF) at Shelton Abbey. In 1984, a plan for rehabilitation of the closed TSF was agreed and financed by the Irish government. The 30-hectare site was remediated by contractors engaged through the Official Receiver to Avoca Mines Ltd.

The TSF was covered in two consecutive phases: firstly, 50cm of shale was taken from an adjacent degraded woodland, and secondly, a 30cm layer of poor-quality topsoil was created from locally available materials. A sheep grazing pasture was established and managed for more than a decade. Then, with the expiry of a state-controlled agricultural lease, a change took place.

The withdrawal of grazing allowed the invasion of a diverse range of plants, in particular, shrubs and low canopy trees – species naturally suitable to the site's semi-waterlogged conditions. The dense, marginal woodland, encircling scrub and open grassland was ingeniously exploited by what ultimately became the Shelton Abbey Shooting Club (SASC) based around what was the ancestral home of the Earl of Wicklow.

The SASC spotted an opportunity presented by the habitats for a managed game-bird rearing programme, and one of the most coveted and respected pheasant shoots in Ireland was born.

Some of the land has since been leased by a syndicate developing a 'driven shoot' known as The Oaks, which is the signature of the Shelton Abbey project and is now believed to contribute over EUR 1.4 million per year to the local economy.

It could be said that the unique combination of woodland, grassland, access roads and paths sets no useful precedent for game-bird projects elsewhere. However, habitat creation, with a specific land use in mind, is a basic feature of land repurposing.

Careful planning works, but so does the exploitation of an opportunity that arises by chance.

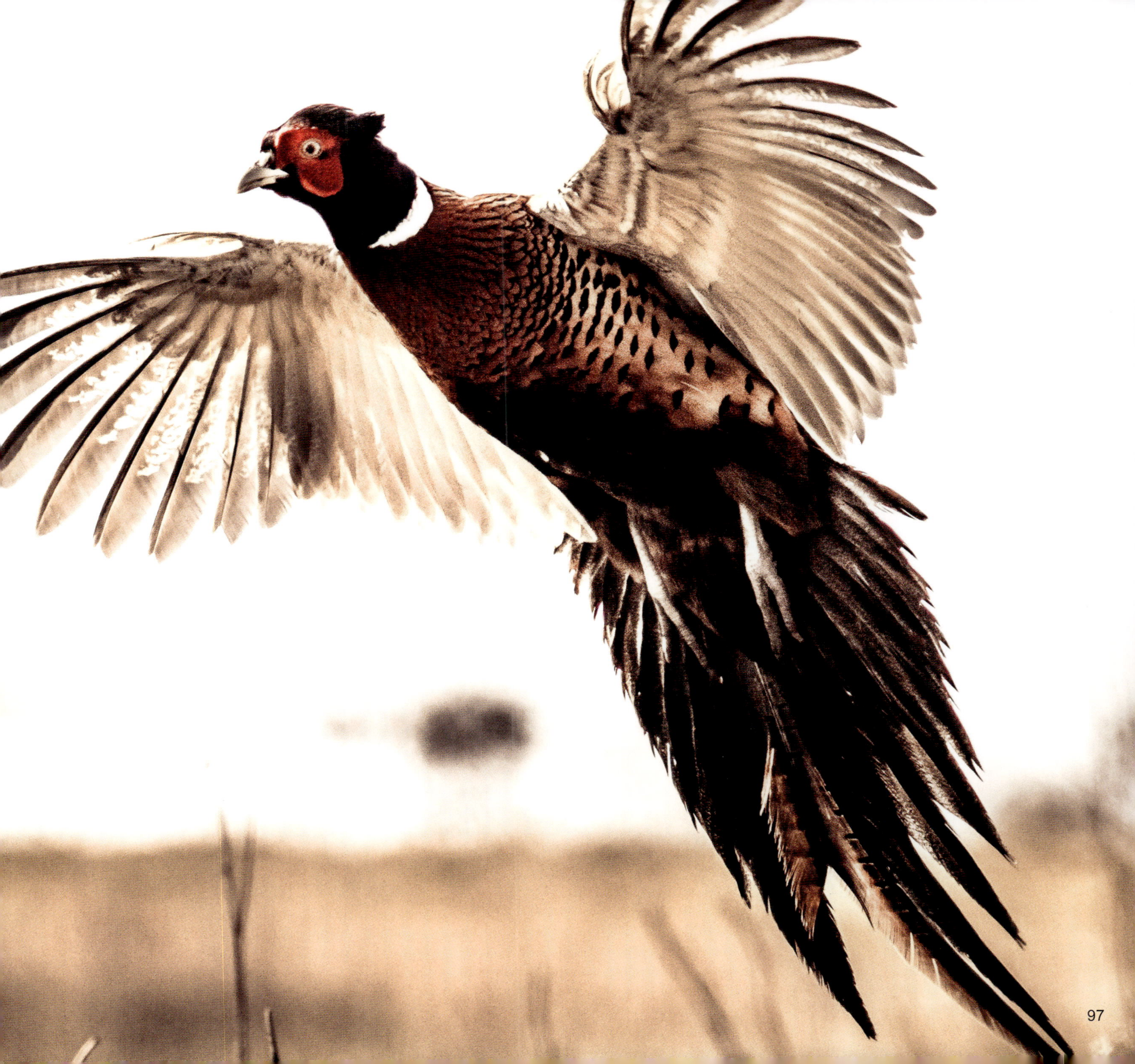

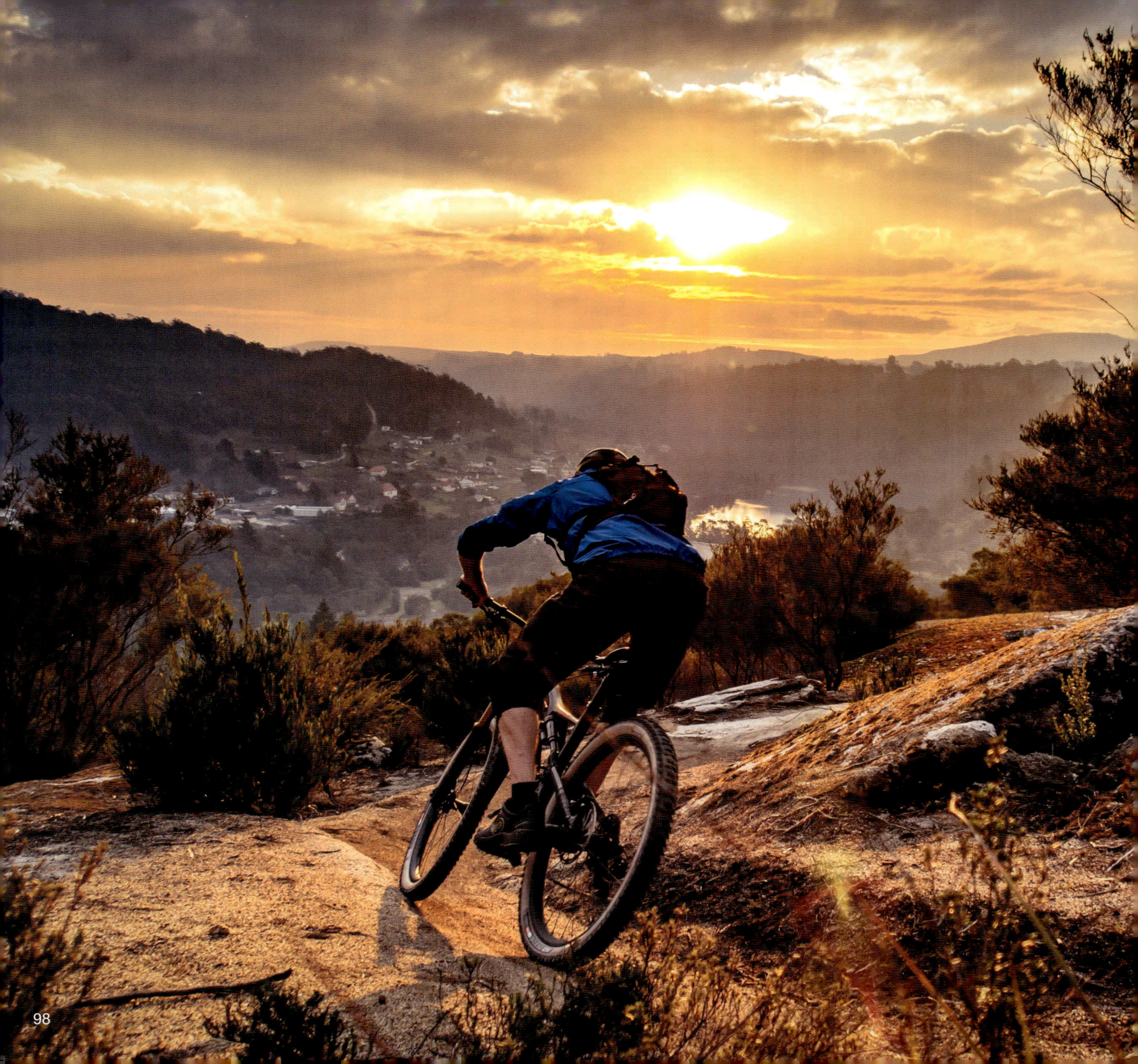

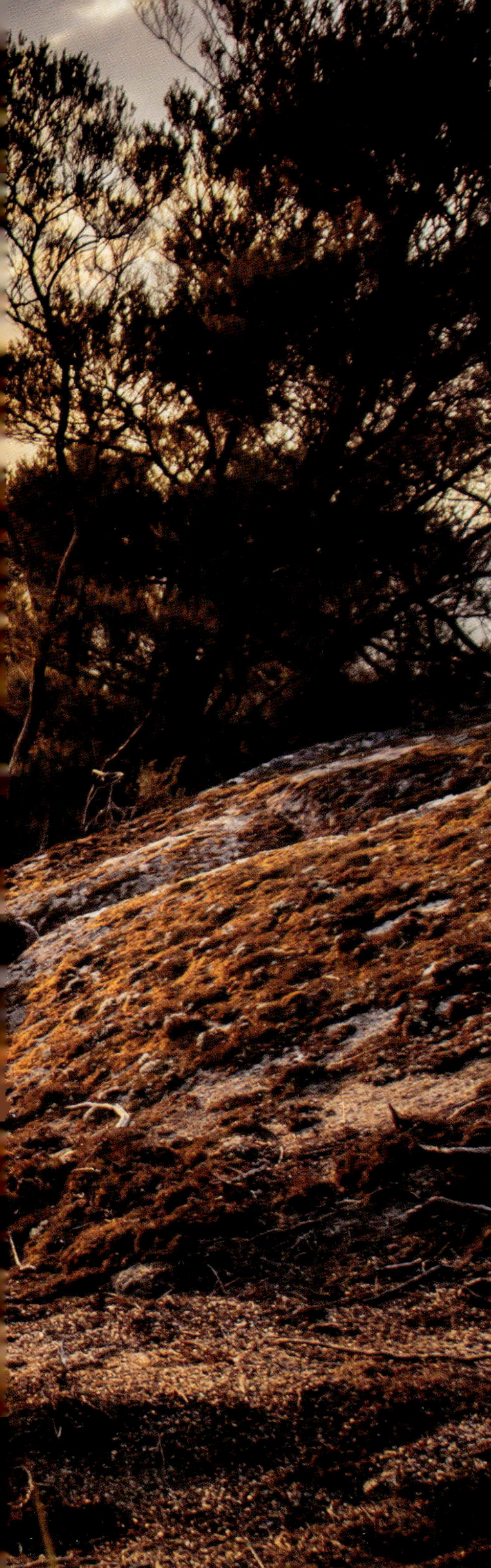

45

Blue Derby

Australia

Aficionados call it 'flow' – a mountain biking (MTB) term describing the harmonious state of mind you experience while riding effortlessly on a trail, with minimal pedalling and braking, when your riding and the terrain become one. Exhilarating and exhausting, with a sprinkling of danger, the MTB trails of Derby, north-east Tasmania, are renowned for the 'flow' they create as you carve through a wilderness that has reclaimed the ravages of miners and loggers, cloaking the remnants of industry with tree-fern groves and gum trees.

Derby, on the banks of the Ringarooma River, is experiencing its third incarnation – first a tin mining township, then a ghost town, and now a burgeoning, world-class MTB mecca known as Blue Derby. Three thousand people lived there at its 19th-century zenith. Disaster struck in 1929 when the river flooded the town and the open pit, killing 14 people. It reopened five years later, but the mine never returned to its previous productivity and closed permanently in 1948. So began a decades-long spiral of decline. Derby became a ghost town as the people left for greener pastures; its buildings became dilapidated, its lots vacant.

The inspiration for Blue Derby came from two local lads who saw potential in the picturesque landscape of this tranquil rural backwater. They connected with Dorset Council when the state government published a strategy to develop bike trails on the island to stimulate rural regeneration. Aware that many other MTB trails were only accessible by car, thus limiting opportunities for community regeneration, they recognised that mountain bikers will cross the world to cycle stunning new locations. They imagined an MTB resort, like a ski resort, where the trails start and end in the town, so it becomes the destination. Opportunities then arise for new accommodation, bike shops, bars and cafés and shops, reversing the historic town's decline.

Facilitated by Dorset Council and the Northern Tasmania Development Corporation, state and federal government grants of AUD 3.25 million helped Blue Derby to build the first 40km of trails. World Trails from Cairns, Queensland, designed and constructed them, and they opened in 2015 as the first major MTB trail network in Tasmania. It was an instant success, attracting more funds, and there are now 125km of trails.

Already well-known in the MTB world, in 2017, assisted by the state government, Derby hosted an international MTB event on the professional Enduro World Series circuit, which had never visited Australia before. Riders were particularly impressed that – unusually – the event was hosted in a town, affording après-ski-style relaxation. As well as trailhead accommodation, the town of Derby sports a microbrewery and distillery, restaurants and bars and even a floating sauna! In both 2017 and 2019, elite riders voted Blue Derby trails as Enduro World Series trail of the year. The World Series returned to race again in 2023, confirming Derby's status as a pre-eminent MTB destination.

For less-than-elite riders, the free-to-ride trails are open year-round. The number and distance of trails means many visitors stay overnight, increasing visitor spend in Derby. Blue Derby attracts over 80,000 visitors annually, mostly for mountain biking, bringing more visitors to this part of Tasmania who generally stay for four to five nights in the region plus a few more nights elsewhere in Tasmania. Annually, this is worth tens of millions of Australian dollars to the Tasmanian economy. About 150 local jobs have been created, in bars, cafés and bike hire businesses. Old businesses are being renovated and new ones are under construction. Strict planning conditions require buildings fit with the traditional architectural vernacular.

Inevitably, there have been some growing pains to this rapid regeneration, such as insufficient visitor-supporting infrastructure, too many visitors at times, a lack of short and long term accommodation – a particular challenge for Derby's workforce – and land supply issues.

The Blue Derby Foundation was set up to promote the area's MTB activities and serve the interests of the community. It fundraises to manage and develop the trails and improve the visitor experience. The council wants to carry out masterplanning for Derby that includes the MTB precinct, reviews of infrastructure and land zone planning, a tourism and marketing plan and a study of the socio-economic impacts.

The first trails generally attracted men; now, the focus is on attracting female riders, and families. To this end, a popular new pump track was recently constructed on mine waste by the riverside, which provides an accessible and fun way to hone riding skills with minimal pedalling. A Blue Derby Off-Road Triathlon is planned for launch to attract a broader sporting audience.

Reassuringly, the town's mining roots are not forgotten. It has retained several charming old buildings and a memorial to those lost in the flood. The Blue Derby Foundation also plans to run guided MTB groups along the trails to explore the post-mining landscape. Some trail names like 'Detonate' and 'Krushka's', (after the pioneering Krushka brothers who, discovered tin here and set up the mine) already reflect the town's mining heritage, and others flow past rusting mining relics half-secreted within the flourishing forest.

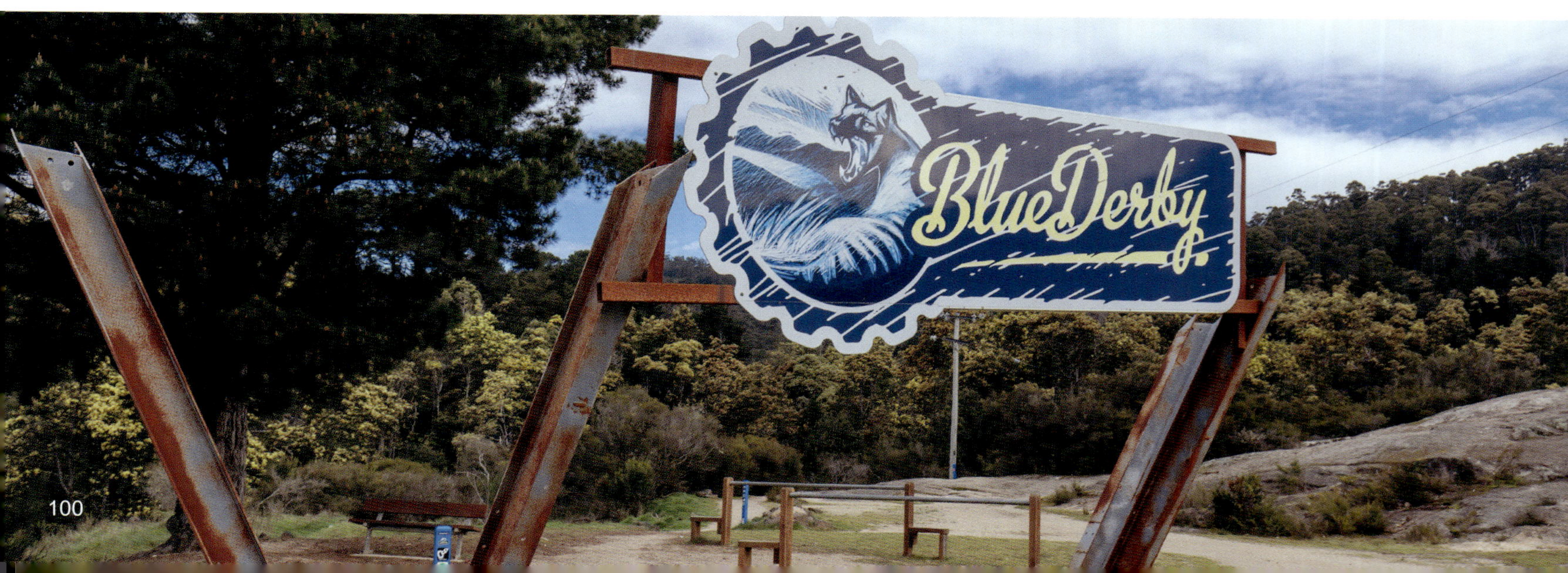

46

Big Sandy Prison

USA

Flat land is at a premium in the beautiful Appalachian Mountain chain. The controversial practice of removing mountaintops to mine coal blights vast expanses of this precious region and backfills narrow valleys with the rubble of decapitated mountains. But the level ground it creates does at least lend itself to all kinds of post-mining development.

The Big Sandy high-security federal prison in Martin County, Kentucky, signifies a growing – but contentious – post-coal economy in central Appalachia. Constructed on a mountaintop removal mining site and a former underground coal mine, the prison houses over 1,400 high-security male prisoners. Notorious inmates have included Chuckie Taylor, son of former Liberian dictator Charles Taylor; Vincent Basciano, acting boss of the Bonanno crime family; and El Sayyid Nosair for his role in the 1993 New York bomb plot.

When it opened in 2003, it was the most expensive federal prison project ever built. Before construction was completed, a perimeter tower sank into an abandoned mineshaft, leading to costly rehabilitation work by the government. This subsidence earned the prison the nickname 'Sink Sink' from local people in a derisive reference to New York's infamous Sing Sing prison.

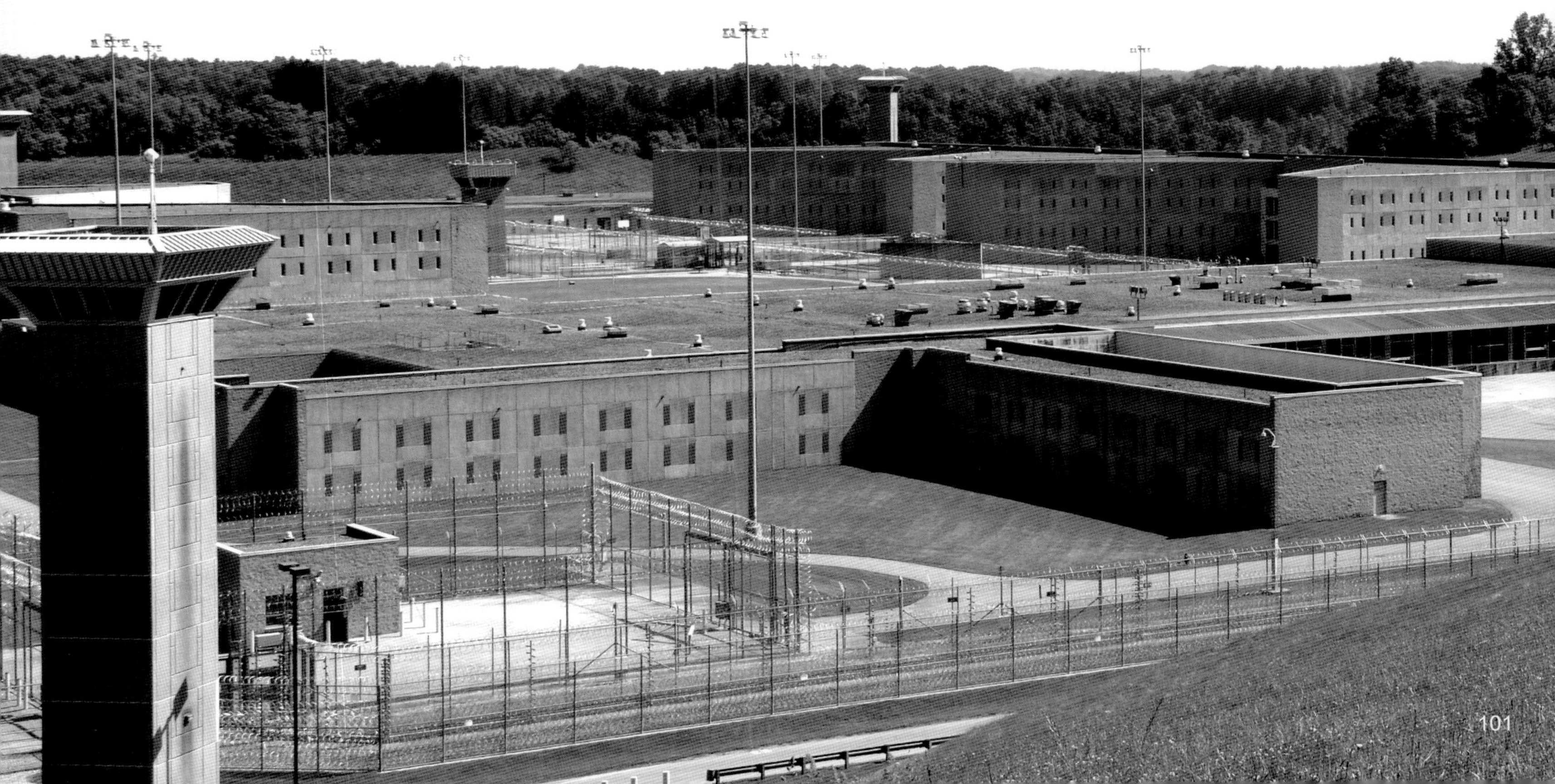

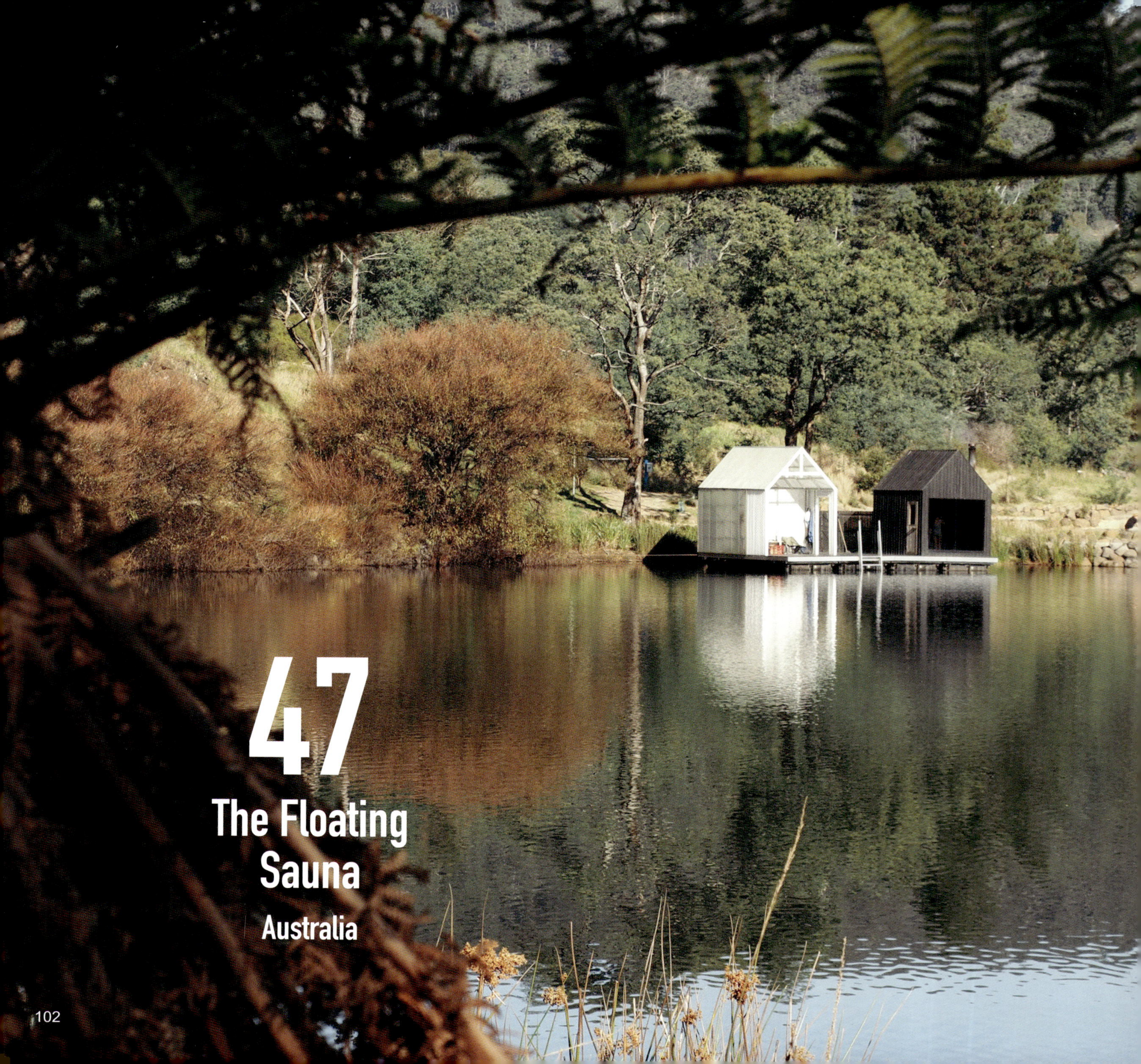

47
The Floating
Sauna
Australia

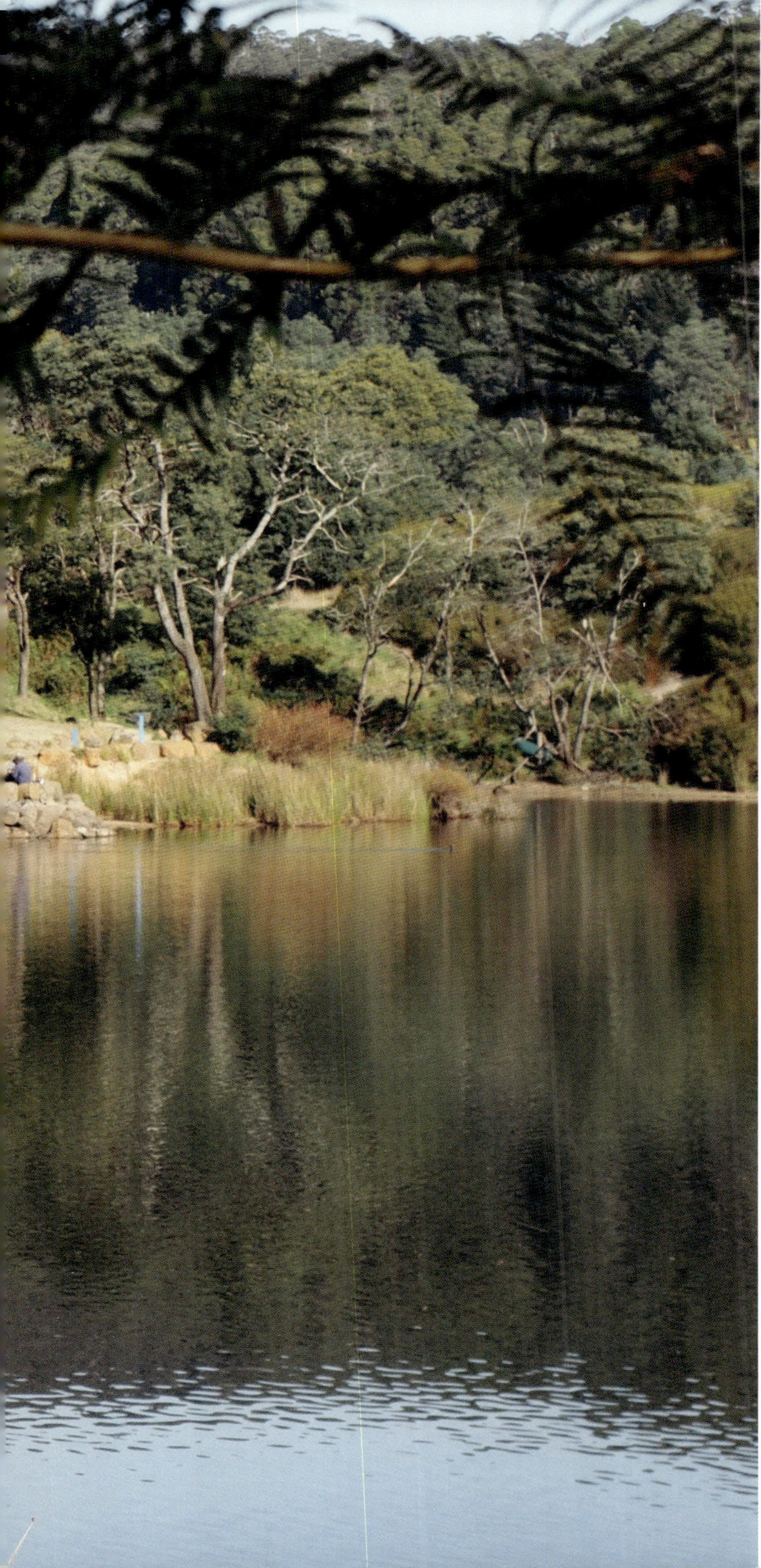

Derby's floating sauna is surprising in several ways. It feels tranquil floating on Lake Derby's calm, blue waters below a steep, craggy hillside cloaked in young Tasmanian rainforest, but this picturesque backdrop was once a denuded, noisy, industrial place. The 65-metre-deep lake was the open pit of Briseis Tin Mine; the craggy hillside was hydraulically mined for ore, and the incongruous angular bank is a waste rock tip.

Today, Briseis Hole, or Lake Derby, is a part of a growing number of recreational activities at the remote township of Derby – a former tin-mining ghost town recently reborn as an international mountain biking Mecca. The sauna's monochromatic sheds – one black, one white – fixed to a pontoon and connected to the shore by a bridge seem initially out of place but gradually make sense. Inspired by the floating saunas common in Scandinavia, Nigel Reeves saw a spin-off business opportunity to accommodate bikers after a long day in the saddle on the famous Blue Derby mountain bike trails.

Designed by Licht Architecture and opened in 2020, Australia's first wood-fired floating sauna took six months to create. The architects received a National Commendation from the National Architecture Awards in 2021 and won the Peter Willmott Award in the Tasmanian equivalent in the same year.

Booked up for weeks in advance, the Floating Sauna has become a destination in its own right. Reeves's aim is to encourage people to reconnect to nature and each other in what feels like natural surroundings with unrestricted views over the lake. The yin-yang of the two buildings is reflected in the counterbalancing experiences of heat from the sauna to the ensuing rejuvenation of a cold-water plunge. And yin-yang might also apply to the repurposing of this former mine site.

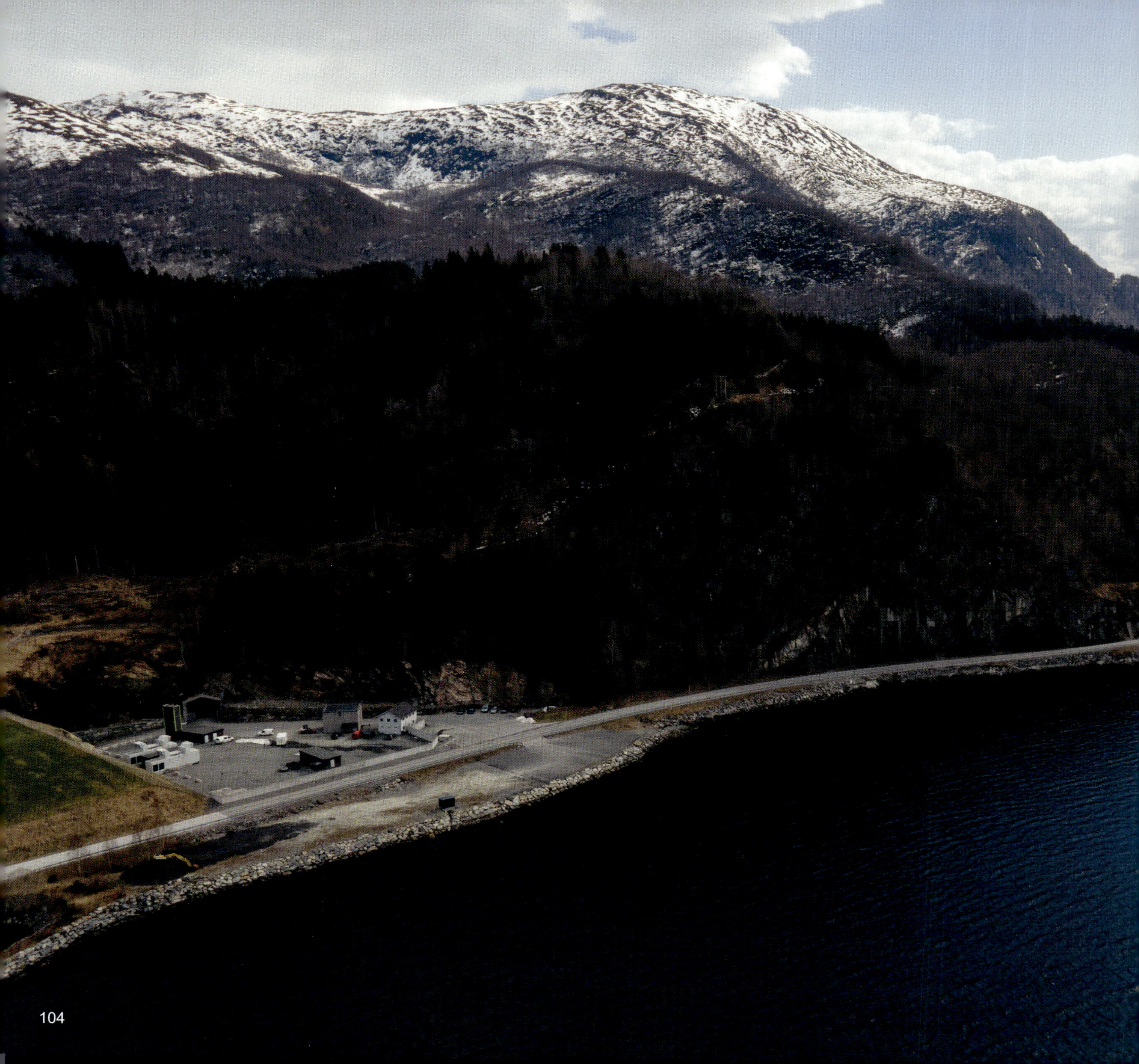

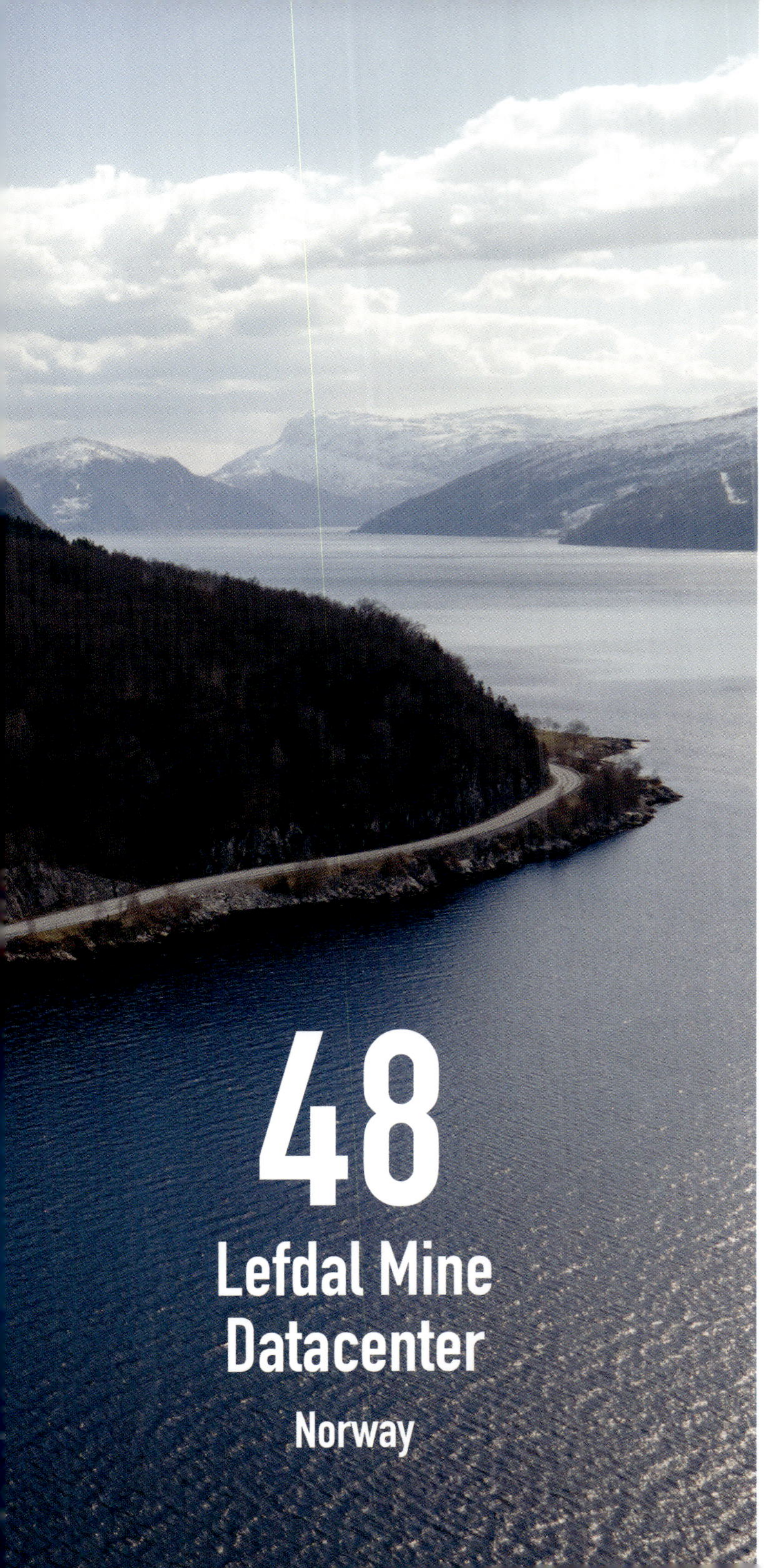

Cloud storage may sound ethereal, but it takes a lot of energy and water to maintain the data centres that make it possible. Collectively, the 8,500-plus data centres generate over 1% of the world's CO_2 emissions. Cryptocurrency mining alone emits over 85 million tonnes of CO_2 every year, and its carbon footprint is increasing – in 2023, the sector produced the same amount of emissions as the entire country of Serbia did in 2019.

Three hundred kilometres northeast of Bergen, the Lefdal Mine Datacenter on the west coast of Norway offers a more sustainable approach. The mine extracted peridot – a yellowish-green stone used mainly for jewellery – from 1971 until it was exhausted in 2009. The idea for the data centre came from the CEO of an IT company, Local Host, whose father had been the mine's works director. The former mine sits beside a fjord and offers 120,000m² of floor space over six levels, with the capacity to expand to 14 levels. Data servers are shipped to the nearby port of Måløy and then transported by truck along the fjord, where they are driven directly inside the mountain.

There are larger data centres, but Lefdal is zero carbon. Around half the energy used by a data centre is for cooling its servers. Lefdal's proximity to the shore means it can use the constant 7°C of the fjord's seawater as a heat exchanger. The closed-loop cooling system pumps fjord water up to the heat exchanger, where it meets the pipes carrying hot water from the facility. The warmer water is fed back into the fjord while the cooled water recirculates around the data centre. This saves not only energy but also water, which would otherwise evaporate via a cooling tower or be discharged as wastewater. Data centres generally use potable water supplied by utility companies for cooling, and only 5% is reused.

As more of the world joins us online, reducing the resources these centres consume is essential. The amount of data doubles every 18 months, but the sector's emissions must drop by 50% to stay on track with Net Zero targets. What Lefdal calls the Norwegian Solution may be the way forward.

49

Tsau ‖Khaeb Sperrgebiet National Park

Namibia

Vast and empty, southern Namibia's almost unheard of Sperrgebiet is one of Africa's last remaining true wildernesses, preserved by restricting access to its 26,000 square kilometres. Its pristine desert and sublime coastline convey isolation, dark skies and a touch of mystery and trepidation, scattered with traces of human endeavour and ingenuity.

The Sperrgebiet's geological wonders include, amongst other things, 1.5-billion-year-old rocks, a 2.5-kilometre-wide meteorite crater – and of course diamonds. Ecologically unique, the Sperrgebiet is also a global biodiversity hotspot – the Succulent Karoo Biome; there are more species of succulent plants here than anywhere else on Earth. But it is also home to rare, hardy animals too, such as the brown hyena. The Orange River that enters the sea near Oranjemund boasts a Ramsar site – one of the country's three globally important coastal wetlands and a metropolis for birds. And along the coast runs Namibia's first Marine Protected Area, home to 35 species of whales and dolphins, two species of seals, and penguins.

While archaeological remains date from at least 300,000 years ago, in the 20th century humans were largely excluded. Colonised by Germany, the area was turned into a restricted zone –'Sperrgebiet' – after coastal diamonds were discovered in 1908. Only company workers could enter this 100-kilometre-wide buffer zone. Small, isolated mining communities were established, connected by railways to the coast, and close to the diamond fields and their processing equipment. As the diamonds ran out, the towns were gradually abandoned.

Traces of railway tracks dissolve into the desert and brown hyenas saunter through mining ghost towns, their wind-scoured doorways and broken windows conceding to desert sands forming dunes in rooms where once people sat. Close by are the rusting hulks of processing plants, desalination plants, power lines and flimsy worker accommodation.

Today, after its creation in 2004, the Tsau ‖Khaeb Sperrgebiet National Park protects this natural and cultural heritage, while still permitting coastal mining. Ironically, the largely untouched nature of this unique wilderness is down to the restrictions that were imposed by the mining industry. The historical mining features are relatively under-exploited tourism assets, though restricted, guided tours are available with the right permits from the Diamond Police. But what of current mining operations? There is a debate as to what extent current mining infrastructure will enrich the area once mining has ceased as it is as much part of the landscape's unique narrative as the abandoned structures and towns from many decades ago.

The national park exemplifies a new type of Namibian protected area, existing to protect biodiversity and cultural heritage, while also contributing to the economy through tourism. So, the national park that you can't visit is changing: tourism concessions are being awarded and management plans are being drawn-up for activities such as desert experiences and ghost town tours. Although it will never be a mass tourism destination, access to the sensitive ecological and cultural features of this land will need to be very carefully managed.

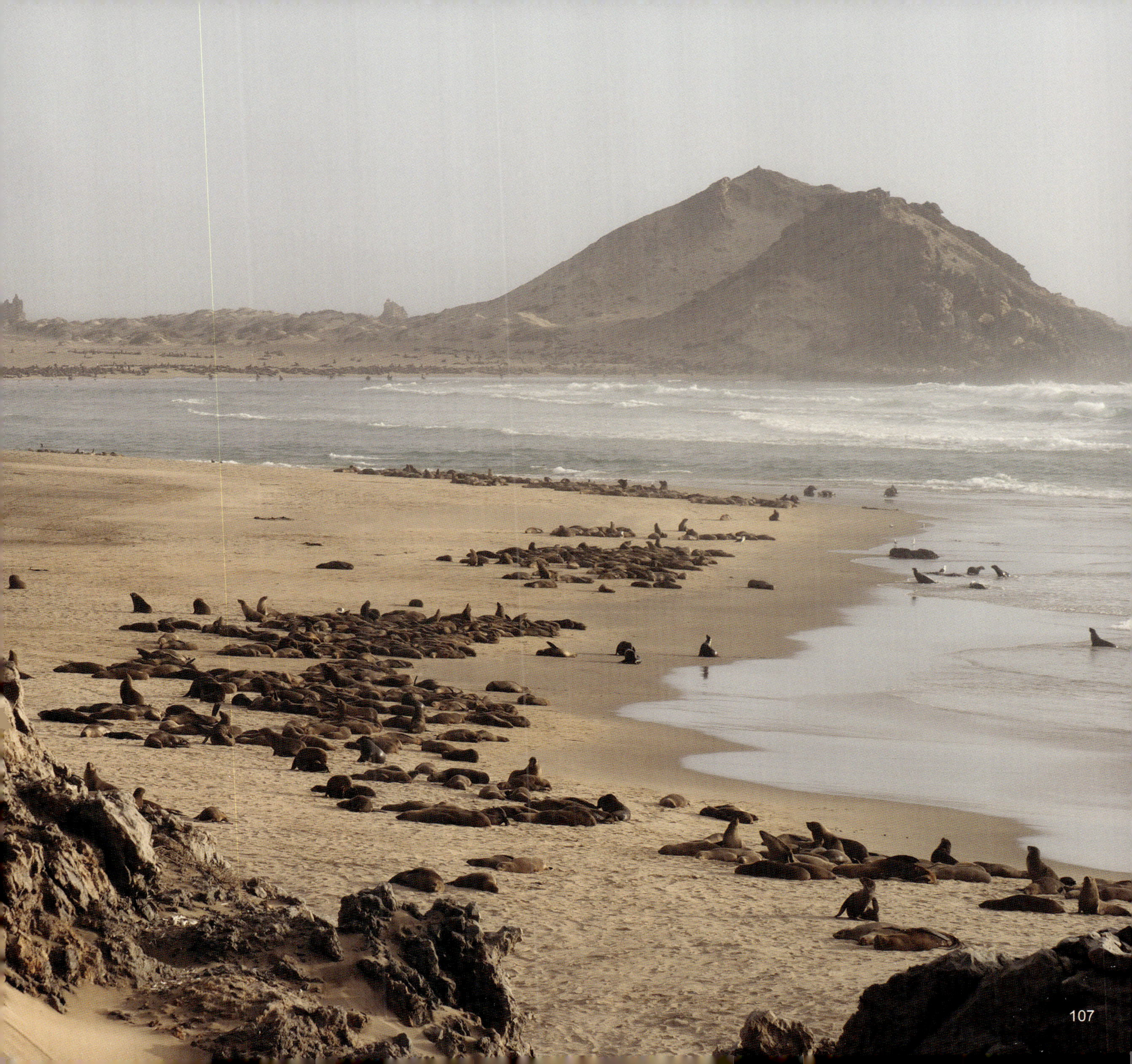

Oranjemund, an angular oasis of low-rise housing, green verges and trees, shimmers like a gem upon a yellow desert, its many trees providing shade and shelter from the sand-laden winds of the mysterious wilderness of Namibia's Sperrgebiet.

Originally a tent camp for alluvial diamond miners on the north bank of the Orange River, Oranjemund was established as a closed mining town in 1936, privately owned by a series of mining companies culminating with Namdeb (a Namibian government/De Beers joint venture). Only company employees and their relatives could enter. Oranjemund was self-sufficient in water and food production, with an abattoir, food-processing and community canteen, though the power station, which served both town and mines, did rely on imported fuel. Housing, schools, medical and recreational facilities were all provided by the company. At its peak, 10,000 people lived there.

In 2007/8, in order to ensure Oranjemund's sustainability, Namdeb and the Namibian Government detached the town's lands from the mining licence area and the Tsau ||Khaeb (Sperrgebiet) National Park (TKNP). It was proclaimed a town in 2011 with the election of a town council a year later.

Research shows that it typically takes 15 years to transform a town like Oranjemund. Three priorities were set. Ownership of services, property and infrastructure was to be completed within three years. Social dependency on Namdeb would be reduced within five years, by which time inhabitants would be aware of their rights and responsibilities in a 'normal' town run by an elected council. Finally, over the next 15 years the local economy would transition from mining to other sources of employment and investment. Namdeb funded a separate legal entity called OMDis (which means 'To make a home' in the local language) to work with others to diversify the economy.

In 2017, restrictions were lifted so that anyone could enter or leave, and road infrastructure was improved between Oranjemund and the rest of the country.

A pilot food nursery has been established as a precursor to

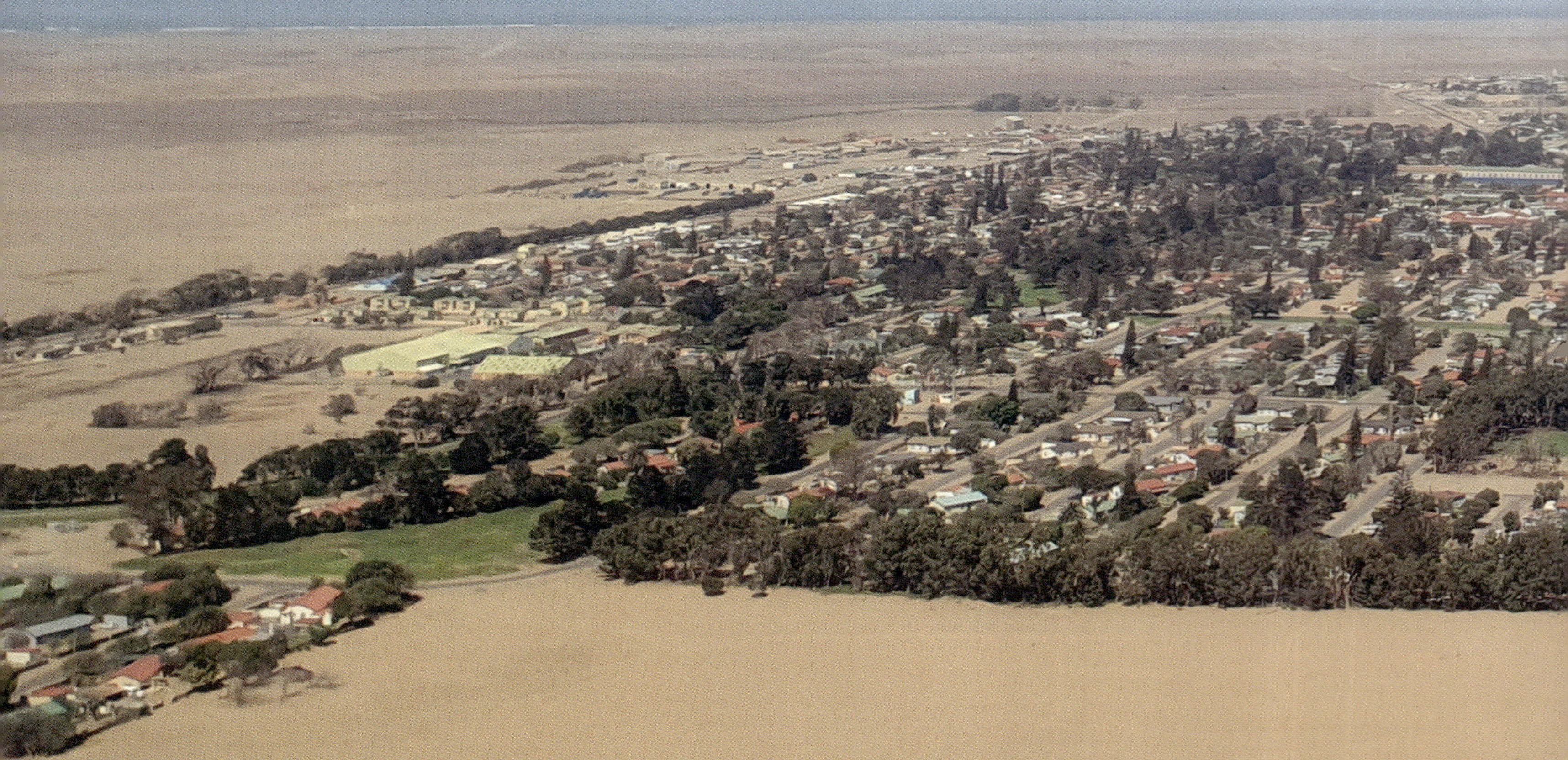

Oranjemund growing its own food once again and even exporting it. Given the almost incessant wind, local wind power is likely to be a major energy source in the future.

Tourism is an obvious priority for Oranjemund, which has an airport, good quality roads, proximity to the Orange River mouth, beaches, protected nature areas and the border crossing with South Africa. But while the region and its wilderness attract adventurous tourists, they generally venture into town only for groceries or fuel. There is little to tempt them to stay longer, but this is beginning to change. Local tour guides are being trained, and car-free routes and e-bike hire are already in place. Artists from across the country have painted vibrant murals on key buildings and installed public sculptures, which have encouraged residents and inspired schools, amongst others, to produce their own public works of art. Serendipitously, the Sperrgebiet has preserved a vast, largely untouched desert wilderness with more than a century-old diamond story to tell. Former Namdeb assets, such as the golf club and yachting club, can be repurposed. The town's Jasper House Museum, which relates the history of the town and its surroundings, has been recently revamped.

OMDis was recently awarded a tourism concession in the TKNP close to Oranjemund and has begun planning around this. A potential future visitor attraction could be the *Bom Jesus* – a 15th-century Portuguese shipwreck unearthed in the Sperrgebiet by Namdeb mining activities, which made headlines around the world. Its cargo included millions of dollars' worth of gold and silver coins and copper ingots. Its new home could be the cavernous old power station in the centre of town.

Namdeb's mines were slated to close at the end of 2022; in October 2021, the life of mine was extended by a further 20 years, but the transformation work was not a futile exercise. People now have a say in the running of their town and can help shape its future with or without mining.

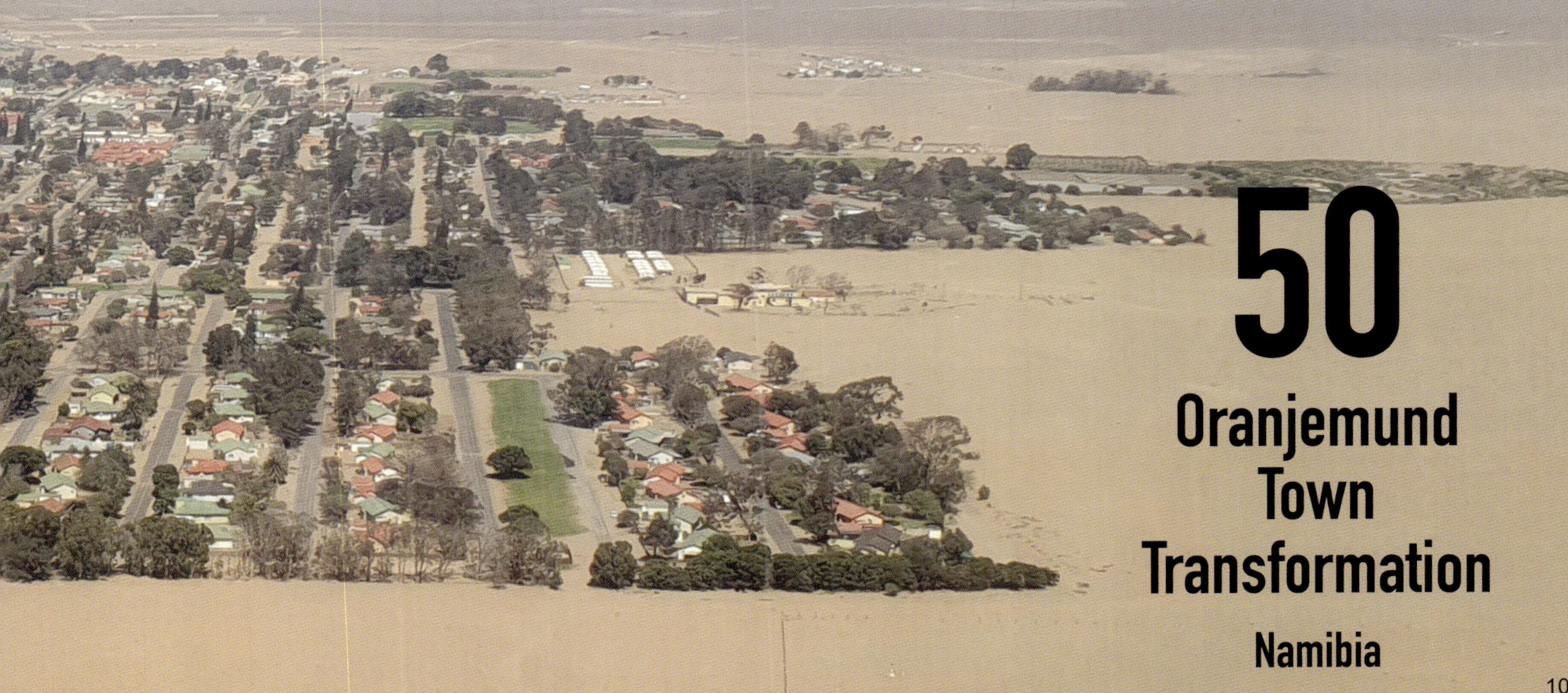

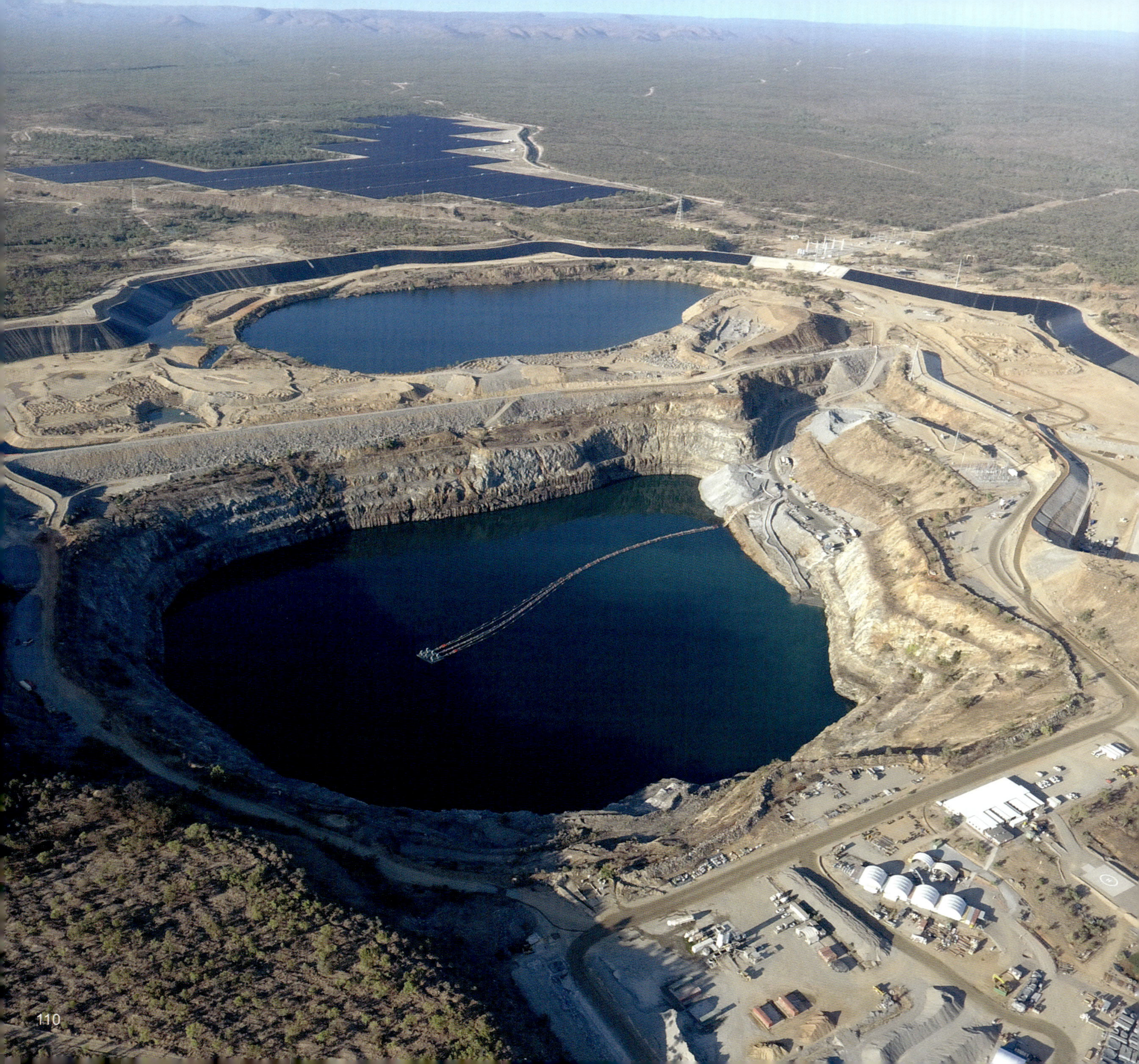

51

Kidston Mine Pumped Storage Hydro Project

Australia

It's a bumpy flight on a tiny plane through towering storm clouds hugging the tops of Queensland's coastal range and on into the Outback. On descending to the dirt airstrip in the middle of nowhere, through the small window are two enormous, flooded open pits and the dark, tessellated rectangles of thousands of solar panels atop a restored tailings facility. The isolation and remoteness are striking — it is a long way from anywhere to build a revolutionary new power station!

Around the world, there are many concepts and plans for reusing mining voids as batteries for storing renewable energy as potential energy in water or compressed air, but in northern Queensland, it's actually happening. A world-first pumped hydro storage scheme is being created from the former Kidston gold mine using the mine's features, such as existing access by road and air, power connectivity, on-site accommodation, amenable geology, water provision, two open pits at different elevations, a flat-topped tailings facility and a community familiar with large scale engineering projects.

Despite its remoteness and harsh climate, Kidston mine and township emerged rapidly during the 1907 Oaks Rush gold rush to become Australia's biggest and richest gold mine until closing in 2001. The site was purchased by Genex Power in 2014 with the intention of turning it into a renewable energy hub and storage facility. First, they installed a 50MW solar farm comprising 540,000 individual panels on the tailings facility — enough to power 26,000 homes. The power generated is transmitted into the grid along the transmission line that originally supplied the mine.

Construction of the unique pumped storage hydro project began in 2021, creating hundreds of construction jobs. It will use the two pits as storage reservoirs – the upper Wises Pit and the lower Eldridge Pit will be connected by two vertical water intake shafts. These will house two 125MW reversible turbines with a start-up time of less than 30 seconds in a purpose-built powerhouse cavern 220 metres underground. The footprint of Wises Pit is being greatly expanded, which includes an awful lot of muck shifting to create a sufficient head of pressure above the turbines to generate electricity. Its sides are also being lined to reduce leakage.

The system will trade daytime solar power for night-time energy: during the day, when power is abundant, the turbines will pump water uphill from Eldridge Pit into Wises Pit, and at night, it will flow back downhill through the turbines to send power to the grid

The multimillion-dollar project was made financially possible by support from the Northern Australia Infrastructure Facility (NAIF) through its loan facility, a grant from the Australian Renewable Energy Agency (ARENA), and from the Queensland State Government through part funding of the new 275kV transmission line.

The pumped hydro and existing solar schemes form the nucleus of what will become the Kidston Clean Energy Hub. Facilitated by the construction of the new transmission line, these power sources are planned to be joined by a 258MW wind farm. This integration of wind, solar and hydro will be globally unique, producing sufficient power for 280,000 homes.

Repurposing the Kidston site in this way enables ongoing benefits to Queensland. The main drawback for Genex, apart from the cost of building the 186-kilometre powerline to connect to the high-power public grid, is continuing compliance with mine-site regulations. The project challenges current rules on how to transfer contingent liabilities from the former site owner, a mining company, to the new owner — a non-mining company that wants to create new environmental and socio-economic values from the site.

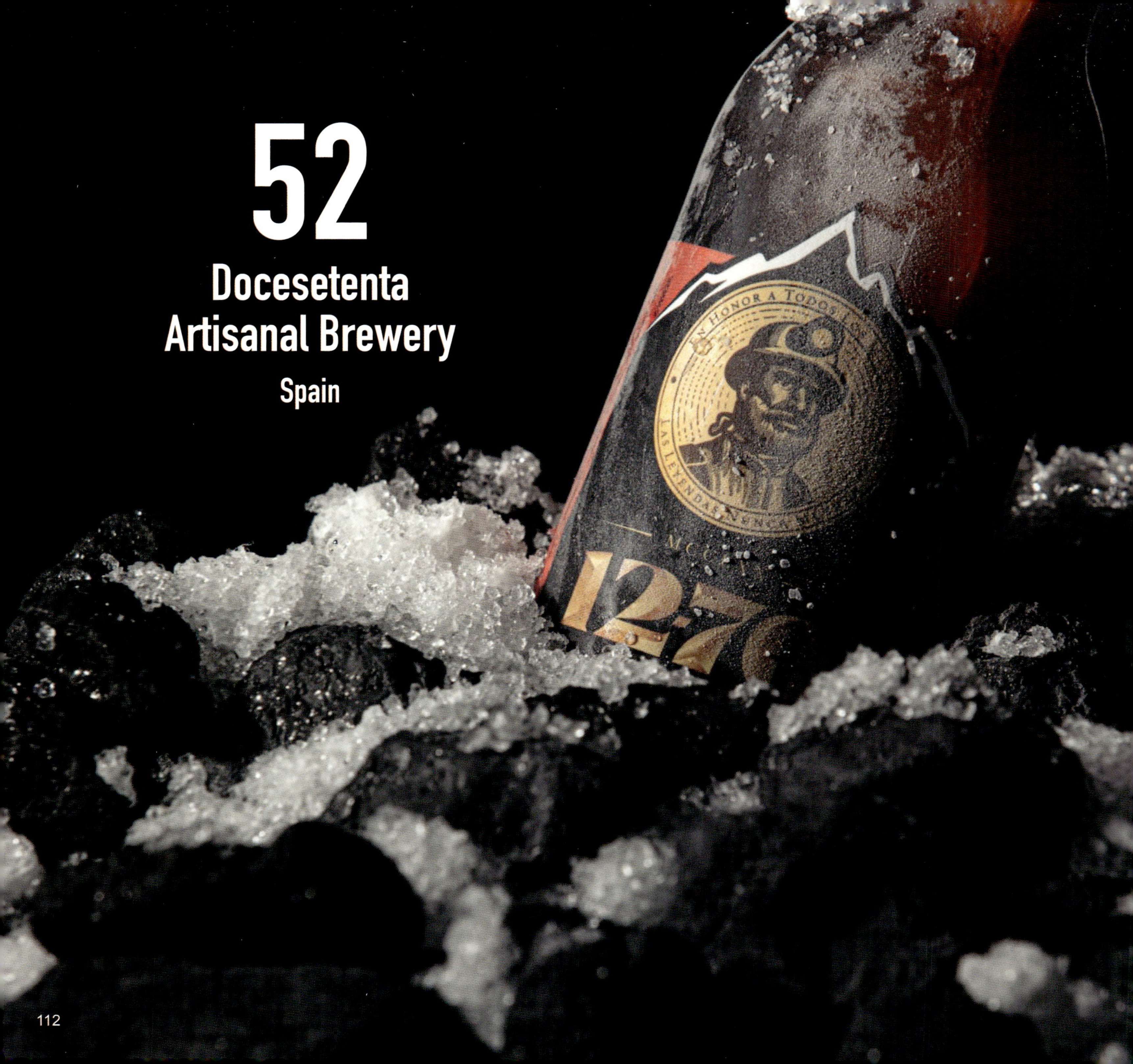

52
Docesetenta
Artisanal Brewery
Spain

The young Juan Jose Villanueva (Juanjo) left his rural coal mining hometown of Villaseca de Laciana in northern Spain to attend university. Meanwhile, the town's Lumajo mine closed in 1997, making his miner father redundant and forcing him to leave to seek work in Oviedo. It was the end of an era; Juanjo's father and grandfathers had toiled underground for decades, and there were few other jobs on offer. Lacania's population plummeted from almost 22,000 to 9,000 today, of whom 6,000 are former mineworkers, with collateral socio-economic decline and erosion of its intrinsic mining culture

In the meantime, Juanjo had been a teacher, a police officer and forensic scientist, and then a successful Panamanian businessman. In 2018, motivated by a strong desire to support the town of his forefathers, Juanjo and his business partner bought the Lumajo mine site, intending to use it to stimulate regeneration in Laciana, but with no firm plan as to how.

Their first project was an artisanal brewery, which they branded 1270, or Docesetenta, recognising the auspicious year that King Alfonso X the Wise granted freedom to the people of the Laciana Valley under the crown's protection. Operating as a social enterprise with a mission to create employment and promote the area's mining heritage, 1270 would, they hoped, bring new life to the dying valley.

With the slogan 'The mine is not dead', beer production began in 2019 in the old miners' changing rooms Juanjo's forebears had used; poignantly, the hooks where the miners used to hang their work clothes still hang from the ceiling. Employing five workers, the brewery makes up to 18,000 litres of beer annually using Leonese hops and local stream water. Two thousand bottles of a novel 1270 beer are fermented 200 metres below ground in old mine workings. And it is an award-winner, too, having won a gold and two bronzes at the prestigious International Beer Challenge in 2023. Brewing and mining entwine even tighter as fermentation research using underground bacteria is underway in old workings 250 metres below ground in partnership with the University of Oviedo.

The mine imparts a unique and strong cultural character to the craft beers and connects the drinker with mining history and culture. The motto' Legends never die' refers to the brave miners who descended into the darkness daily, some never to return. The bottle labels display a miner's head, and even the labels' colours represent mining: red for the blood spilt, black for the coal and gold for the beer's colour.

Despite the beautiful scenery and wildlife of the Laciana Valley (a UNESCO Biosphere Reserve) and the rugged Cantabrian Mountains, there has been relatively little tourism; but growing awareness of the brewery has attracted more visitors from outside. There is no other brewery in the area, so 1270 has connected tourism with culture and history. In 2020, an underground experience was opened as the only mining visitor attraction in the region. After a tour of the brewery and a beer-tasting, visitors are taken into a short section of the 34 kilometres of underground mine workings where the miners' work, culture and history are portrayed 'to ensure that all the sacrifice and blood that was spilt in those tunnels are not forgotten'. In 2023, 16,000 people visited the mine – a record far beyond Juanjo's expectations when he started the project. The timing of the tours is carefully choreographed to begin at 12.00 and 16.00. A visit takes approximately two hours, and there are no on-site catering facilities, so when visitors leave the site at lunchtime and dinnertime, respectively, they are directed to the local restaurants in the town, thus sharing opportunities with other local businesses.

All of 1270's developments have been privately funded, with parts of the site due to be offered to other businesses to attract them into the area. The initiative has already generated an additional 15 jobs in the area, with a plan to increase this to 75 in the medium term. To his earlier diverse career path, Juanjo can now add 'brewing and tourism regeneration entrepreneur'!

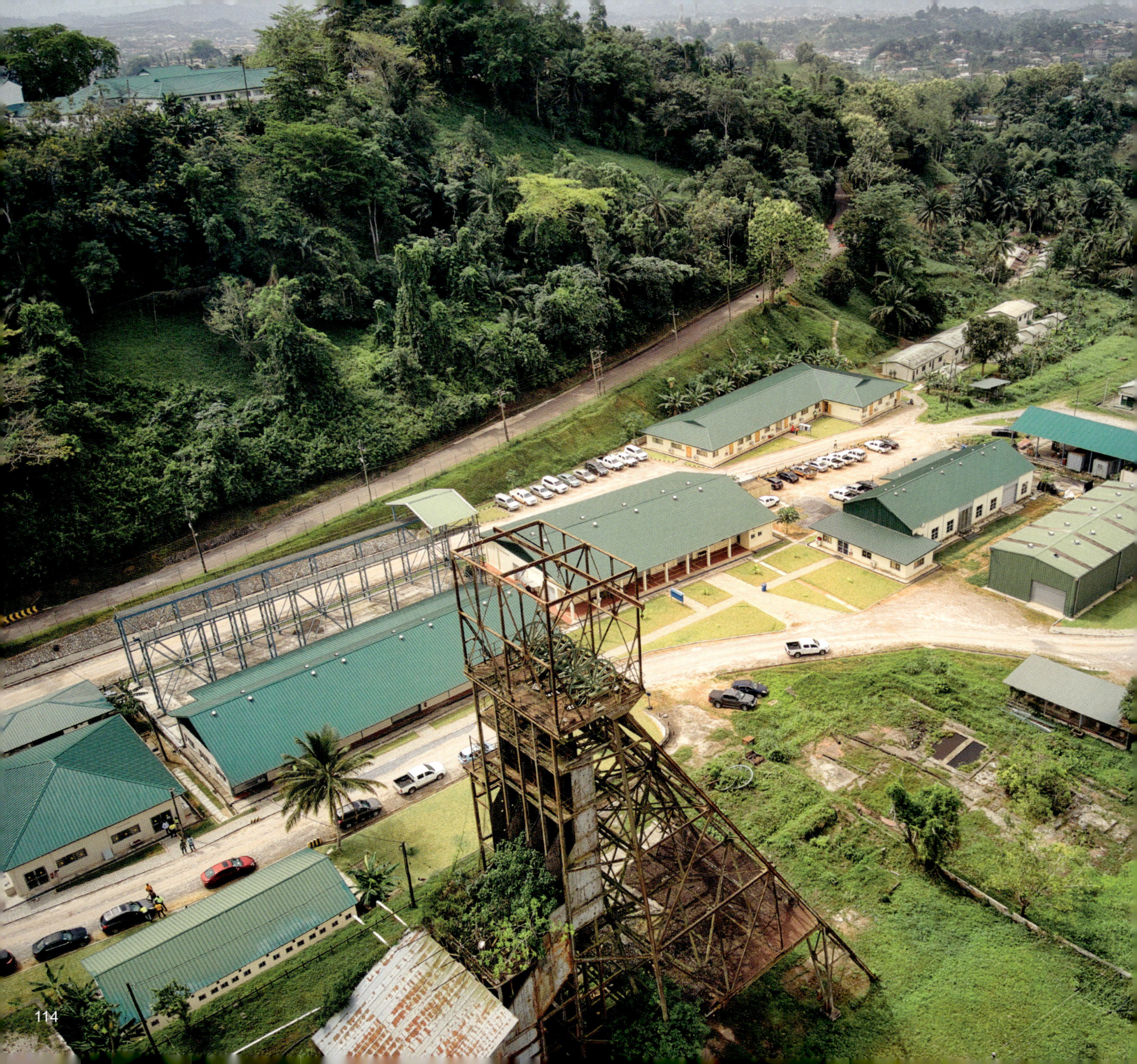

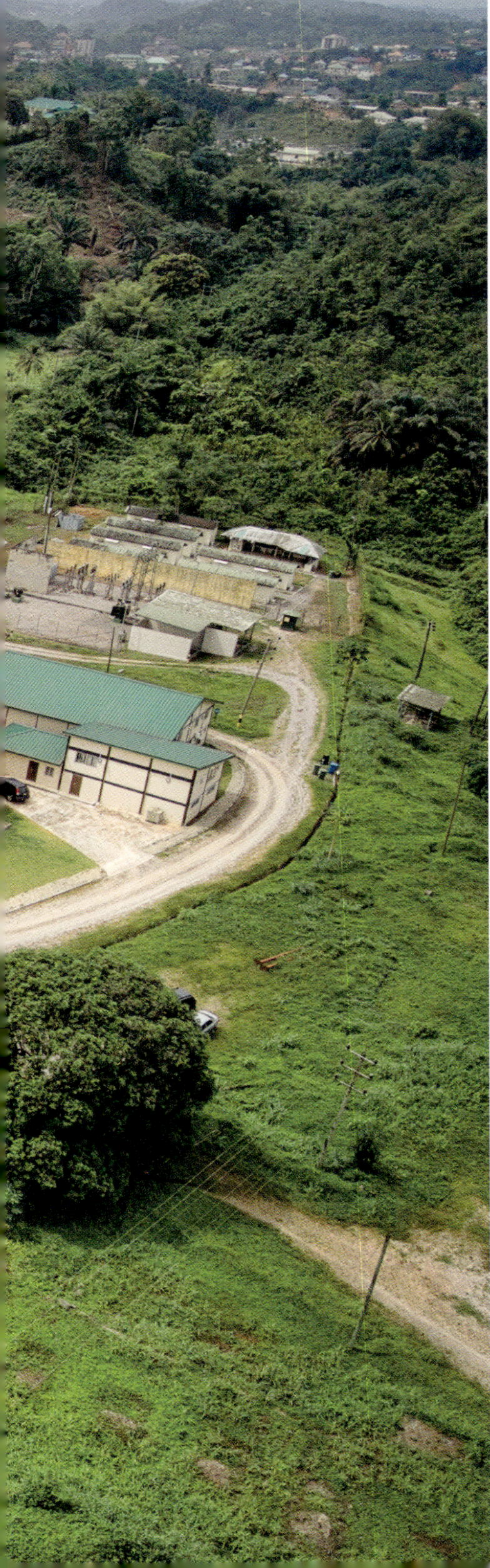

53

Tarkwa Gold Mine Learning and Development Centre

Ghana

The new trainees' beaming smiles tell you all you need to know about the importance of the Gold Fields mining company's training and development centre at Tarkwa Mine, Ghana. The students' two-year course covers all the engineering skills and knowledge required to work on a mine.

The centre itself is an asset to the mine and to the country. Its repurposed buildings sit below the impressive, fading headframe of the unused Akoon Vertical Shaft, now greening with vines. Tarkwa Mine is renowned across Africa, and its headframe, one of only two remaining in Ghana, symbolises the historical development of gold mining in the country. In 1990 the underground mine workings were decommissioned, and the mid-20th century headframe and run-down single-storey buildings around it were due to be flattened and the land reforested.

The potential of the headframe as a symbol of learning was recognised by the mine's project manager. A trained architect, he developed the idea for a model training centre and mining heritage facility before demolition could proceed. This reduced waste, saved land, improved amenities, enhanced aesthetics and ultimately saved money.

Although the creation of this tertiary training facility is a work in progress, it has become essential due to the limited number of aligned technical training institutions. A centre of excellence, it provides hands-on training for employees, contractors, communities, artisans, mechanics, electricians and other trades relevant to the operational needs of a mining or other industrial company and includes virtual reality truck driving simulators.

The second phase will refurbish the headframe into a museum for educational and heritage purposes, offering a glimpse into Tarkwa's rich mining past.

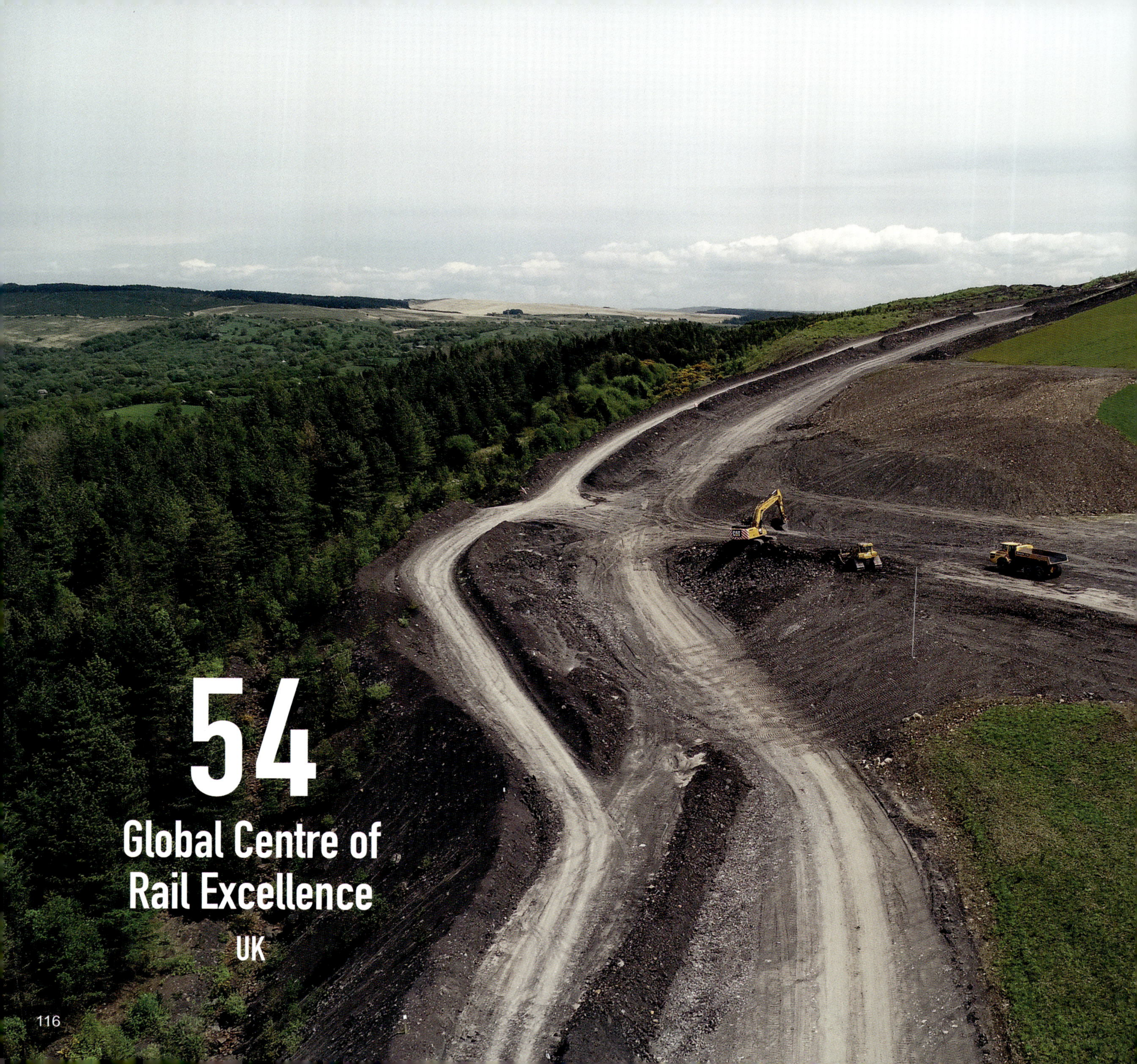

54
Global Centre of
Rail Excellence
UK

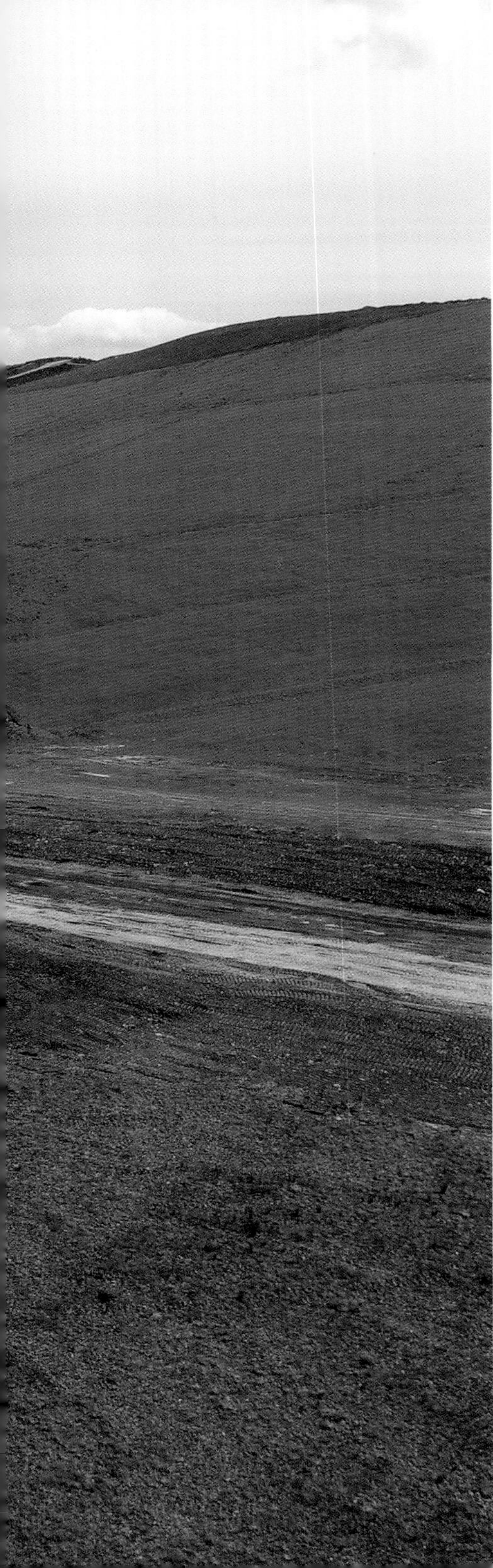

The Global Centre of Rail Excellence (GRCE) sits on high land at the head of the Dulais and Swansea valleys, with breathtaking views over the Welsh valleys and the hills of Bannau Brycheiniog National Park. Its 700-hectare site is on the former Nant Helen opencast mine and Onllywn Washery in an area where coal was mined for more than 200 years until 2021.

It acquired the site from mining company Celtic Energy and will be a European first – a one-stop-shop for testing new rolling stock, carrying out world-class research and development, certifying rail infrastructure and technologies and training to support the rail transport technologies of the future. Key features will include two electrified rail loops, an electrified seven-kilometre line for high-speed rolling stock and another seven-kilometre electrified track for testing rail infrastructure, including a two-platform station. GCRE will also include visitor and conference facilities, a business park, and a hotel for researchers and potentially tourists visiting the adjacent national park. An existing branch line that used to transport coal from the mine to the port some 25 kilometres away was a major plus. Its refurbishment will allow rolling stock and passengers to be brought to the site by rail, rather than road, as well as people.

The concept had been kicking around for years until 2021, when the Welsh government formed GCRE Ltd as a special-purpose vehicle. The GBP 400 million project has so far received GBP 70 million of pump-priming funds from the Welsh and UK governments, with the remainder to be met by private investment.

GCRE will create an estimated 100 to 150 direct, high-quality jobs and should facilitate other major investments and R&D in the area after decades of post-mining decline. The development will also be supported by higher education, with rail research universities using the GCRE site for academic R&D. The research is beginning; in early 2023, 24 projects were successful in the first round of research funding for the 'Innovation in Railway Construction' competition. Potential European customers are showing interest, and a number of high-profile clients have signed up to use the site for research and development, including rolling stock manufacturer Hitachi, Transport for Wales (the public transport arm of the Welsh government), Thales, Ricardo and Connected Places Catapult.

It's early days, but preparatory on-site works have already begun. Up to three million cubic metres of spoil are being used to fashion final post-mining landforms and create test track routes, while common grazing lands are being restored in between. The exposed nature of the site makes it ideal for wind and solar energy, and discussions are underway to make the project self-sufficient in green energy. GCRE aims to open in 2027 as the UK's first net zero railway.

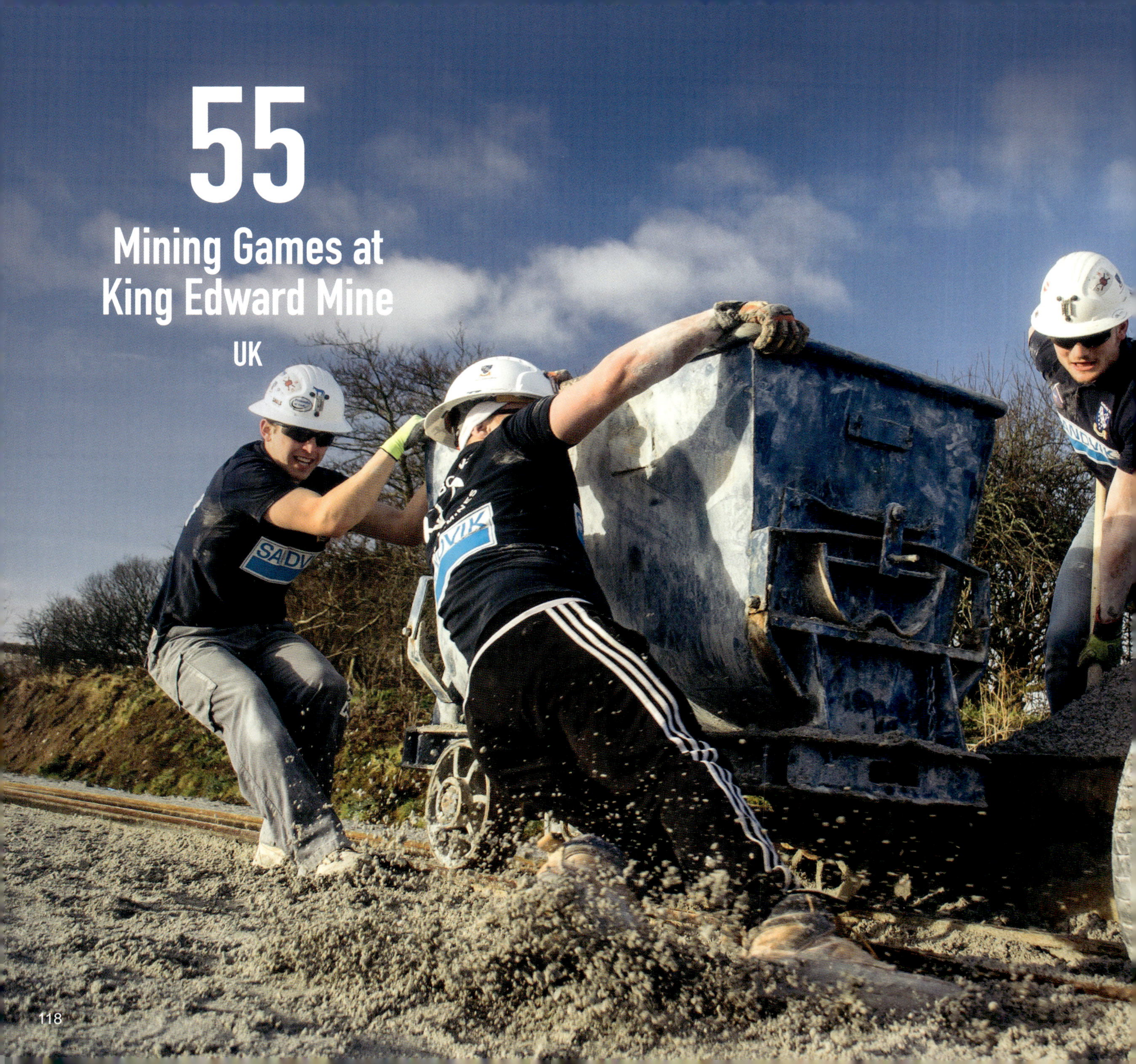

55
Mining Games at
King Edward Mine
UK

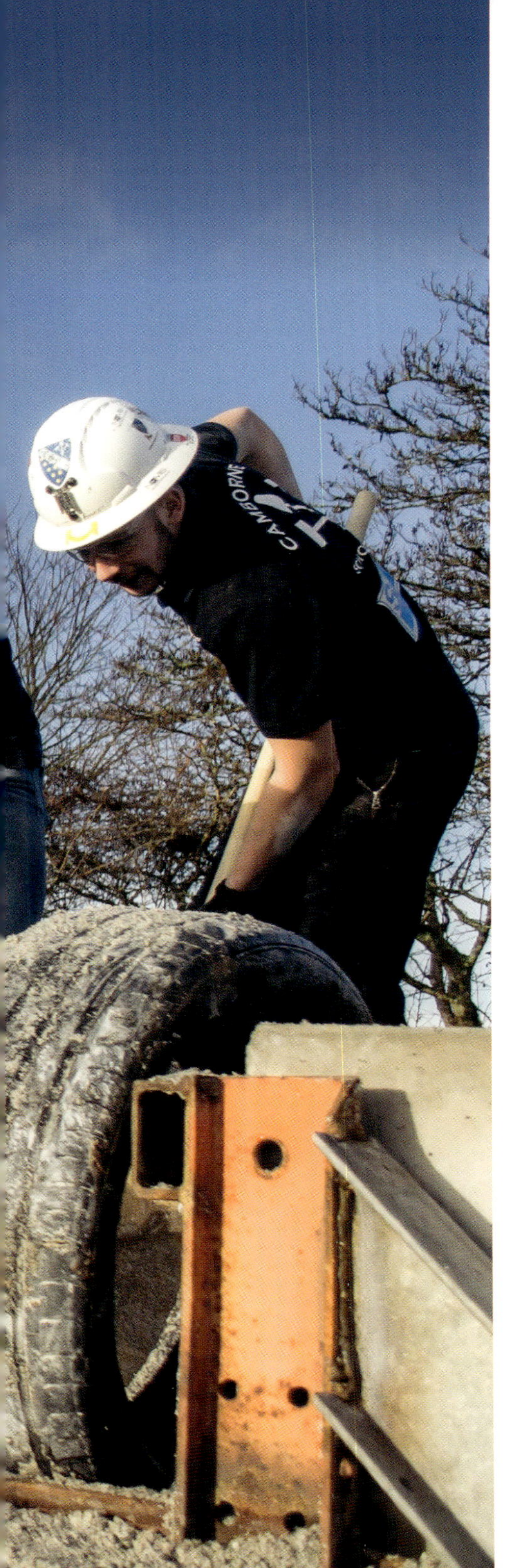

In the shadow of King Edward Mine's (KEM's) ruined Cornish engine houses, beneath the hardhats, it's all straining sinews, grimaces and perspiration for the competitors of the 2018 International Intercollegiate Mining Competition, aka the International Mining Games (IMG). The games were established in the late 1970s to honour the 91 miners who died in the 1972 Sunshine Mine disaster in Idaho, USA, and all mineworkers killed at work. The student-run competition includes seven gruelling events: jackleg drilling, track laying, hand mucking, Swede saw, gold panning, hand drilling and surveying. The IMG keeps old-school mining skills alive while reminding future miners of their heritage and encouraging teamwork and comradeship, which are crucial for working in the industry.

For over a century, KEM was the fount of education for generations of Camborne School of Mines (CSM) students. In response to local industry needs, CSM opened its doors in 1887. In the late 19th century, Cornwall's hard rock metal mining industry was in decline, so CSM acquired, developed and equipped some abandoned underground workings as a teaching mine, which became known as King Edward Mine in 1901. Its students received high-quality, hands-on mining and processing training combined with academic classroom teaching. The tin ore they mined and processed was sold to cover teaching costs. Mining at KEM ended in 1920, but the surface buildings were used for teaching well into the 21st century and are still used by CSM's men's and women's IMG teams as a training ground. CSM is now part of the University of Exeter, and the site is owned by Cornwall Council.

King Edward Mine is independently run as a museum by the dedicated volunteers of the KEM charity incorporated organisation, hosting visitors interested in Cornwall's mining heritage and hundreds of school children on organised visits every year. It is an integral part of the Cornwall and West Devon Mining Landscape World Heritage Site, inscribed by UNESCO in 2006, and comprises Cornwall's oldest complete set of mine buildings. The site is Grade II*-listed, with renovation and maintenance supported by admission charges, donations, the CSM Trust and, till recently, EU public funding.

CSM attracted students from across the world, and still does; KEM will host future IMGs, attracting teams of international mining students to compete again in its historic setting. Inspired by the 40th IMG held in Cornwall in 2018, local schools developed their own version, known as the Cornwall Schools Mining Games, also hosted by KEM and supported by local minerals and mining companies. Young school students, bedecked in Hi-Viz vests, compete against each other in practical, mining-related, team activities, in the haunting presence of KEM's engine houses. The lunchtime break is an opportunity for local experts to tell of mining history using the surrounding landscape as a teaching aid. A recent event involved 14 Cornish schools comprising over 160 students in 22 teams. It aimed to excite the young competitors about earth sciences, engineering and mining so that perhaps some of them will be able to capitalise on the recent surge in interest in mining in Cornwall – and beyond – when they leave school.

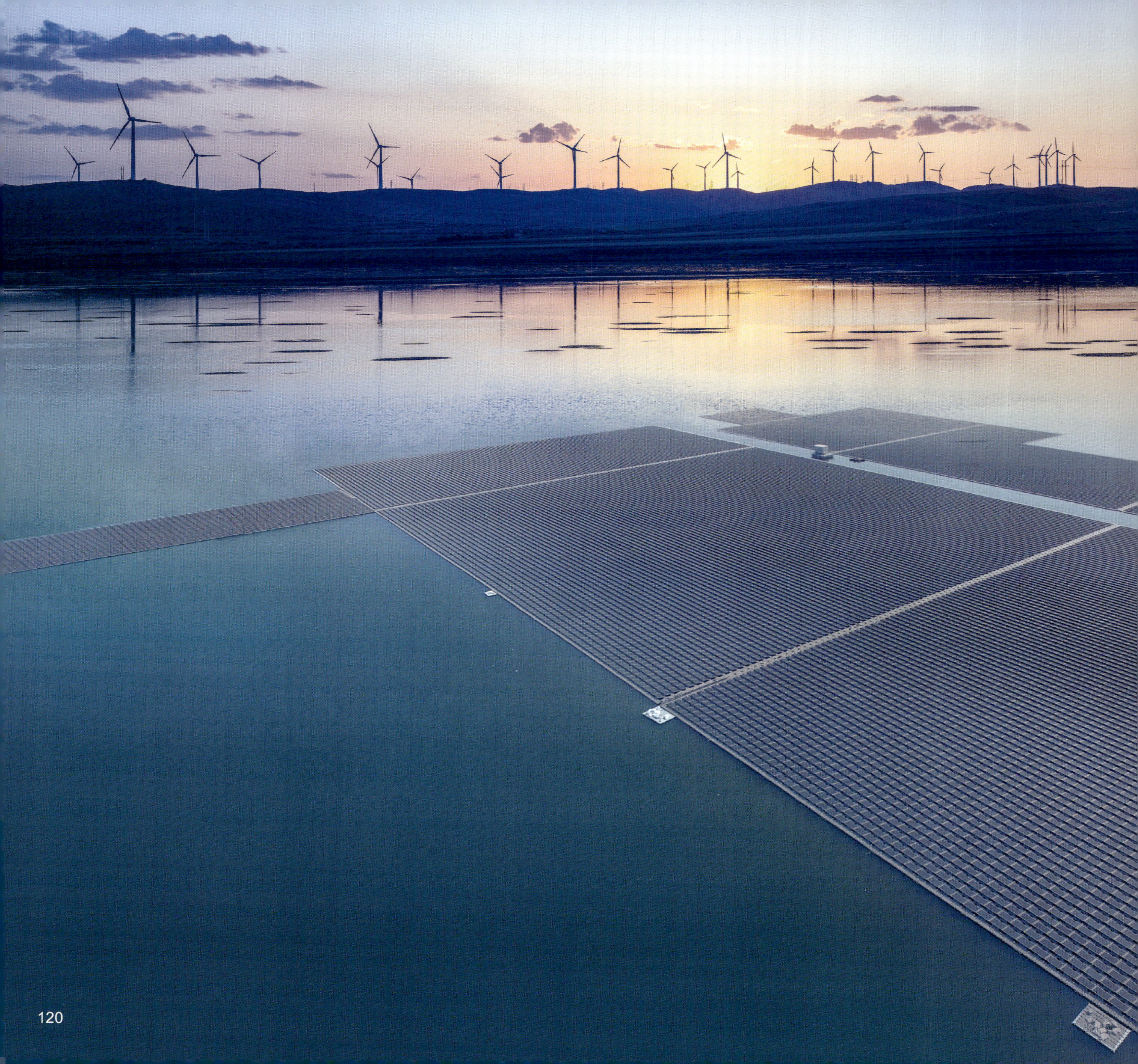

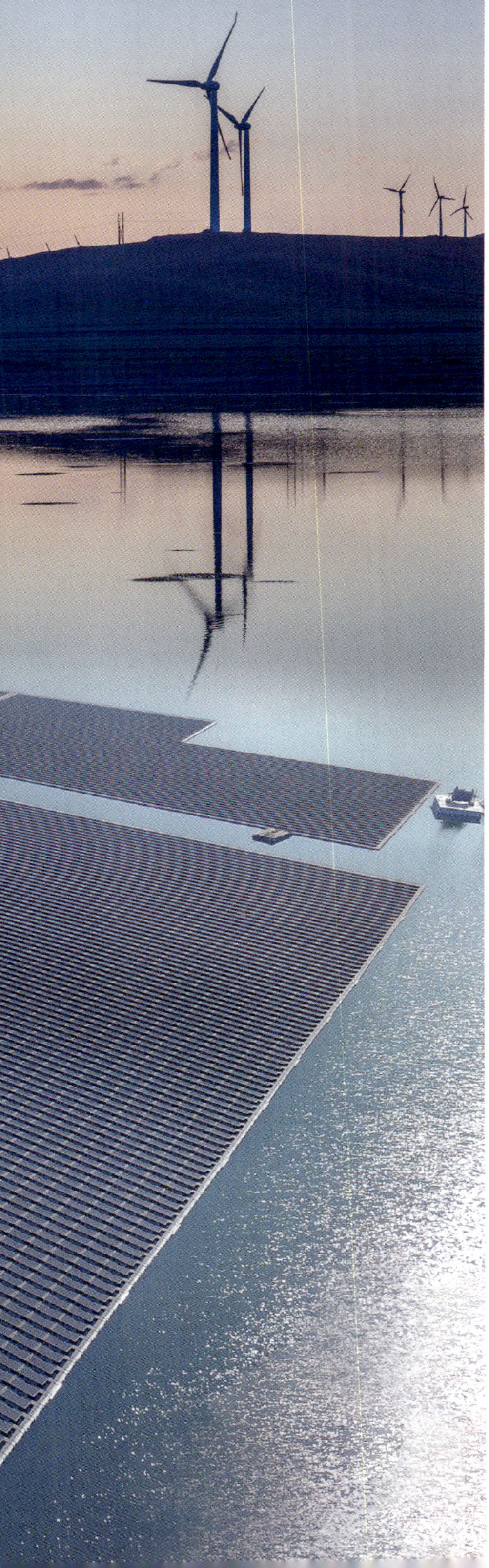

56
Three Gorges Floating Solar Farm

China

Huainan City in the Chinese province of Anhui was founded in 1950 as a major coal mining centre. Since the 1960s, excessive underground mining has led to the catastrophic collapse of the overlying strata leading to huge tracts of land at the surface being irretrievably damaged by subsidence. Simultaneously, when a mine closes and its pumps are turned off, the rebounding groundwater may gradually flood the subsided areas, which also accumulate surface water run-off, ultimately forming new lakes. Subsidence around Huainan City has badly affected 27 townships and 311,000 residents, covering an area of 205 square kilometres of which about 70% have become flooded as lakes.

In response to the Chinese government's policy of achieving carbon neutrality by 2060, some renewable energy companies have seen the subsidence lakes as an opportunity to install large-scale, floating solar photovoltaics. Floating solar installations are booming worldwide; they have advantages over land-based installations because they do not compete for high-value agricultural or development land. Solar panels work more efficiently at lower temperatures, so the water's cooling effects make floating solar more productive. And in drought-prone areas the panels reduce evaporation, so conserving water, and shading can reduce blooms of toxic blue-green algae. The aquatic conditions reduce dust settlement on the panels, but conversely also encourage the growth of algae upon them. On the downside, a floating installation is more costly to install than land-based solar and requires anchoring to the lakebed. There are also concerns that aquatic life, including commercial lake fish, may be affected.

One such company, the China Three Gorges Renewables Group, has installed the world's largest floating solar farm on a subsidence lake in the Huainan coal district. Financed by low-interest loans from central government, with a contribution from local government, the around a billion yuan project was commissioned in 2018. Covering 320 hectares, its photovoltaic panels have a total capacity of 150 megawatts – enough to power 94,000 homes. The installation supports about 100 part-time jobs in the local community. Other companies have also installed similar systems in the area. The rapidly growing green energy provision in this former coal mining heartland has obvious benefits for human health, air quality and the supply of cheap, clean energy.

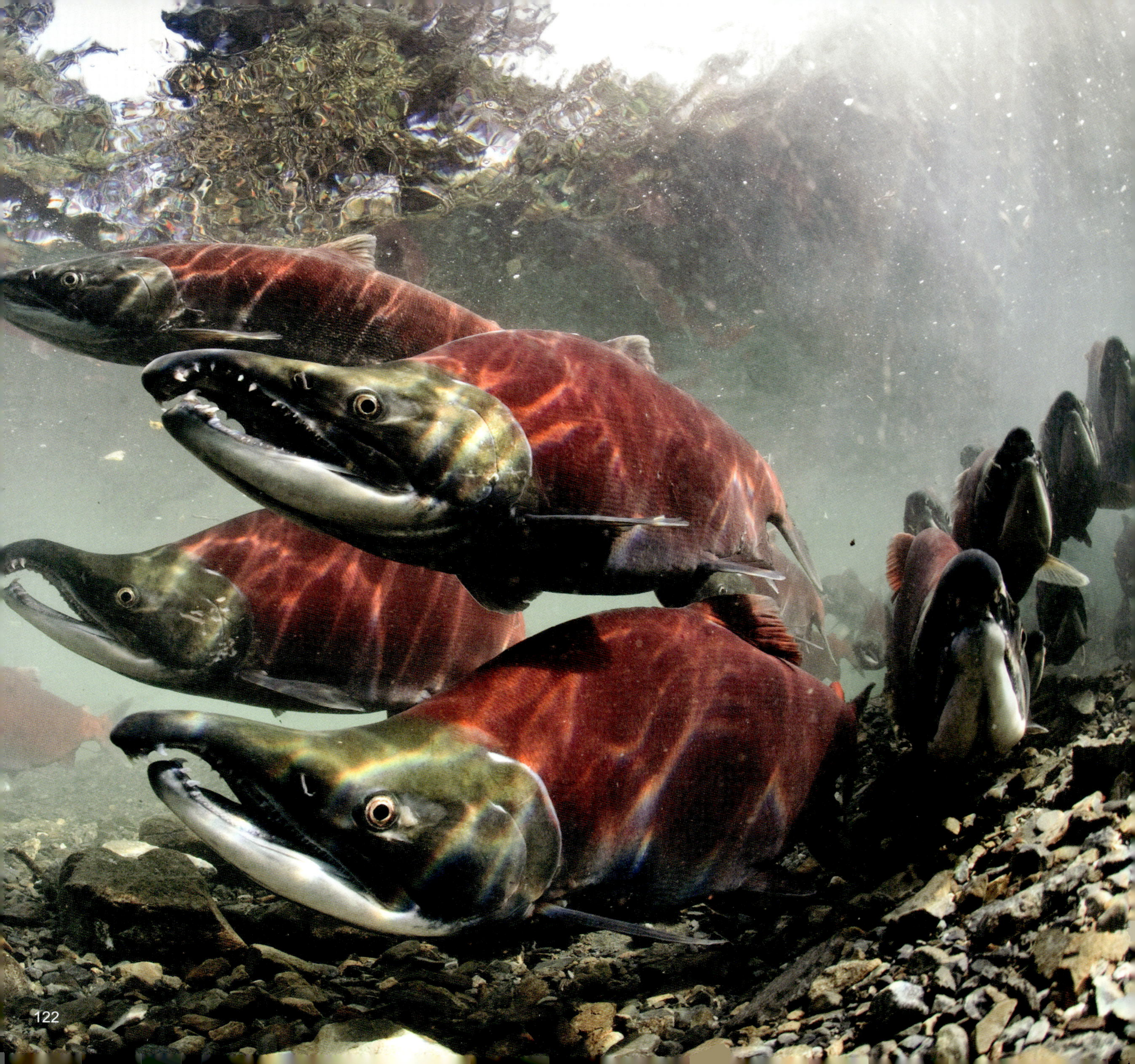

57

Salmon Gold

USA

As Mark Twain almost said, 'There's gold in them thar rivers – still!' Mining of the rich placer (streambed alluvial) gold deposits that sparked the late-19th century gold rushes in the wilds of Alaska, the Yukon Territory and British Columbia continues because gold still hides in those previously mined sediments.

The rivers of this remote region are home to globally significant populations of salmon, grayling and other fish species that migrate up from the sea to spawn, many of which are culturally significant to indigenous peoples. They are deeply concerned about the disruption caused to these streams and rivers by historic – and current – placer mining activities and the resulting dwindling fish populations.

Flying over the expansive alluvial gold fields that cross the Alaska/Yukon border, Stephen D'Esposito, the CEO of NGO RESOLVE, noted the extensive damage to rivers and riparian habitats that has persisted for decades; the tailings piles, disturbed riparian areas and disrupted stream flows of shallow riffles and multiple channels combine to diminish the available habitat for salmon migration and spawning.

In response RESOLVE launched Salmon Gold in 2018, using its own unrestricted funds, primary funding from Apple and Tiffany & Co., and support from Newmont to test the concept of mining residual gold in disturbed riverine habitats while simultaneously restoring them for salmon and other species. Salmon Gold raises funds to repair damaged rivers and their surrounding lands to produce gold with a positive story in the marketplace, and that demonstrates the value of restoration, working in partnership with responsible placer miners, ecological restoration organisations, regulators and communities, indigenous groups and manufacturers. The gold won from these sediments is then sold to companies that require responsibly sourced gold for their supply chains.

The concept was proved at Jack Wade Creek in Alaska's Fortymile Wild and Scenic River system. Salmon Gold worked with the placer mining father and son team of Dean and Chris Race to extract residual gold and restore a stream section and an area of abandoned mineland. The gold is in the supply chains of Tiffany & Co., Mejuri and Apple. Tiffany & Co. helped launch the project, while Mejuri is the new jewellery partner as of 2023. Blockchain ensures the traceability of gold from miner to refiner in Apple's supply chain.

The Race Family Mining Operation received the national 2019 Hardrock Mineral Small Operator Award from the US Bureau of Land Management for their progressive and pioneering environmental work. Similar Salmon Gold projects are in development at Gold Creek, also in Alaska, and Sulphur Creek in the Yukon.

The success of the Jack Wade Creek demonstration led to Salmon Gold becoming an integral project of the RESOLVE spin-off company, Regeneration Enterprises. While Salmon Gold's aim is to scale up activities across the former gold fields of Alaska, the Yukon and British Columbia, Regeneration is focused on re-mining and restoration across all industrial, hard rock mining, especially for critical minerals and copper. The aim is to create a dependable revenue stream to fund restoration in tandem with re-mining. Established as a B Corp with support from Rio Tinto, Apple, Mejuri and others, Regeneration Enterprises will reclaim mine wastes while securing minerals for the energy transition, green tech, and sustainable brands. It has a pipeline of 70 potential sites on which to apply this blueprint and growing interest from regulators, local communities and indigenous peoples.

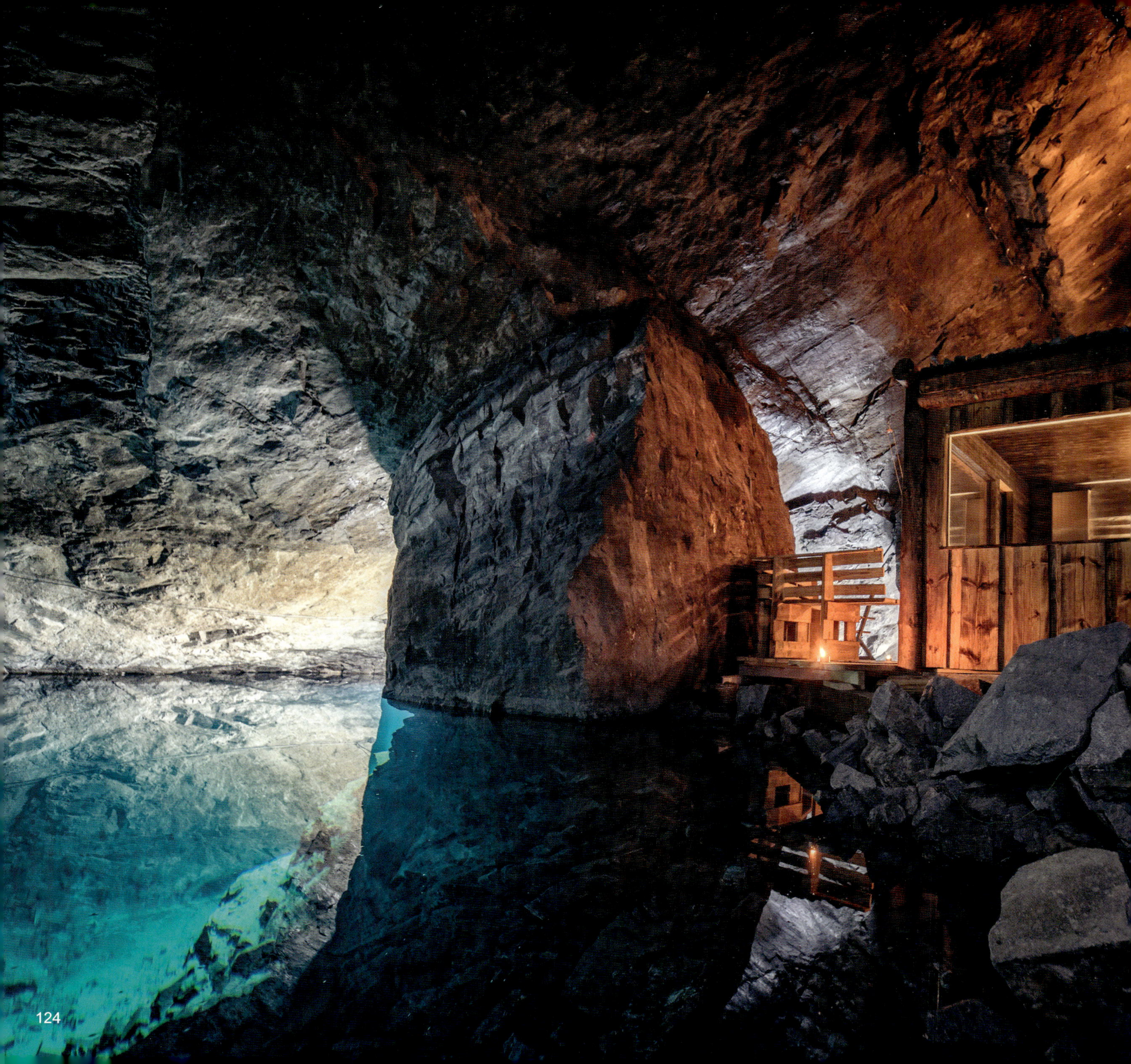

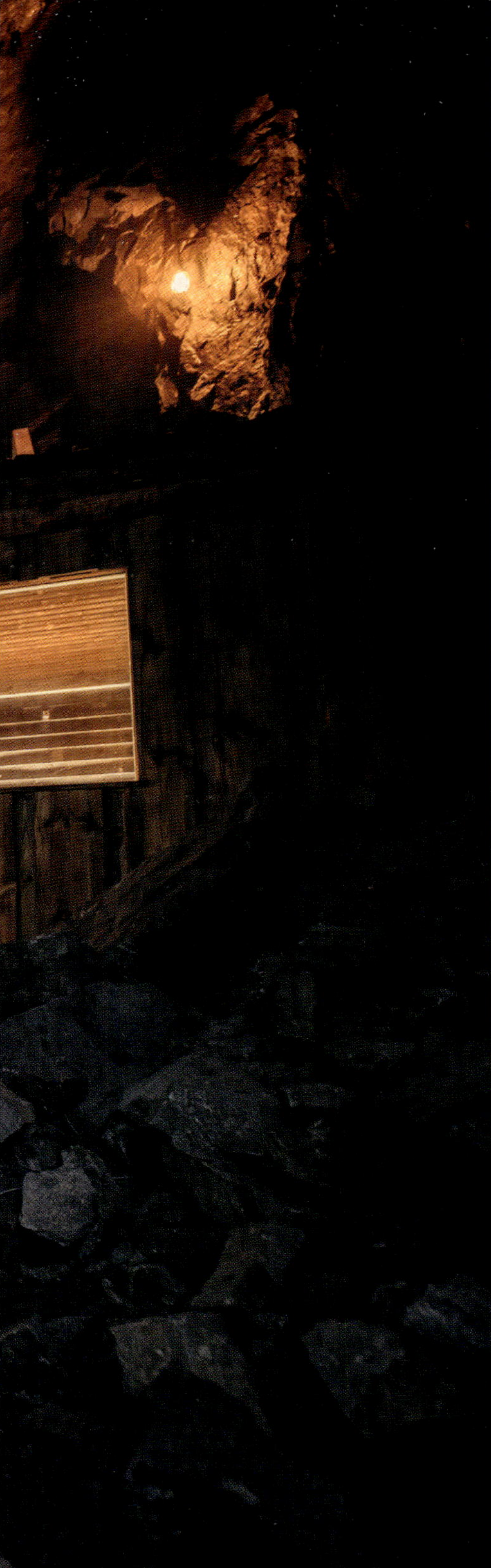

58

Adventuremine

Sweden

Adventuremine was born from a real adventure. In the late nineties, three school friends climbed over the fence guarding the old Tuna Hästberg iron mine and went exploring, spending days underground. Initially, they considered keeping it a secret between the three of them, but they decided that it was too good not to share, and a few years later, they began taking visitors into the depths. After a few more years, a verbal agreement was reached with the owner, which was followed around ten years later by a lease, and finally, the site was purchased.

Adventuremine is now on a commercial footing thanks to years of support from local volunteers and has over 5,000 visitors a year, providing the community with additional income from overnight stays in the area. They train police and fire department personnel and the Swedish Armed Forces. They still offer guided tours, but they have expanded to include conferences, performances and diving. The mine has around 30km of flooded passageways,

and over a hundred divers are certified there every year. Emerging from long dives in the chilly water, the friends started joking about installing a sauna down there. They contacted several companies, but the response was the same – conditions in the mine would make it impossible to build and operate a sauna.

So, they decided to have a go themselves. Timber came from the forest above, power cables were laid for lighting and ventilation, and incredibly heavy anti-fogging windows were procured and somehow installed.

It should have just been a matter of ordering sauna stones, too. Unfortunately, there was a shortage thanks to a home sauna construction boom during the coronavirus pandemic. Resourceful as ever, they made their own from the mine's basalt-like dolerite and iron ore, and they converted a cement mixer into a stone tumbler to smooth them down.

It takes around 45 minutes to get to the sauna through the old tunnels. The sauna experience includes traditional cold baths in the surprisingly pure mine water as well as storytelling and music, and visitors have the option to dine in the old dynamite store. If navigating your way to a sauna that is a hundred metres below the surface doesn't sound adventurous enough, there are plans to create a Shower of Hell – 120 litres of cold water dropped on your head from a height of four metres.

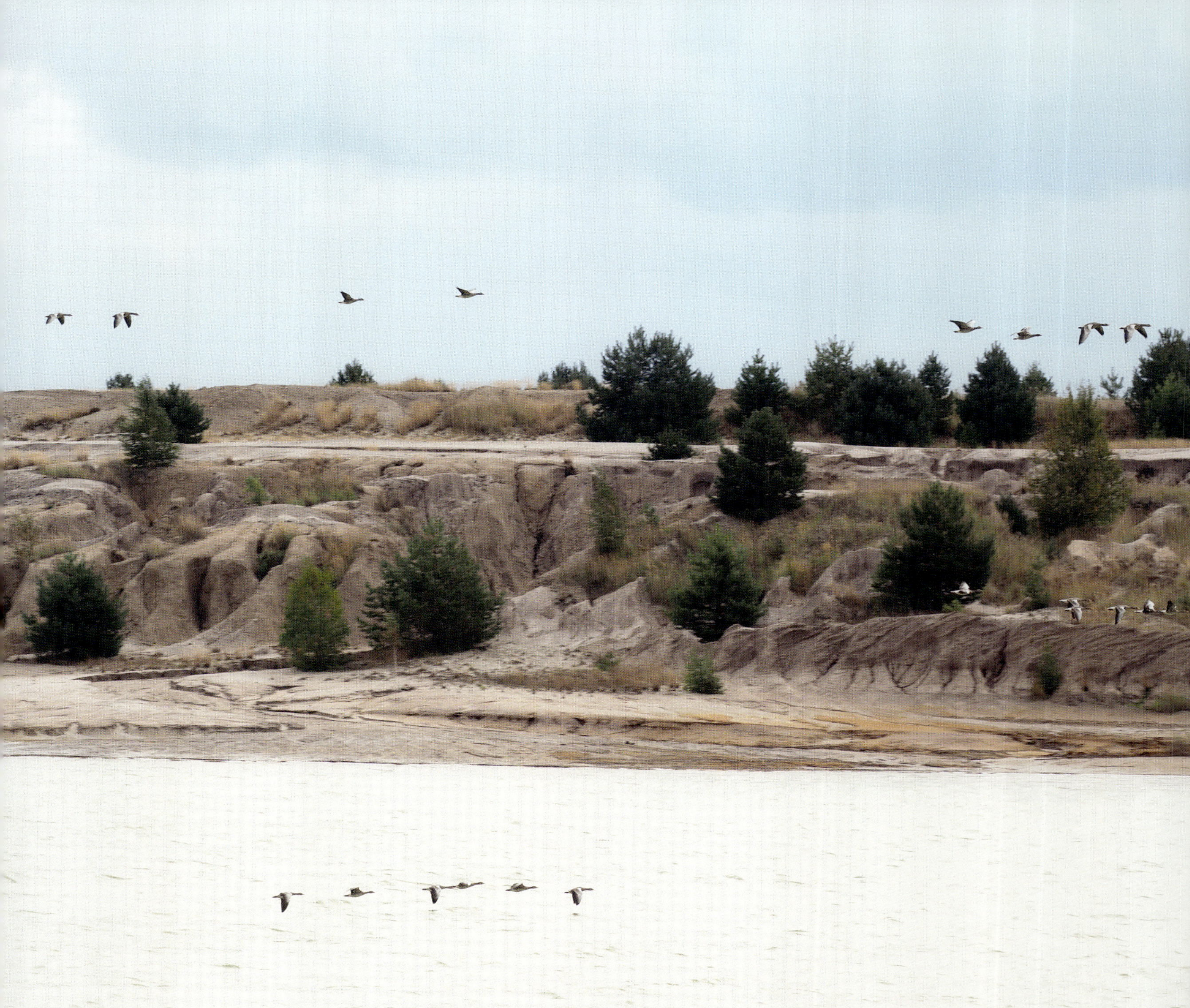

59
Sielmann's Natural Landscape Wanninchen

Germany

Beneath a leaden sky, V-formations of honking geese cross the monochrome moonscape before gliding down to the lake's surface with its crumbling banks and its sand dunes. This post-mining landscape of extremes offers no shelter from sun or wind, extreme heat and cold, scarce nutrients, unstable land and soil and polluted lake water, but its young trees testify that life can persist. This is Sielmann's Natural Landscape Wanninchen in eastern Germany, and despite the name there is simultaneously nothing natural about the origins of this man-made place, but everything natural about how it is being allowed to recover.

The late Heinz Sielmann was a renowned German adventurer and natural history filmmaker who became an activist and conservationist after witnessing the growing destruction of the natural world during his travels. In 1994, he established the Heinz Sielmann Foundation to encourage people to engage with nature and to conserve threatened habitats.

The Sielmann Natural Landscape Wanninchen sits on land that until 1991 was part of the vast Schlabendorf-Süd opencast lignite mine. The mine closed shortly after German reunification. LMBV, the government-owned company responsible for rehabilitating former state-owned mining company land, moved in and engineered the landforms to promote stability and create small islands, a geographically diverse lakeshore and tree plantations. Unlike most similar sites in the region, the 40-metre-deep pit was not flooded but was allowed to fill slowly with groundwater, which began its inexorable rise after the mine closed and its pumps were turned off. Lake formation will take many years as ridges, hollows and trenches become imperceptibly flooded creating dynamic new wildlife habitats. This groundwater is acidic and high in dissolved iron, necessitating applications of lime to improve lake water quality.

The Natural Landscape is surrounded by the Lower Lusatian Ridge Nature Park, which forms part of the Lower Lusatian Ridge – a 50-kilometre-long range of low moraine hills that mark the final extent of Ice Age ice sheets of 200,000 years ago (the Wolstonian Stage). It is now a recovering landscape of quiet, rural land uses, conservation and recreation.

The Heinz Sielmann Foundation bought more than 3,300 hectares of this surreal desert landscape from LMBV to create its natural landscape reserve. Recognising that its vast areas of continuous, unbroken landscape, lack of disturbance, topographic variety and diverse and nutrient-poor soil conditions are ideal for many unusual and specialist plants and animals, they have left the land to recover by making not just space for nature but giving it time too. Antithetically, despite the skeletal landscape's brutal appearance, its biodiversity is particularly sensitive to human interference. This contradicts the tendency (and regulations in many jurisdictions) that post-mining land must be restored as soon as possible and that any barren areas are frowned upon and seen as a a problem to fix.

The approach has been vindicated by the presence of numerous rare, endangered, unusual and specialist species of plants and animals. Many of these are small and nondescript, but two flagship species have proven the concept: each autumn, the site provides a stop-over for thousands of majestic common cranes on their annual migration, as well as impressive flocks of thousands of migrating greylag geese. And wolves frequent the area, taking advantage of the relative lack of people.

The Wanninchen Nature Experience Centre opened in 2002 to provide information about the site and its wildlife to visitors, promote environmental education and encourage adventure in the natural world. Tracks along the water's edge and wildlife-watching hides allow visitors to explore and overlook the new world on the opposing bank in the hope of seeing a crane, or even a wolf.

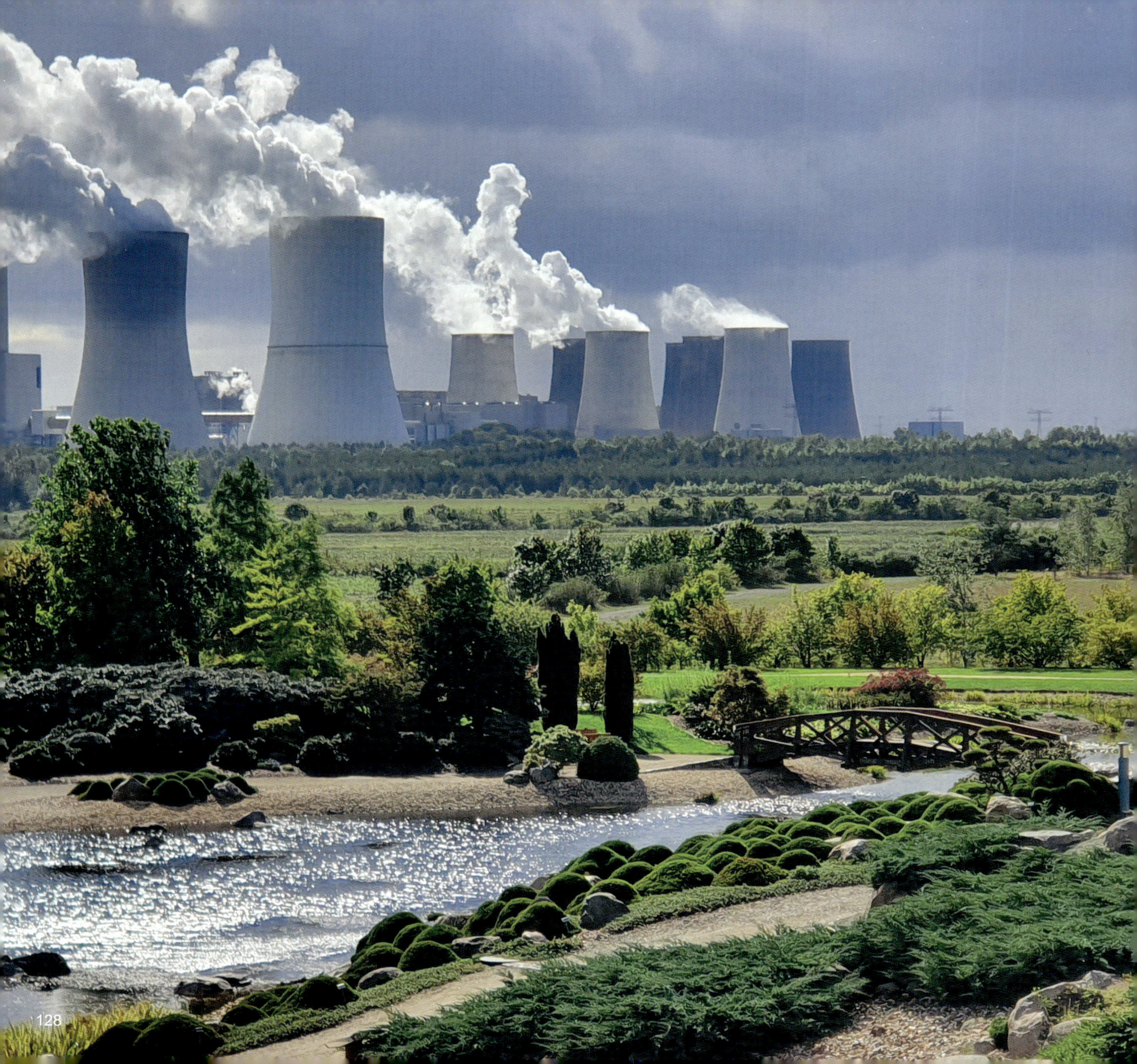

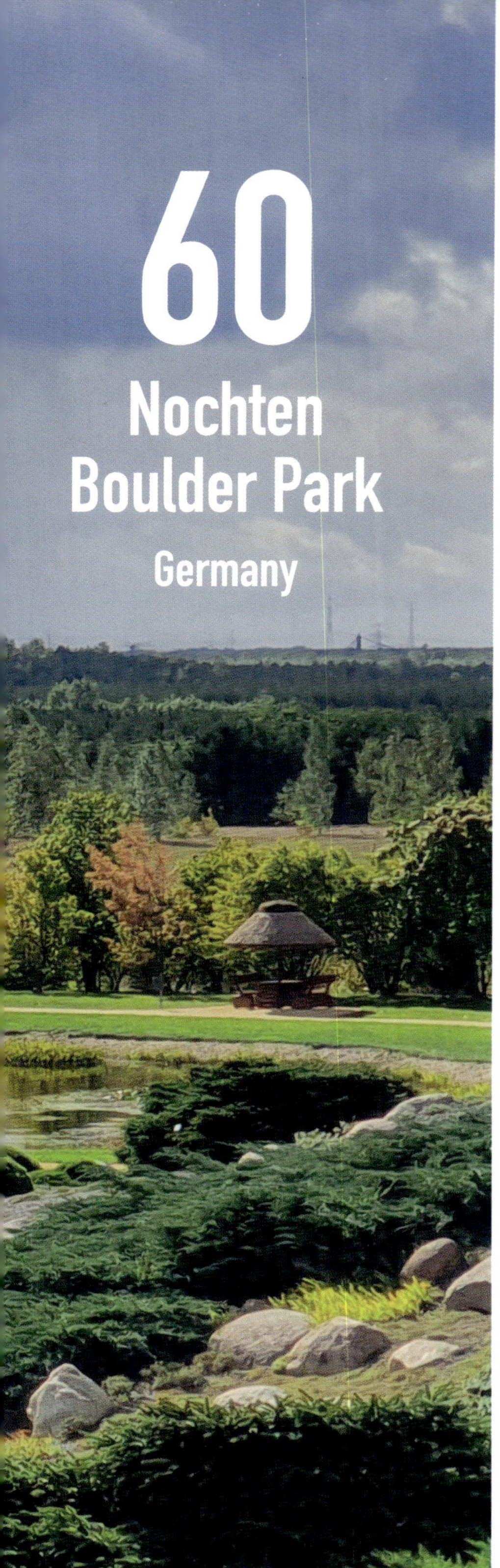

60

Nochten Boulder Park

Germany

Parks and gardens are common reuses for old mine sites, but beneath the clouds billowing from the cooling towers of the Boxberg lignite power station sits a garden with a difference: at the Nochten Boulder Park, fashioned from the scars of mined land, itinerant boulders rather than plants are the real rock stars.

As the excavators of Saxony's vast Nochten opencast mine consumed overburden to access the underlying lignite, they unearthed exotic boulders that had been carried hundreds of kilometres from their Scandinavian homelands by a super-slow ice train during the last Ice Age. As the frozen blanket melted, the geological travellers took up permanent residence in a new land. And there they stayed, for millennia, until modern mining revealed them.

The original concept emerged in the late 1990s within the mining company LAUBAG – then the operators of the mine, which is now operated by LEAG. Planting began in 2000; now, the garden is mature, with extensive beds of vibrant plant colour and bright young trees.

Its low hills, ponds and babbling streams were sculpted from a mound of overburden. Among the 100,000 plants, 7,000 glacial erratics take centre stage. Each one is a fragment of natural history with a deep-time story to tell, given that many are well over a billion years old. The diverse rocks differ in colour, pattern and texture – some are surprisingly beautiful; their unique appearances and compositions enabled geologists to pinpoint the precise Scandinavian locations from which they emanated, and these are marked on a giant map of the region on the side of an artificial hill using the boulders themselves as markers. Interestingly, knowing their origins and final resting places makes it possible to determine the flow directions of their ancient ice conveyors.

The park attracts tens of thousands of visitors annually and incorporates facilities for formal meetings and events, a café, an information area and a souvenir shop. Several kilometres of winding gravel paths and trails encourage exploration of the 20-hectare park's itinerant geology, monuments and planted areas. The latter grade from more formal gardens, through planted heathland, merging into a more natural heath area populated by plants found in the local area, including a heather moor, which includes rarities and small trees rescued as the mines took their land. This transitions into the natural heath and forest areas across the park boundary.

There are many profoundly interesting aspects to the Nochten Boulder Park; in the shadow of the powerplant, one ponders deep time and the juxtaposition of beauty and ruined land. From atop a low hillock, if you divert your gaze away from the mesmerising cloud factory, you may discern lumbering behemoths looming over the reforested mine lands in the distance, still devouring earth in their opencast home.

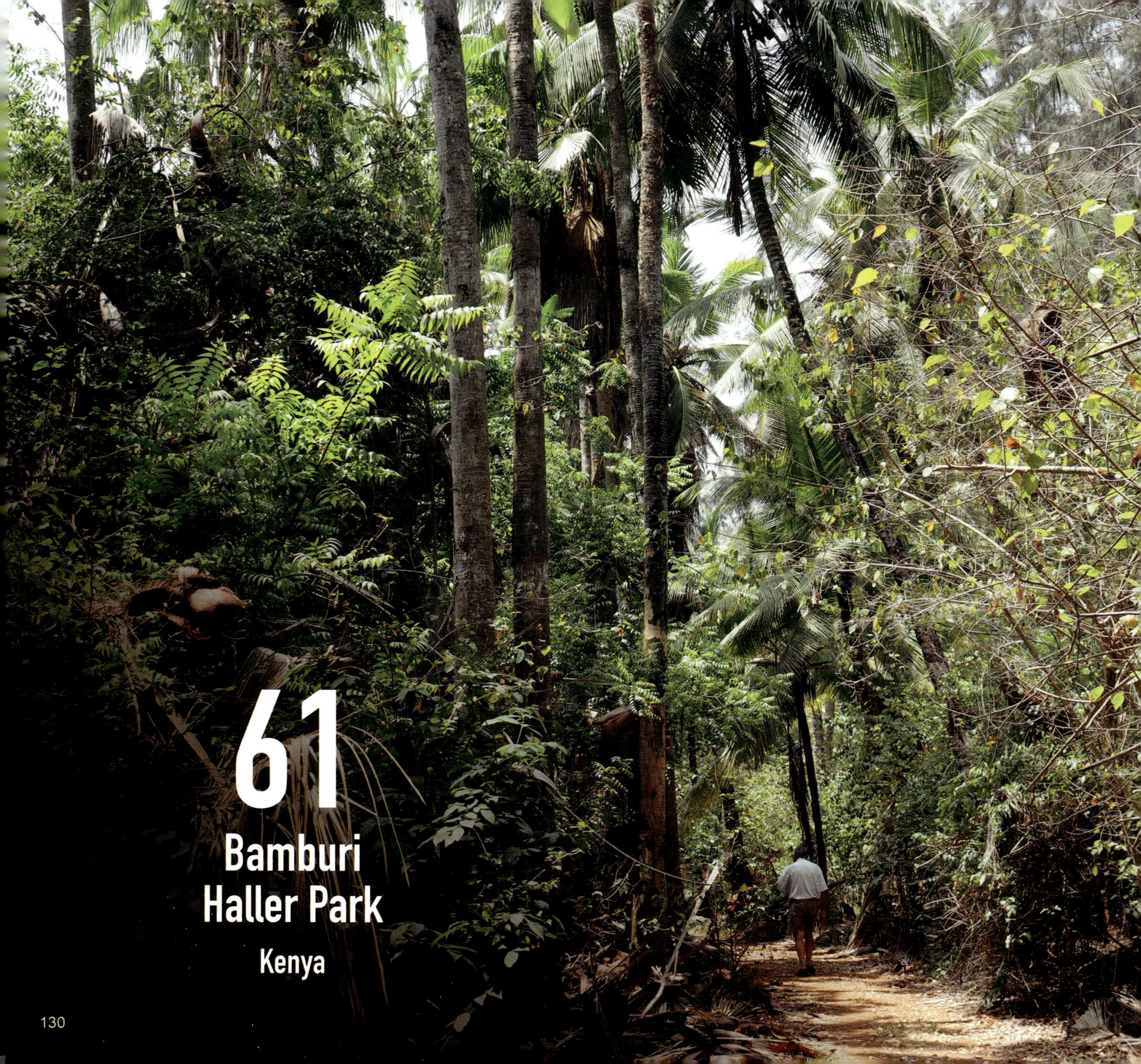

61
Bamburi
Haller Park

Kenya

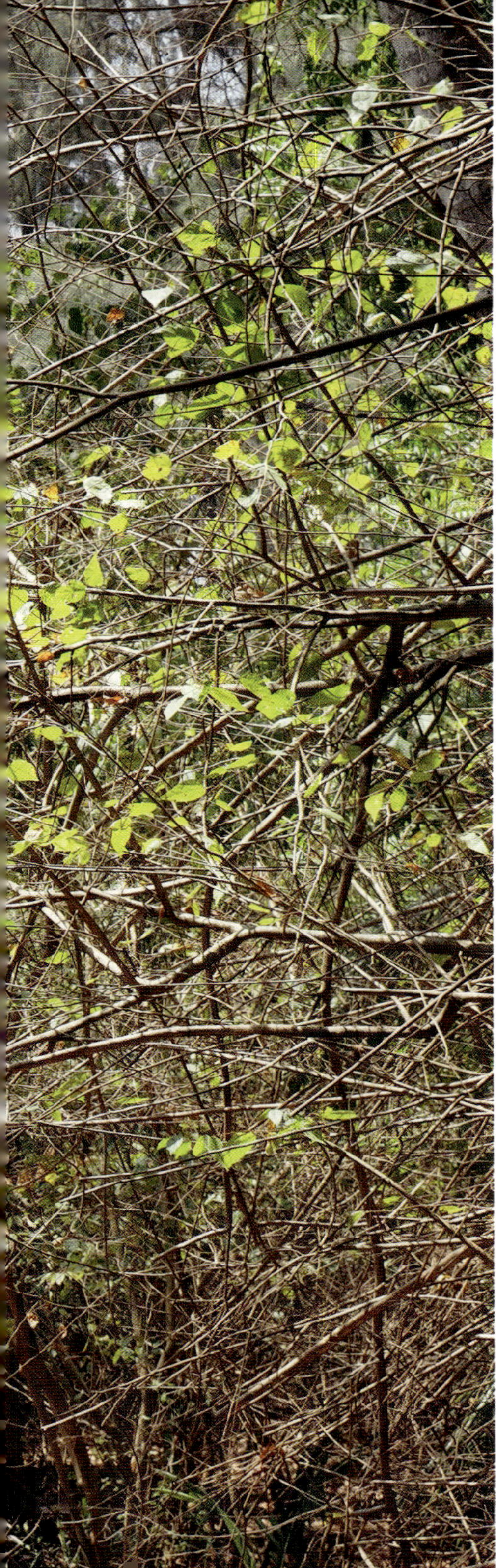

Wandering in the dappled shade through coastal Kenya's tropical forest with Dr René Haller is an education in vision and passionate endeavour. The trees drip with life, and the calls of unseen birds pierce the humid air as he digs a small hole with his trowel to proudly show a few centimetres of substrate darkened with organic matter – the result of more than 50 years of ecological process.

The renowned story of Kenya's Bamburi Haller Park is the result of Haller's determination and decades of applied research. In 1959, Swiss-born Haller, a naturalist, horticulturist, landscape designer and tropical agronomist, began work at Mombasa's Bamburi Cement Company with the brief to produce food for the workforce on the company's land. Ten years later, he persuaded the company to let him restore their enormous limestone quarries close to the Mombasa coastline. These were challenging sites, covering several square kilometres of infertile, sun-baked and compacted rocky surface with saline water just below the surface.

Before ecological restoration practices on mine sites were scientifically established, as head of the company's Garden Department, Haller worked from first principles, adapting conventional horticultural techniques to establish trees on this barren land by trialling 26 species planted in pits arduously hacked out of the concrete-like substrate. Haller admits with hindsight that this approach was too elaborate and wrong – a rethink was needed to kickstart ecological processes, and critical to this was choosing the best species for the circumstances and alleviating substrate compaction. For Haller, cracking the soil code was key to driving holistic change for both 'ecology and economy'.

The trial showed that only one species would readily cope with the harsh conditions – the casuarina tree native to Australia but naturalised along the East African coast. It resembles a feathery conifer and is admirably adapted to cope with drought and heat with its tightly bunched leaves resembling pine needles, reducing their surface area to moisture loss. Despite their tenacity, however, the young casuarinas began to wither, prompting further investigation. Haller observed that casuarina tree roots on local sand dunes were intimately associated with symbiotic microbes. One critical symbiosis is with bacteria that convert atmospheric nitrogen in soils to nitrates, a natural fertiliser which is then absorbed through the tree's roots. He transferred some of this soil to the saplings suffering in the quarry, and they began to flourish.

The high tannin content of casuarina needles makes the resulting leaf litter resistant to decomposition, which builds up on the surface, stymieing soil formation. This time, Haller turned his enquiring mind to his beachside garden, where he found red-legged millipedes munching the tough casuarina leaf litter into a fine organic tilth, which ultimately becomes humus – the key organic constituent of a soil. Haller translocated hundreds of millipedes to the quarry, where they got to work on soil formation. Casuarina became his go-to restoration primer to create soil.

Compaction remained a challenge; digging tree planting pits was time-consuming work, and as the trees grew, their roots were confined by the compacted soil, leading to water and nutrient stress and instability. Haller instituted a regime of substrate ripping, with the saplings planted into the ripping lines. As these trees grew, they cast shade, reducing temperature extremes and increasing soil organic matter, which enhances water retention and fertility. Other species of animals and plants moved in and quickly became part of this emerging novel ecosystem.

A million trees were planted within ten years. Despite the extreme conditions, the short-lived casuarinas grew quickly, reaching 30 metres in that time. Dense stands of casuarinas were thinned out and then sold for timber or firewood, which enabled the planting of other tree species, including commercial species like African mahogany. Other species colonised the area naturally through wind and animal dispersal, so today, large areas of the park are floristically diverse and self-sustaining. To some extent, the re-greened quarry has become a biodiversity sanctuary, with over 30 species on the endangered species Red List.

Animals such as vervet monkeys and birds entered the system naturally, and through trial and error further animals were introduced, each with a role in assisting ecosystem recovery. These included ants to assist with nutrient cycling; bushpigs to disturb the undergrowth and aerate the soil as they snuffle for invertebrates; giraffes to convert higher-level tree leaves into faeces, augmenting soil fertility and distributing seeds; a herd of oryx to graze on poor quality grasses – they can survive in harsh environments, with little water; hippos keep the small,

brackish lakes alive; and even Aldabra giant tortoises bulldozing through the undergrowth. Small deer and antelopes were trialled as potential bushmeat. Accumulating animal dung caused a plague of flies, so dung beetles were tasked with industriously collecting, dispersing and burying faeces to speed up decomposition.

Interconnected waterbodies and wetlands meander through the landscape as envisioned in Haller's original master plan. The ground is only centimetres above saline groundwater, which is linked to the sea through the porous limestone, rising and falling with the tides. The brackish lake margins and wetlands offer ideal conditions for impressive mangrove ferns, aerially rooted pandanus and mangrove trees.

In line with his original remit to feed this company's workforce, he grew fruit and vegetable crops, including coconuts and bananas. He also selected eland antelopes for domestication because their milk's antibiotic properties keep it fresh for months without refrigeration, and as native livestock, they are resistant to many pests and diseases.

Given the hydrodynamics of the site, Haller developed a commercially viable, integrated aquaculture system consisting of a fish farm, a crocodile farm and a biological water treatment area of Nile cabbage and rice paddy fields for water purification. The aim was to use all organic wastes and wastewater streams as a resource for further food production in a closed-loop system. Set up at the same time as the reforestation project, the famous Haller Park fish farm was based on a system of concrete tanks of flowing water enabling the fish to swim

against a current. It produced up to 40 tonnes of fish per year.

The park was Haller's launchpad for a lifetime of research and development of low-cost initiatives in food production, conservation, health, education and business development to help people in Kenya through the Haller Foundation and his Baobab Trust. He has received numerous awards in recognition of his achievements, including a prestigious UN Global 500 Roll of Honour award, previous recipients of which include luminaries like President Jimmy Carter, Sir David Attenborough and Gro Harlem Brundtland.

A unique hybrid ecosystem, now managed by Lafarge Ecosystems Ltd, exists where once was a barren wasteland. Over 150,000 visitors annually explore the two main parts of its vast area. The first is Bamburi Haller Park, with a focus on tourism and conservation education based on the tropical forest gardens, reptile and butterfly displays, and an orphanage for rescued animals. Its conference and event venues sit beneath the trees, including an impressive auditorium with a beautifully thatched roof, designed by students and built in the 1980s from locally sourced natural materials. The second part is the 16km of Bamburi Forest Trails, which encourage exploration by bike or on foot beneath the cooling, dappled shade of palm trees swaying in the coastal breeze. Both combine in a breathtaking 50-year incarnation of Haller's vision as he wanders ahead into his forest.

Previous page: René Haller wanders in his tropical forest, which was once a barren quarry.

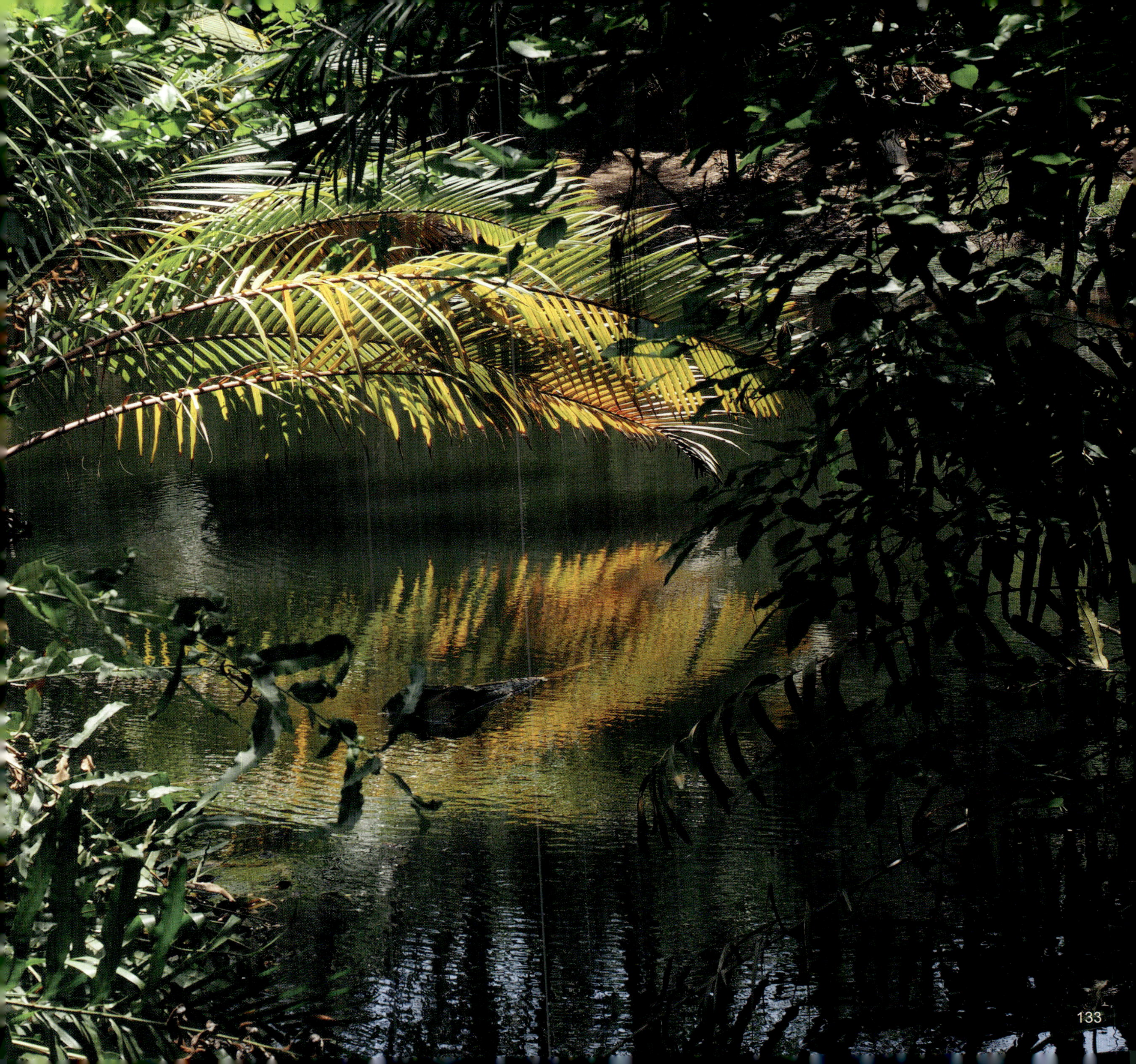

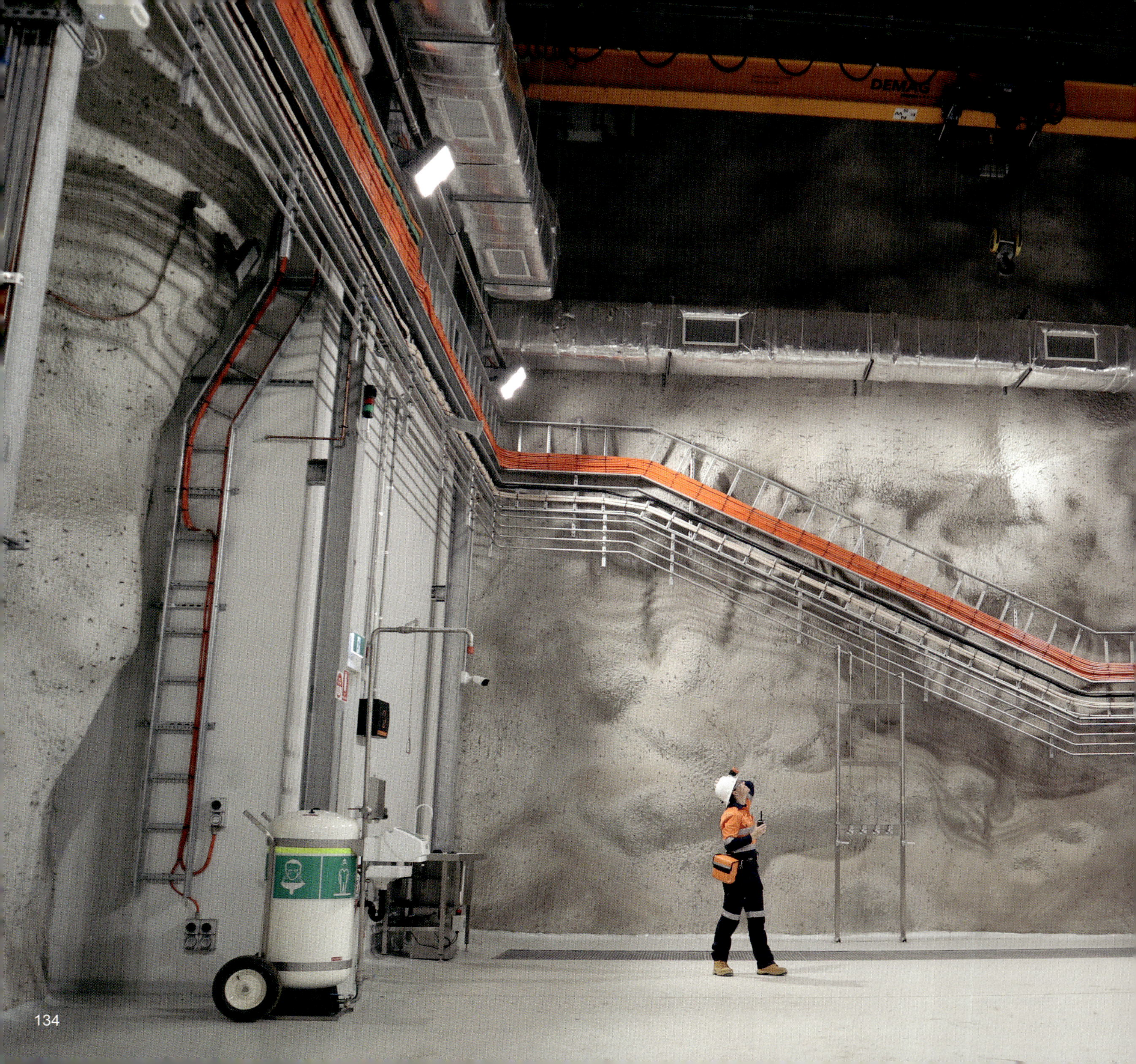

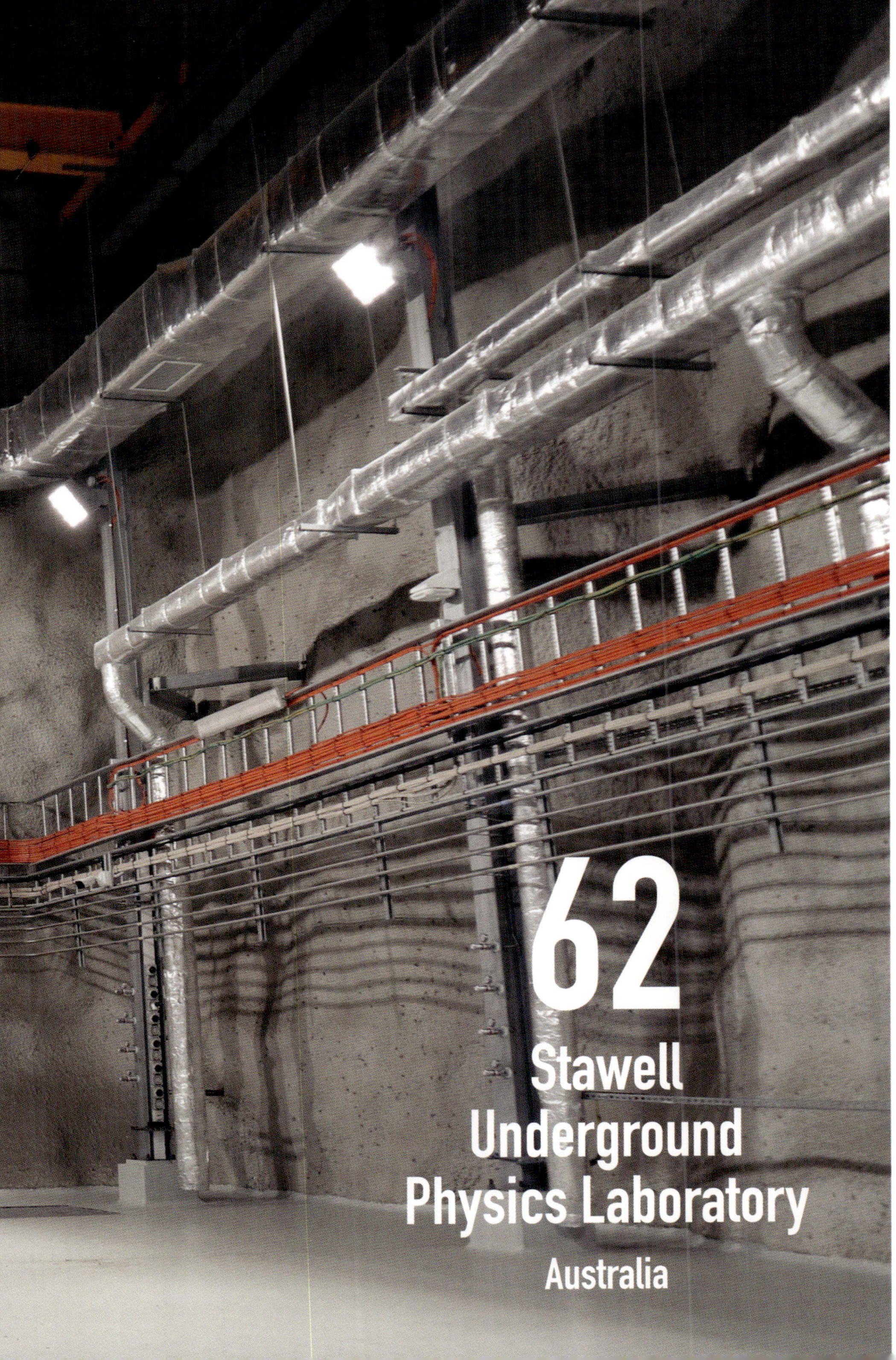

This vast void hewn from solid rock, with its bright white lights, gleaming metalwork and immaculate appearance, feels like a sterile outer space experience rather than anything to do with mining. It's a 25-minute, 10-kilometre downhill drive into the underworld to reach the Stawell Underground Physics Laboratory (SUPL), over a kilometre below the remote gold mining town of Stawell, Victoria, Australia.

There are several underground physics laboratories in mines across the world, but SUPL is the first in the southern hemisphere. It aims to unlock one of the biggest mysteries of the universe – the nature of dark matter. There is about five times the mass of dark matter in the universe than ordinary matter. It pervades everything including you and this book, but it is benign and interacts with barely anything except gravity, which makes it virtually impossible to detect and so far its existence has only been inferred by physicists through its gravitational effects on other objects in the universe. No one knows what this elusive material consists of – a popular working theory is that dark matter is made from subatomic particles, known as WIMPs – weakly interacting massive particles.

For over 20 years, tantalising glimpses of what might be dark matter have been made by the DAMA/LIBRA detector deep inside an Italian mountain. It has reported the same predictable annual oscillating signal, so a southern hemisphere counterweight, SABRE (Sodium Iodide with Active Background Rejection Experiment), was deemed necessary to remove any influence of more mundane, Earth-bound seasonality on physical effects.

Out of a population of about 6,200 people, Stawell Gold Mine (SGM) employs around 300. However, from 2012 the company announced it was winding down operations, which led to mothballing in 2016, raising major concerns in the community about what would happen to the local economy post-mining. Local leaders embarked on a long process of finding a new economic path, stimulating ideas and work to diversify the town's economy, and even utilising a copy of the original *101 Things to Do with a Hole in the Ground* book in the process! The local MP, mayor and council worked tirelessly with other branches of government and stakeholders while the search for a suitable site for a southern hemisphere dark matter laboratory was underway in Australia. The stars aligned and the decision was made to create Australia's first underground physics laboratory at Stawell. The mine restarted operations in 2019 and, in the same year, the construction of SUPL's underground world began using local expertise and workers wherever possible and supported by SGM in providing tunnelling expertise and underground services. SUPL Ltd, a new company, was set up to take the project forward.

The dark matter detector will be so sensitive that all potential sources of extraneous background radiation have to be minimised. Locating the detector deep underground shields the facility against cosmic rays from space and the sun – potential major sources of interference. Prior to construction an Australia-wide search was conducted to identify ultra-low radiation materials for the concrete for SUPL. Ultimately, aggregate from South Australia, cement from Queensland and sand from a local supplier were used. The steel used in the void was also subject to similar radiation checks. Shotcrete (concrete sprayed onto the cavern's stone walls) was then coated with a special paint to shield against radon (a common radioactive gas) leakage into the cavern from the surrounding rock. The air-conditioned cavern is held at a positive air pressure to minimise the entry of dust, further reducing radiation and keeping things spotless.

SUPL is set to be the epicentre of dark matter research in Australia. It arose as a collaboration between state and federal governments (the federal minister for education was a local MP) with funding from both, and was heavily supported by the mayor and local council and SGM. The Australian Government and the Victorian State Government each provided AUD 5 million to the build which was supplemented by major support from the Stawell Gold Mine and AUD 1.9 million from the University of Melbourne. The Australian Research Council awarded a consortium of universities AUD 35 million to establish a Centre of Excellence for Dark Matter Particle Physics, which included a key experimental research theme for experiments in SUPL. And SUPL is already putting the town on the map – and not just in Australia: several prestigious national and international universities are collaborating on the venture.

SUPL's SABRE south detector will be a big beast, measuring 4m^3 and weighing 120 tonnes. Transporting this into the cavern along 10km of underground decline tunnel will be a serious logistical challenge. The detector's heart will consist of seven ultra-pure, flawless sodium iodide crystals. A subatomic particle passing through a crystal will generate a flash of light, or scintillation, when it hits an atomic nucleus, which will be recorded by highly sensitive light detectors. Interferences from cosmic and gamma rays from other sources will be combatted by a range of measures, including surrounding the crystals with a 12-tonne tank of a scintillating liquid, which emits light when a gamma ray passes through it.

If it works, SUPL will have a unique place in the history of physics for its ground-breaking contribution to our understanding of the formation of the universe. But if dark matter remains elusive, this superb low-radiation research facility can be used in other technical disciplines, such as researching ultra-sensitive detectors for a range of particle and quantum detection devices, materials and pharmaceuticals, producing a host of highly trained scientists and engineers ready to enter leading-edge technology industries. And, encouragingly, the buzz associated with the search for dark matter in a nearby underground cavern seems to have increased local schoolchildren's interest in STEM subjects.

63
Ópera de Arame
Brazil

The Ópera de Arame, or 'Wire Opera House', in Curitiba is an architectural marvel that has repurposed an old quarry into one of Latin America's premier cultural stages. Nestled within the Parque das Pedreiras alongside the Paulo Leminski Cultural Space, the opera house rests on an artificial lake. Seating around 1,500 and hosting a diverse array of shows from popular to classical, the venue has seen performances by world-renowned artists.

Located to the north of Curitiba's city centre in the state of Paraná, South Brazil, this exemplary building is fashioned entirely from glass and steel tubes, its distinctive shape and structure drawing in music and architecture enthusiasts from around the globe. The creation of Curitiba's former mayor Jaime Lerner and architect Domingos Bongestabs, the steel structure was erected in just 75 days and opened on 18 March 1992.

A bridge offers visitors a panoramic view of the tubular construction as they make their way to the opera house. Inside, the glass roof and walls afford a breathtaking view of the natural surroundings, embracing lush vegetation, the quarry's rocky outcrops and cascading waterfalls.

Beyond its original cultural purpose, the Wire Opera House offers art exhibitions and restaurants, as well as ecological trails in the surrounding park. In 2013, the opera house underwent a renovation with more comfortable seats, and improved acoustics and hydraulic systems.

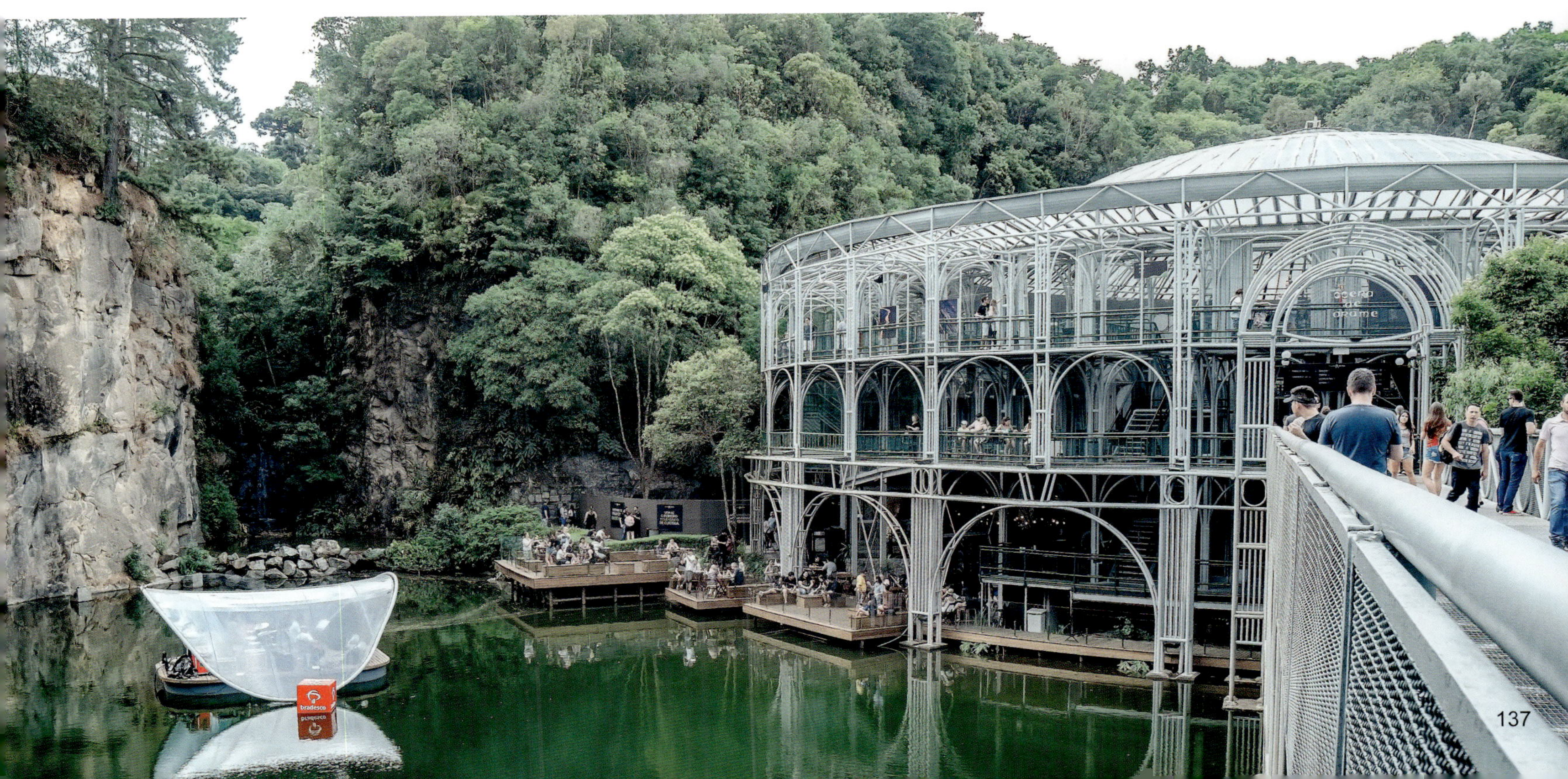

64

Ffos Las Racecourse

UK

Coal mining in the Trimsaran area of Carmarthenshire, west Wales, extends back centuries. Celtic Energy opened its Ffos Las mine in 1983, which became one of Europe's deepest opencast mines.

Named after the farm that originally occupied the land, *ffos las* is Welsh for 'Blue Ditch'. The mine closed in 1997.

Surrounded by the tranquil countryside of the Gwendraeth Valley and with views over Carmarthen Bay, the 240-hectare brownfield site was acquired in 2003 by Dai Walters, one of Wales' best-known entrepreneurs, with a passion for horse racing. The self-made businessman grew from humble beginnings as an apprentice at a South Wales coal mine in the 1970s to the owner of plant hire and major construction businesses.

After several years of planning, the council approved the site's use as a racecourse, housing and a hotel in 2006. The GBP 20 million project took only two years to build and opened in 2009.

Ffos Las was the first National Hunt racecourse built in the UK since 1927 and allows a dual purpose with flat race meetings in the summer. It is a left-handed course, about 2.4 kilometres long, and incorporates a six-furlong straight course too. It hosts 23 racing fixtures per year, including the annual Welsh Champion Hurdle race. The site is a 90-minute drive from the Welsh ferry port of Fishguard, providing easy access for the Irish horse racing community. The racecourse also offers conference facilities and venue hire. The surrounding reclaimed mine land sports housing estates and mixed industrial and retail facilities, a solar farm, areas for nature, and surface water storage lagoons. In 2018, Dai Walters sold Ffos Las to the Arena Racing Company.

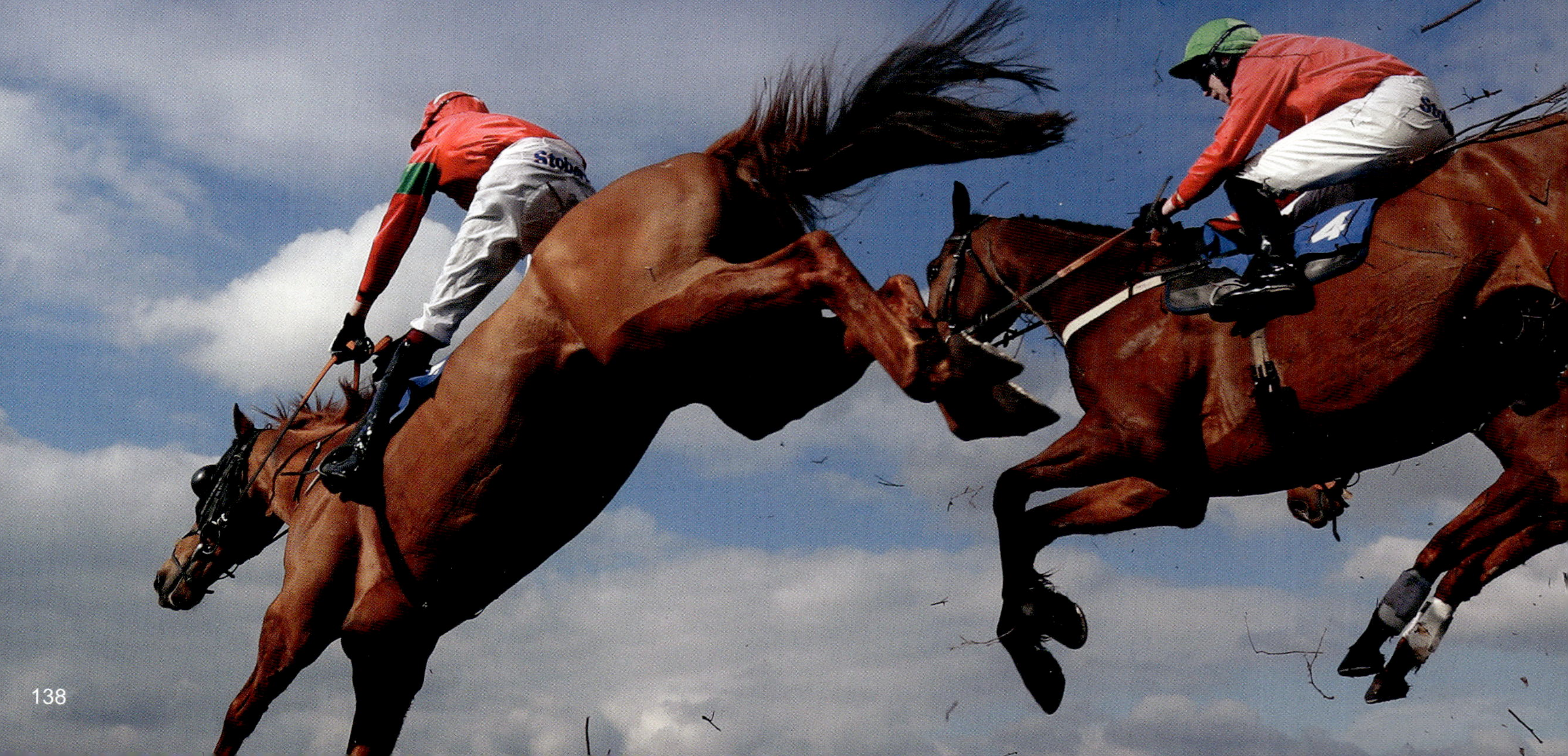

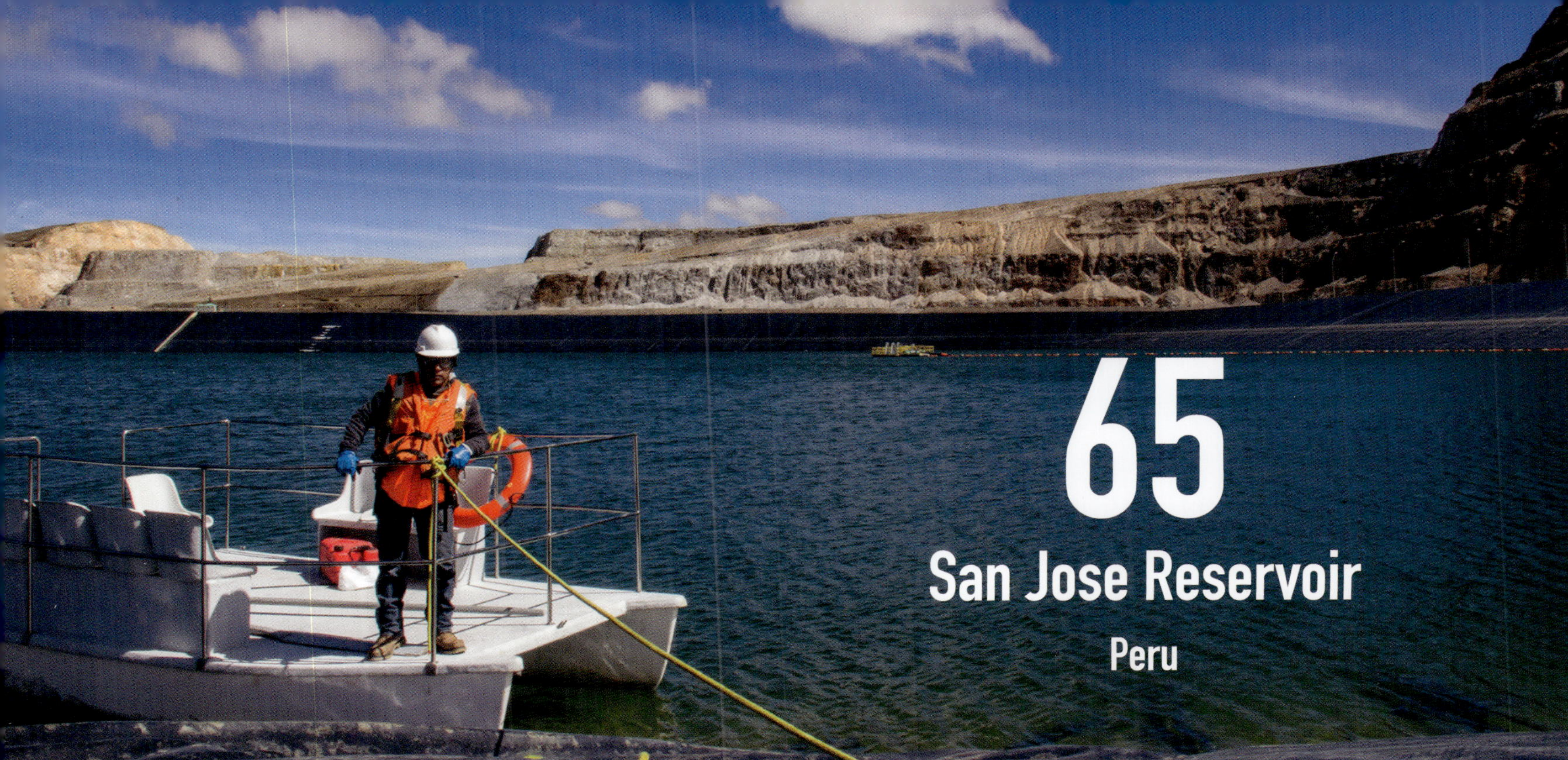

Four thousand metres up in the Peruvian Andes of Cajamarca, Newmont's expansive Yanacocha gold mining complex is the largest in South America. There is also a long history of high-altitude farming that is adapted to the region's marked wet and dry seasons. Cajamarca's ancient irrigation canal systems are a testament to the engineering skills of thousands of years ago. Rural *campesino* farmers downstream of the mining complex still rely on this canal network for collecting and distributing water from natural seeps and springs to allow them to grow crops, such as potatoes, corn and alfalfa through the dry season.

The development of the San Jose mine, one of the mines in the Yanacocha complex, and other facilities, affected both the canals' structures and water availability, leading to heightened community tensions over the sustainability of water supplies. The company agreed to pay compensation for the lost water and to identify other water source options.

In response, in 2007 after mining in the San Jose pit had ended, the company transformed it into a reservoir by re-engineering the pit to allow a geosynthetic liner to be installed to store water for the local agricultural communities. Water from the San Jose Reservoir now flows into the canal network, supplying year-round irrigation to almost 1,000 hectares of agricultural land, which supports 1,000 *campesino* families. Yanacocha's water treatment plant recharges the reservoir, and after the complex closes, water treatment and reservoir storage will continue.

Although initially tense due to the effects of the mine on the canals, relations between community and company have got better since Newmont's commitment to ensure year-round water supplies and its support of other community development projects. *Campesino* farmers that receive water from San Jose no longer completely depend on the vagaries of seasonal rainfall for their crop and livestock yields, resulting in improved certainty, greater productivity and a better quality of life.

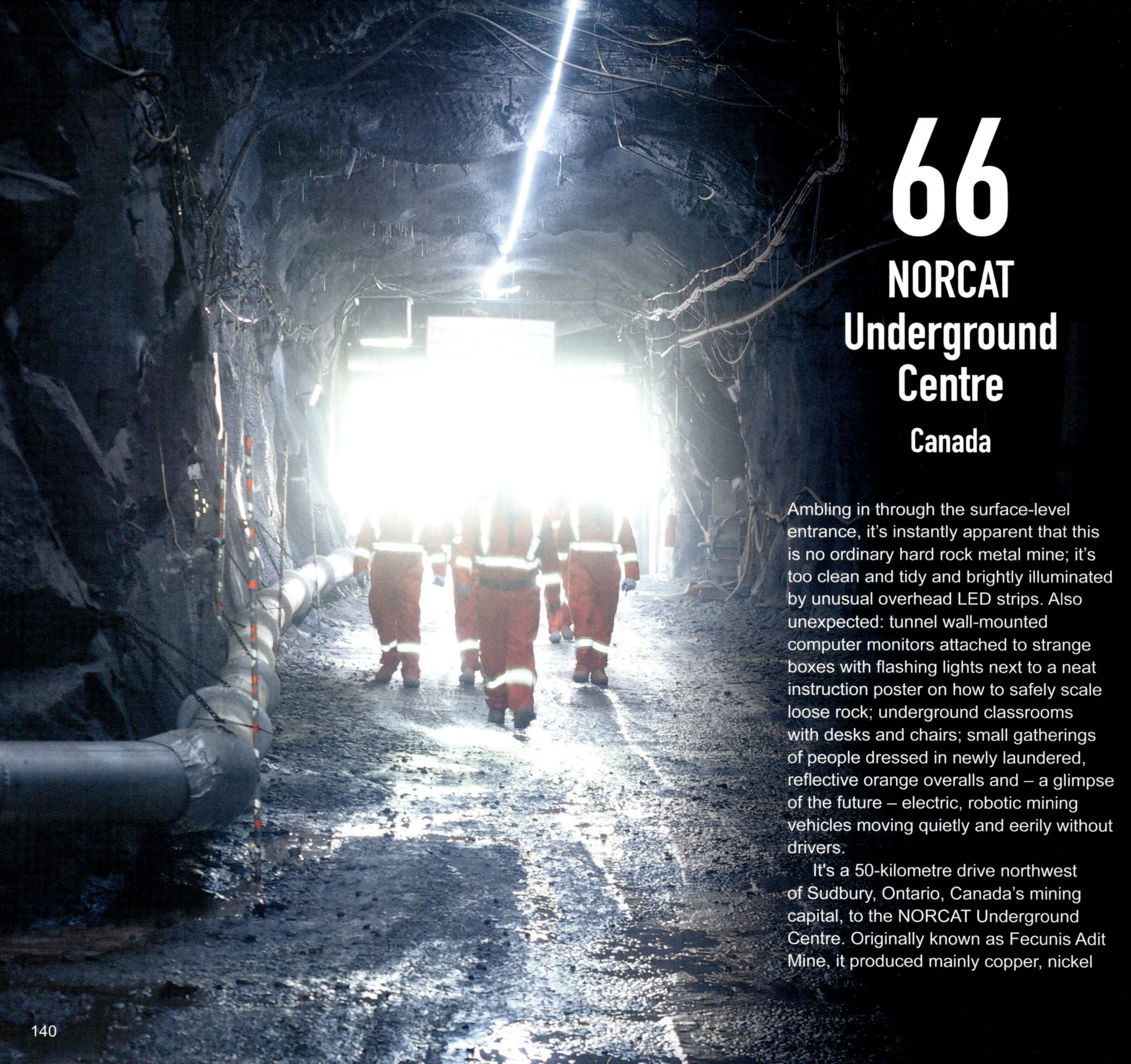

NORCAT Underground Centre

Canada

Ambling in through the surface-level entrance, it's instantly apparent that this is no ordinary hard rock metal mine; it's too clean and tidy and brightly illuminated by unusual overhead LED strips. Also unexpected: tunnel wall-mounted computer monitors attached to strange boxes with flashing lights next to a neat instruction poster on how to safely scale loose rock; underground classrooms with desks and chairs; small gatherings of people dressed in newly laundered, reflective orange overalls and – a glimpse of the future – electric, robotic mining vehicles moving quietly and eerily without drivers.

It's a 50-kilometre drive northwest of Sudbury, Ontario, Canada's mining capital, to the NORCAT Underground Centre. Originally known as Fecunis Adit Mine, it produced mainly copper, nickel and platinum group metals. Formerly owned by Falconbridge (now Glencore), it opened in the 1950s and was slated for closure 40 years later. In the early 2000s, NORCAT assumed operational control of the mine based on a land lease with Glencore.

Founded in 1995, NORCAT was set up by northern Ontario's business and academic sectors to provide experiential, hands-on skills development for new and existing miners and to advance mining technologies for long-term regional and national benefit. It is the only centre globally that owns and operates an underground mine for the primary purposes of providing training and – uniquely – a stepping stone for tech companies to test new technologies in a controlled, operational mining environment, which can be problematic in a large-scale commercial mine.

With over three kilometres of tunnels accessible, the Underground Centre takes about 1,000 trainees per year, mainly from Northern Ontario. Courses include underground operator training, hard rock mining methods and diamond drilling. NORCAT provides accredited qualifications with a focus on safety and electrification (battery electric vehicles). The students are the mine's workforce and the ore they extract is bought and processed by Glencore.

NORCAT is a critical part of Sudbury's world-leading mining technology cluster and a regional hub for Ontario's Tech Network (a provincial centre of excellence). Its Underground Centre is a global one-stop shop for mining's emerging transformational technologies. Every year, it supports about 50 mining technology projects and hosts, on average, one overseas mining delegation per week. They tour the facilities and experience the technologies in action. The facility also enhances Canada's international reputation as a leading mining nation, with the potential to attract further high-tech mining investment into Sudbury and Ontario.

Just a few of the wide range of technologies that have been tested, developed and demonstrated include ruggedised wifi infrastructure to enable a subterranean 'internet of things'; envelope-pushing cellphone networking technology; new types of ventilation systems; robotic vehicles and drone mapping and surveying platforms. Customers range from tech giants like Hexagon Mining and Rogers Communication to small local start-ups and SMEs.

In 2022, NORCAT hosted 'Mining Transformed' – the first industry exhibition held in a working underground mine. The event attracted 250 delegates to view 50 emerging technology installations from 15 countries. NORCAT brokers constructive interactions between mining technology companies ('innovation builders') and mining companies ('innovation buyers'), giving the 'builders' the opportunity to present their equipment in real mine conditions, which is not possible at a conventional tradeshow, thus answering the inevitable buyer's question: 'Does this work underground?' The event's success was derived from the 'spirit of the event rooted in the place'. There are plans to run it again.

The Centre also provides entrepreneurial assistance through its programmes, facilities and services in support of its mining technology 'ecosystem', including business start-up mentoring, marketing, business incubator hub and workspace, related training events and access to public and private funding opportunities.

Since the Covid pandemic, over 250 new jobs have been created in companies that went through NORCAT. The organisation itself employs about 95 full-time equivalents, consisting of 108 employees, 15 of which support training and innovation in the Underground Centre

NORCAT is a non-profit organisation and doesn't have investment interests in the companies it works with. It receives limited operational government funding and operates on an earned-revenue model, charging fees for its facilities and services that support training and the innovation projects. It does receive grants from the government for capital developments like buildings. The Underground Centre was recently expanded with an innovation hub to improve facilities. Funded by CAD 4.5 million from government and industry, as a measure of success its ten new offices were all leased out before it even opened.

67

Life Ribermine

Spain and Portugal

Mineral waste tips are often the most visible impact of a mine's activities and may persist for decades, centuries and even millennia after abandonment or closure. The conventional practice of geometric benched tips using a steep gradient or terraced approach often exacerbates erosion, forming runoff gullies. These cause knock-on impacts on land, water and air quality from the denuded surfaces. Such highly eroded lands limit options for post-mining reuse while remaining visually intrusive for those who live nearby and diminishing the landscape character.

However, geomorphic reclamation (aka geomorphic rehabilitation) involves sculpting, building and replicating aesthetically appealing, culturally appropriate landforms that offer a better canvas for a more vibrant ecology.

The EUR 3 million, EU-funded Life Ribermine project, under Professor José Martin Duque of Madrid's Complutense University and others, is a pioneering geomorphic reclamation project that is working its magic to heal mining-damaged river ecosystems on two challenging abandoned mine sites –

the Santa Engracia china clay quarry in Spain and Portugal's Lousal pyrite mine.

In its central Spanish setting of deep green forests, picturesque gorges and turquoise rivers, the piercingly white, visually jarring scar of the abandoned Santa Engracia quarry outside Peñalén village in Guadalajara Province eats into the *mesa* overlooking the upper Tagus (or Tajo) river – the longest on the Iberian Peninsula. Its freely eroding surfaces used to supply hundreds of tonnes of milky sediments annually into the formerly pristine rivers of the protected Alto Tajo Natural Park and UNESCO-designated geopark. Abandoned in 1990, the quarry was a complexity of unconsolidated mineral wastes, hard rock high walls and an unstable open pit, with several features that promoted severe gully erosion and precluded vegetation establishment that would otherwise help to bind the substrate and soften the scar.

The innovative Life Ribermine demonstration project applies three landscape design software packages to one site: GeoFluv – Natural Regrade for remodelling unconsolidated waste dumps; Talus Royal to mimic natural cliffs on the quarry high wall; and SIBERIA to model and predict erosion. Across 30 hectares of the site, an erosion-resistant topography of stable cliffs and gentle hills, ridges and meandering streams was designed and constructed to promote a self-sustaining vegetation cover.

Geomorphological input data was meticulously collected by scientists from unblemished landscapes in the site's vicinity that have, over millennia, reached a natural equilibrium between

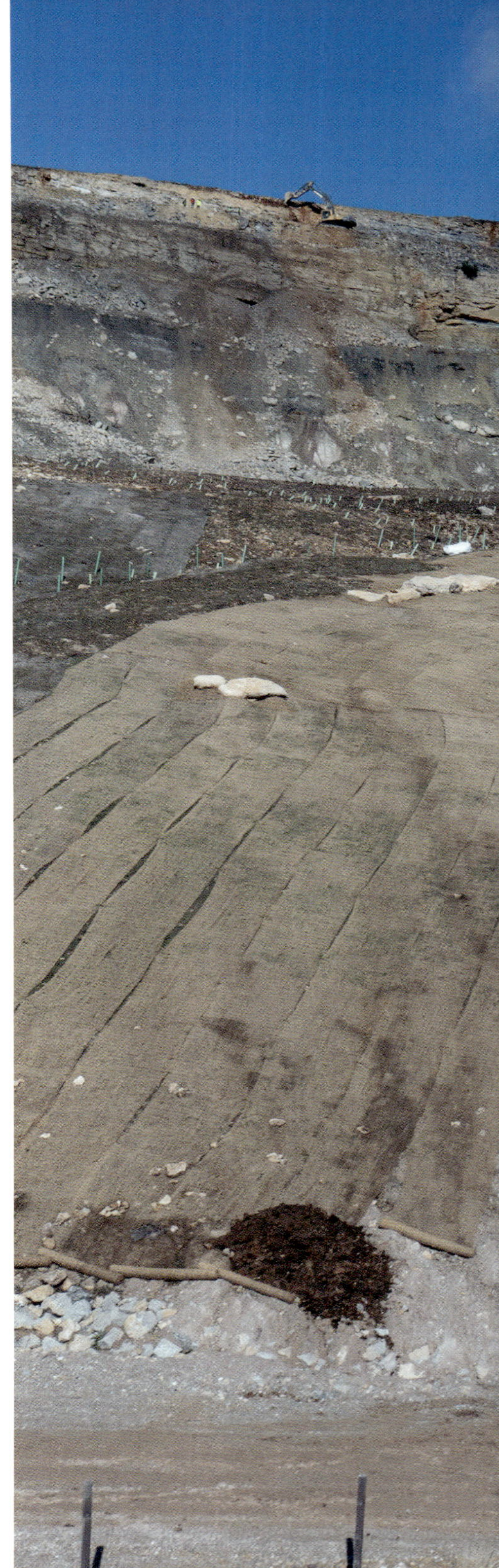

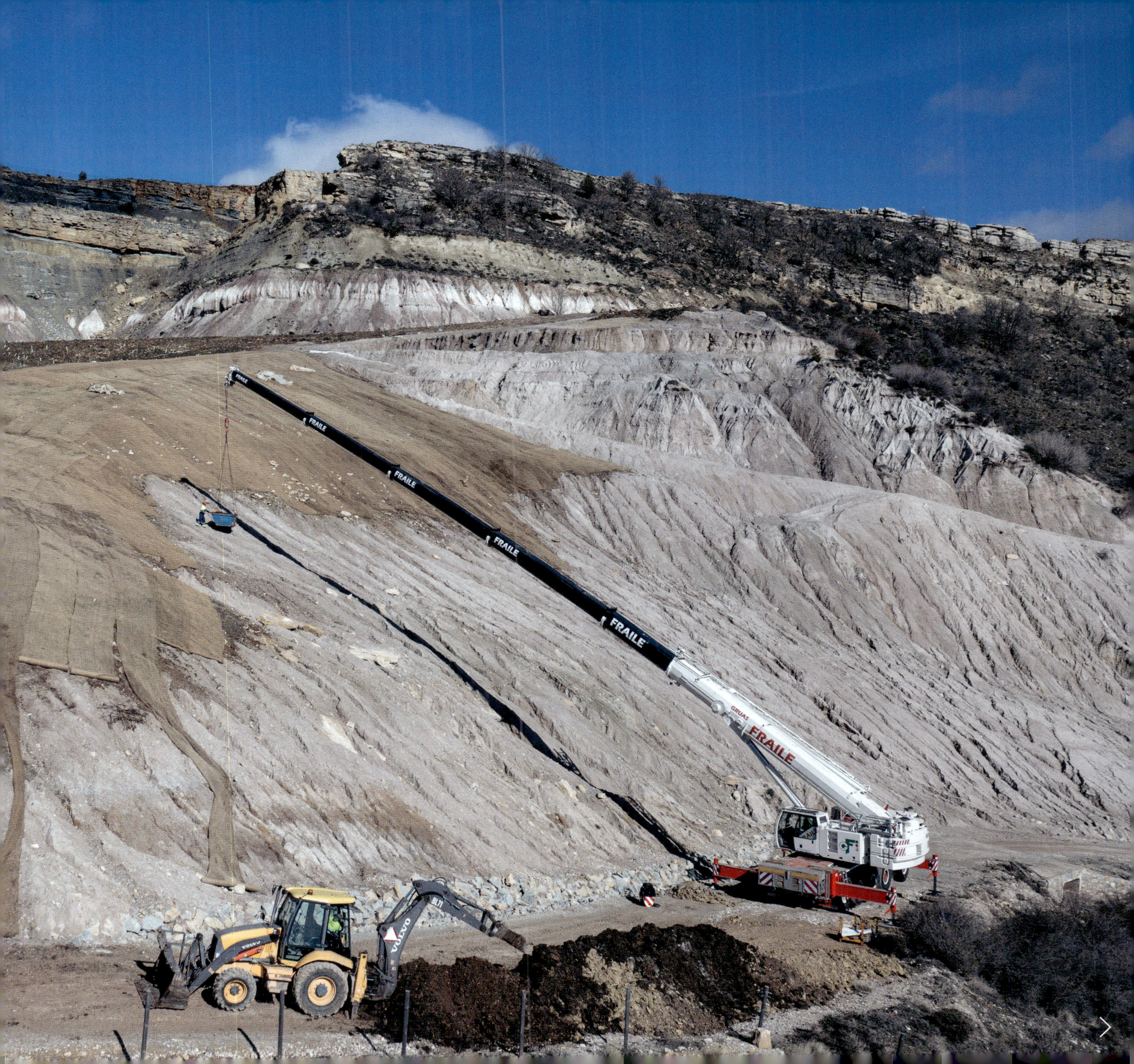

geology, weather and ecology. The Natural Regrade software then models the post-mining landforms that would form naturally over time in the interplay between unconsolidated material and prevailing weather. Drainage basins are its fundamental planning units, with bespoke slopes and channels that allow water to flow in a dynamic yet stable, landscape-friendly way and that 'plumb' into the surrounding drainage patterns.

The tools of the 'catchment plumbers' are yellow earth-moving machines that sculpt the plan's complex slopes and channels after surveyors have intricately pegged out the detail of the computer model. The trick is to find machinery operators who relish the challenge of learning a new approach and appreciate their role in creating a positive legacy. Their machines dance to the designers' tune while orchestrating the mosaic of small hills and ridges, S-shaped slopes and sinuous channels that – once completed – is functional, resilient and integrates visually with the surrounding terrain. The scale of topographic variation from millimetres and centimetres to metres and hundreds of metres creates diverse environmental conditions, more ecological niches and, ultimately, richer biodiversity.

Vegetation is critical in reducing erosion and visually and ecologically knits the site into the surrounding landscape while enhancing nature and ecosystem services. Depending on the slope properties and the ecological goals, Life Ribermine demonstrates several revegetation innovations to accelerate ecological succession towards the surrounding natural pine-oak forest ecological community.

This is Talus Royal's first mine site application; it was originally developed in France to model road cuttings. It digitally remodelled the quarry's visually intrusive high walls by, essentially, compressing time to create cliff profiles that would take an age to form under natural conditions. Spectacular explosions – known as restoration blasting – then produced an assembly of naturalistic vertical faces, narrow ledges, chutes, spurs and scree slopes reminiscent of surrounding natural cliff formations. The final details were skilfully applied by a digger operator who intricately sculpted the blasted boulders for the desired effect. This geomorphic variety aids moisture retention, improves vegetation establishment and encourages wildlife colonisation.

The third piece of the computer modelling jigsaw was to understand the landscape's progression over time in response to erosion. Developed in Australia and used by Life Ribermine in another European first, SIBERIA models landscape evolution from decades to millennia based on local environmental data to forecast erosion and investigate potential flaws in the post-mining design.

Dealing with a mine site's physical challenges is one thing, but what of the tricky geochemical problems faced at many sites? The highly erodible, sulphidic wastes of Portugal's abandoned Lousal pyrite mine have oozed heavy metal-laden acid drainage into local watercourses for decades. So, in another innovation, Life Ribermine combined an erosion-controlling GeoFluv geomorphic design with a soil cover system comprising an acid-neutralising subsoil layer covering the sulphidic waste. This was overlaid with a surface cover to direct runoff into naturalistic drainage channels armoured with acid-neutralising limestone. The cover system decreases water and air percolation into the underlying wastes to reduce acidic subsurface flows. The new landscape is integrated with a wetland system as the final neutralising step. Although now physically and geochemically stabilised, as with many acid mine drainage systems, the site will require ongoing maintenance, such as replenishing the limestone aggregate channel linings, potentially for decades.

Controlling erosion on post-mining sites is a prerequisite, but it is often not enough; without the inherent landform landscape integration of geomorphic design, mine rehabilitation will continue to be functionally and visually limited and long-term erosion control will be challenging. The goal of geomorphic reclamation is to reduce post-closure risk, expense and liabilities; indeed, it has been formally recognised as best practice in the EU's compendium of best available techniques for managing mineral wastes. Life Ribermine demonstrates that this is possible and, alongside similar projects, intimates that costs are comparable to more conventional methods. Three-years' worth of initial monitoring at both the Peñalén and Lousal sites shows the significant improvement in water quality and healing river systems.

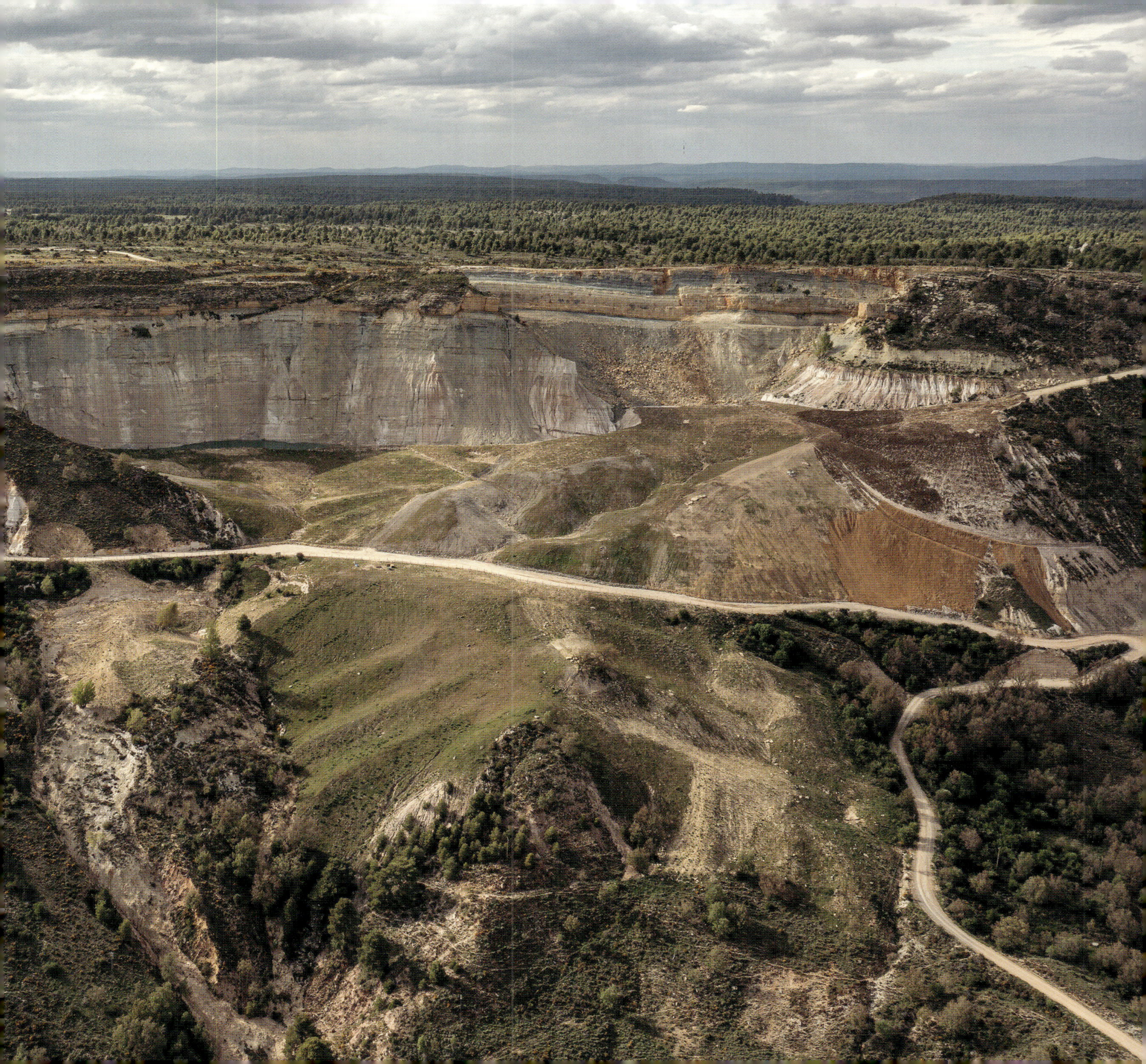

68
Pike County Airport
USA

Famed for its swathes of rich forest and ridge and valley terrain, Appalachia has geological bones that make it one of the world's great coalfields. Despite the region's economic importance, its beautiful but rugged geography has confined railways and, till recently, roads, to the narrow valley floors. There is also limited flat land available for development. Pike County, Kentucky, is a case in point: despite being the state's largest county, only 2% of its land was flat enough for development, of which half was floodplains.

Mountaintop removal mining literally blasts mountain tops apart to reveal the coal within, while its waste fills the valleys. Its controversial impacts are well-known, but one benefit is the creation of rare expanses of flat land ripe for economic development. A number of community airports have been constructed at strategic locations in the region on these manmade upland plateaus.

Demand for an airport in Pikeville County grew to support the 1970s' burgeoning coal economy, stimulate economic diversification and create jobs at a time when the road infrastructure was inferior to the impressive transport engineering feats visible across Appalachia today.

In 1976 the county judge, Wayne T. Rutherford, set up the Pike County Airport Board with USD 1.3 million of coal severance tax (a tax imposed by the state on the extraction of non-renewable resources) with the mission to find a site suitable for an airport. They selected a promising mountaintop site near Cowpen Creek, close to a major highway and at the notional hub of a surrounding – if dispersed – population of a few hundred thousand people.

The land was donated by the coal company, which had mined about 350 million cubic metres of rock to access coal within the erstwhile mountain, fashioning a broad upland plateau – an ideal geography for a small airport just a few kilometres from Pikeville's business district. Establishing a regional airport was risky, given no guarantee of state or federal financial support, but the board persevered and contracted civil engineers R.D. Zande & Associates to construct the facility because of their experience of designing airports on similar sites.

Today, publicly owned Pike County Airport, also known as Hatcher Field (James Hatcher was a famed local coal industry figure), boasts a 1,676-metre runway and associated infrastructure, prospective development areas, proximity to US Highway 23 and 200 hectares of additional land. In recent years, it has averaged 28 flights per day, more or less evenly split between local and transient, including a few military flights.

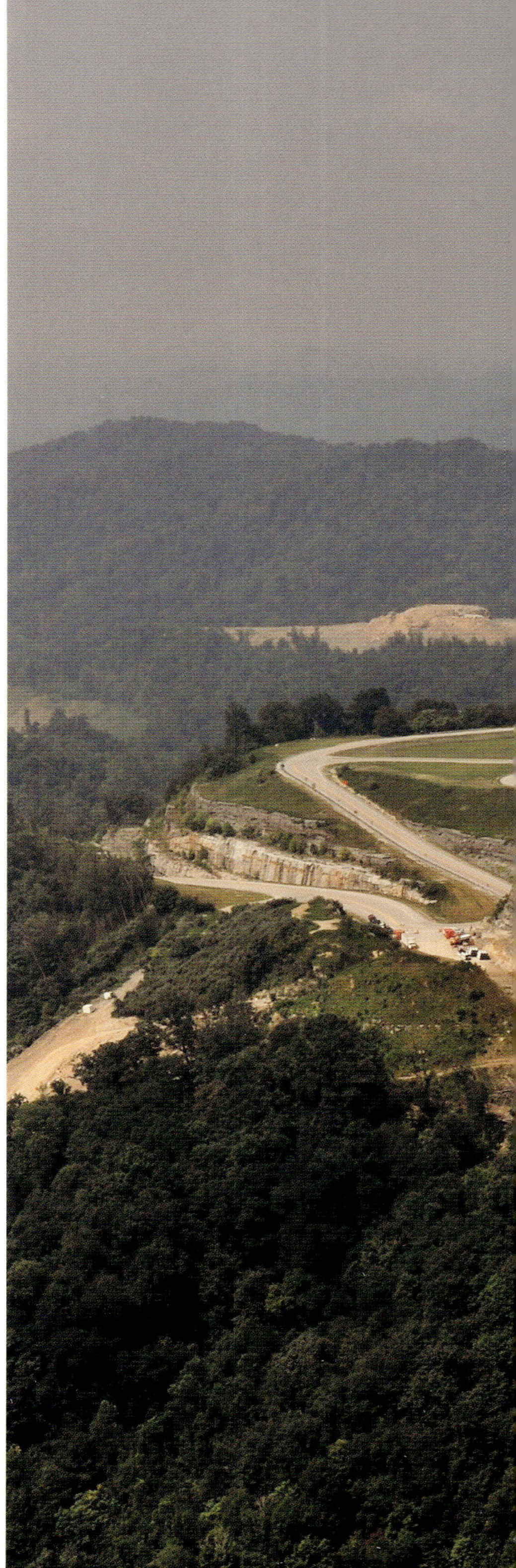

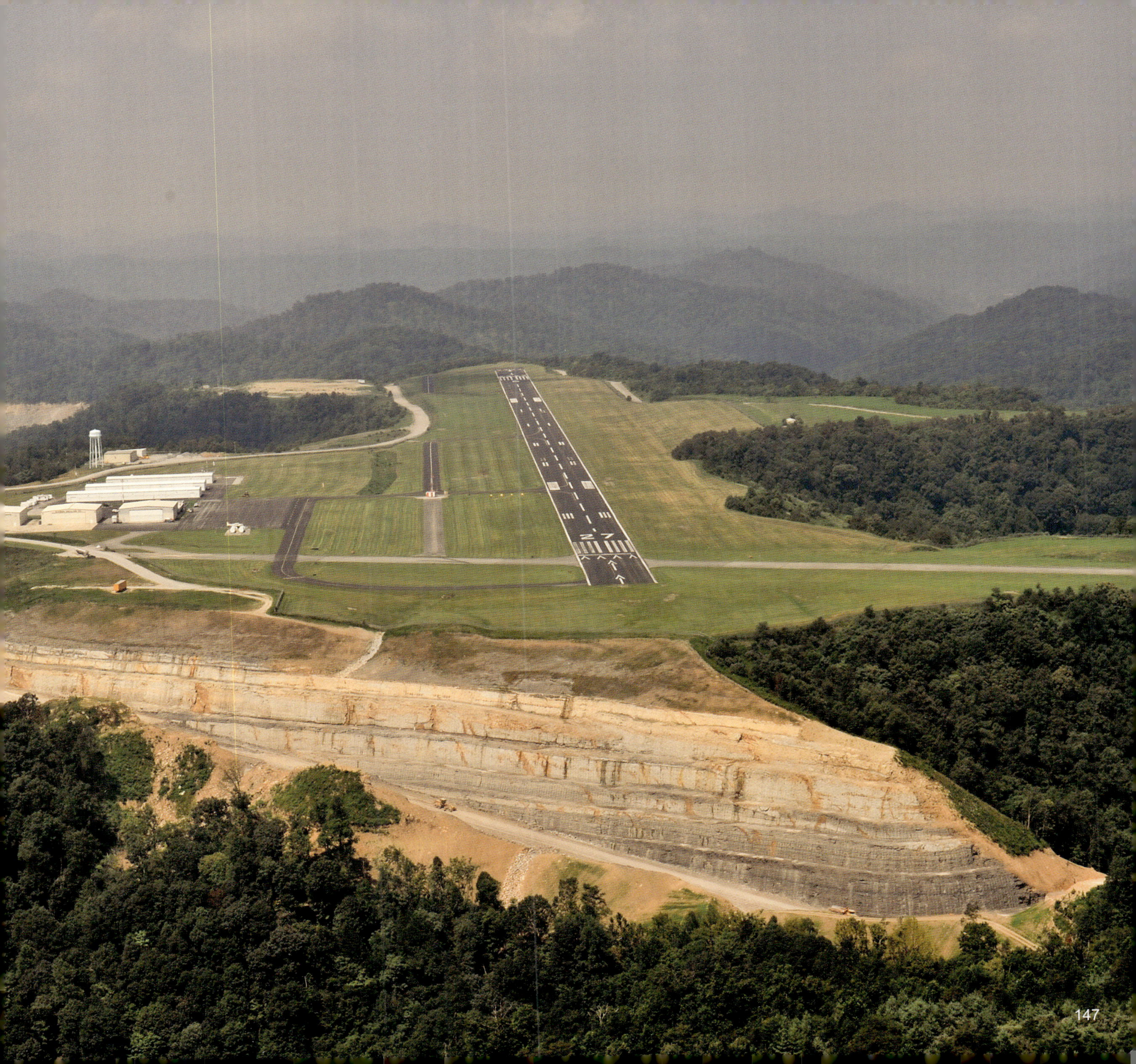

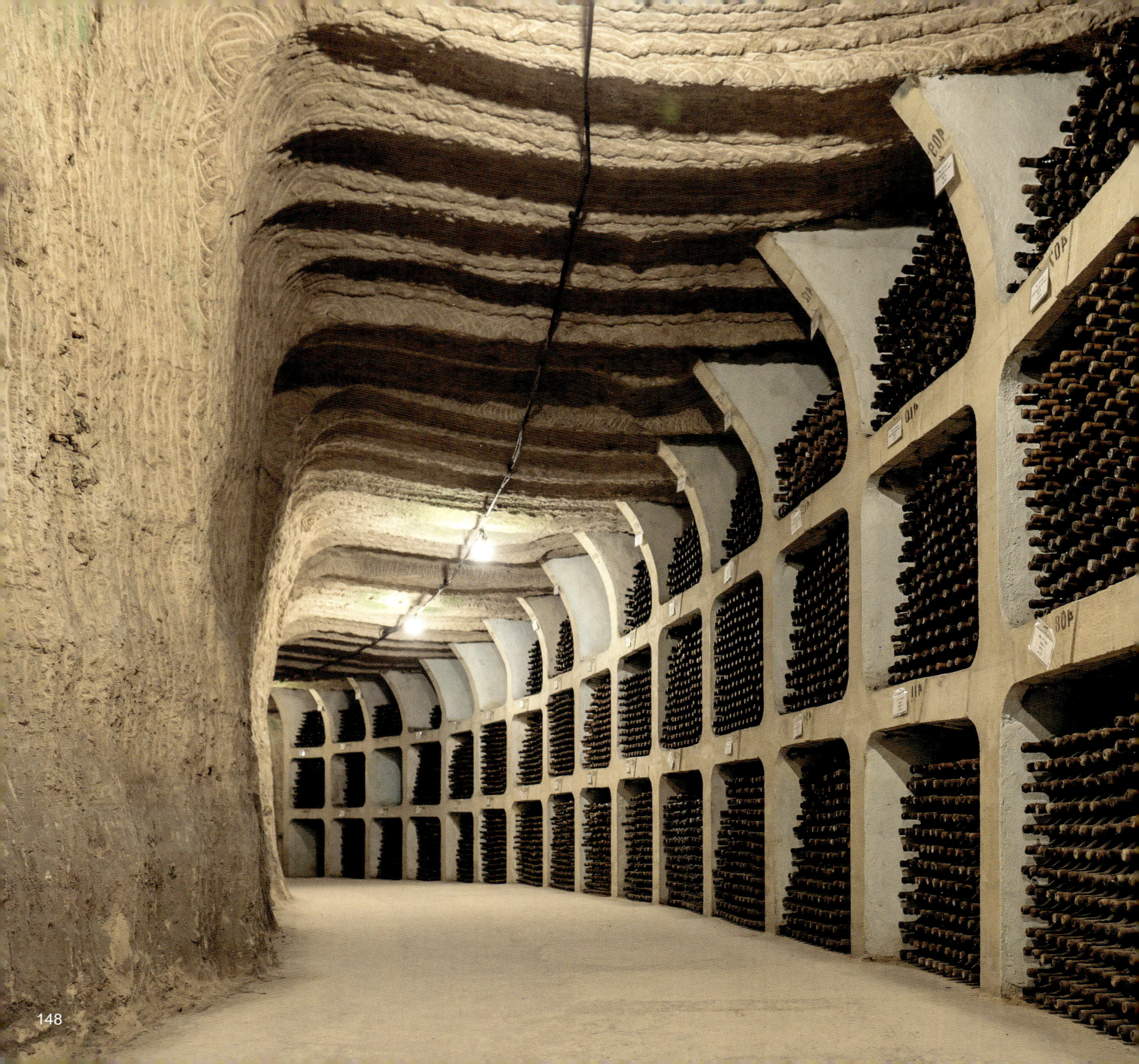

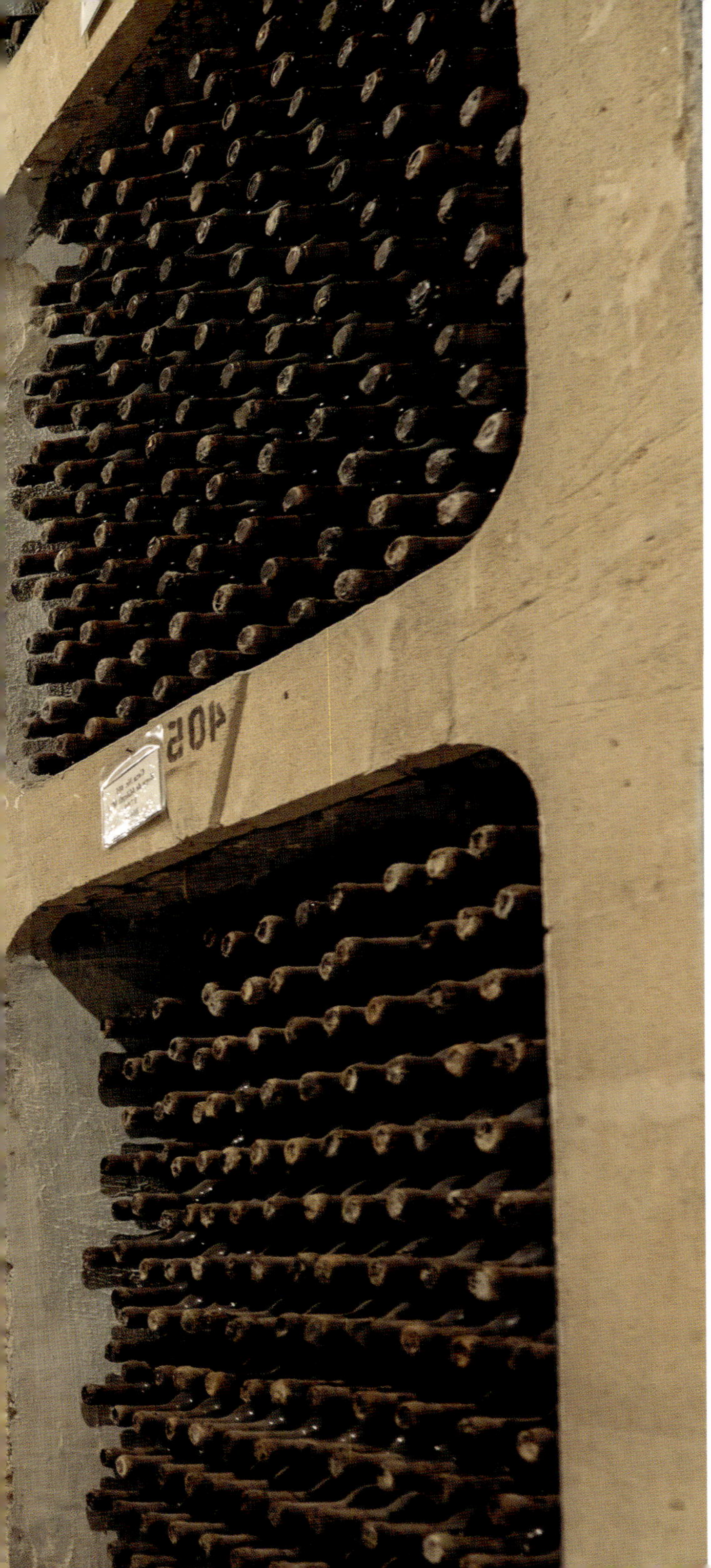

69
Mileștii Mici Wine Cellar
Moldova

The Republic of Moldova is one of the world's largest wine producers by volume. Only slightly larger than Belgium, Moldova uses more of its land proportionally for growing grapes than any other wine-producing nation – over 100,000 hectares at different scales from vineyards to smallholdings. Grapes for sale on the roadside are a familiar sight.

Fittingly, Moldova also has the world's largest wine cellar, according to the *Guinness Book of World Records*. Sixteen kilometres south of the capital, Chișinău, a former limestone mine, holds the most bottles of wine in the world.

Mileștii Mici has been a state-owned and operated vineyard since it opened in 1968. Much of its production went to its Soviet neighbours and later to their successors. But following a Russian embargo in response to trade with the EU, increasing amounts are now sold to Europe.

In addition to the vineyard itself, Mileștii Mici comprises over 220km of tunnels, of which 55km are in use. At its heart is the 'Golden Collection': two million bottles of wine held in a honeycomb structure of alcoves. There is also a 'secret' room within the Golden Collection, created for the most valuable wines.

The distances involved require a vehicle. Headlights illuminate rows of vast casks as you drive through 'streets' named after wine varieties. The main tasting room is deep inside the mine, which puts a limit on what you can drink if you drive yourself, but there is an electric trolley service for those who want to make the most of the experience.

70

Elliot Lake Retirement Community

Canada

Elliot Lake was born quickly in the 1950s' 'Uranium Rush' after the metal's discovery among northern Ontario's forests and lakes on the traditional territory of the Serpent River First Nation. Uranium was in high demand for America's nuclear weapons programme, and within just five years, nine mining companies ran 12 mines and 11 mills, extracting and processing millions of tonnes of ore. The government-planned town carved from the wilderness accommodated 25,000 people and was called the Uranium Capital of the World.

American demand ended abruptly in 1959, and the population plummeted below 6,000 by the 1970s. An upsurge in nuclear power generation drove a short-lived revival, and the town boomed again to around 20,000 in the mid-90s. The development of Saskatchewan's much richer deposits and the cancellation of Crown Corporation contracts led to the closure of the remaining mines later that decade, with Denison Mines and Rio Algom (now BHP) operating the last mines. This led to a rapid population decline to around 10,000 at the century's end. So many out-of-work people in the middle of nowhere required profound social adjustment.

Context is everything to post-mining success. A favourable political environment allowed a cushion of employment for a few years as the mines wound down, providing time for

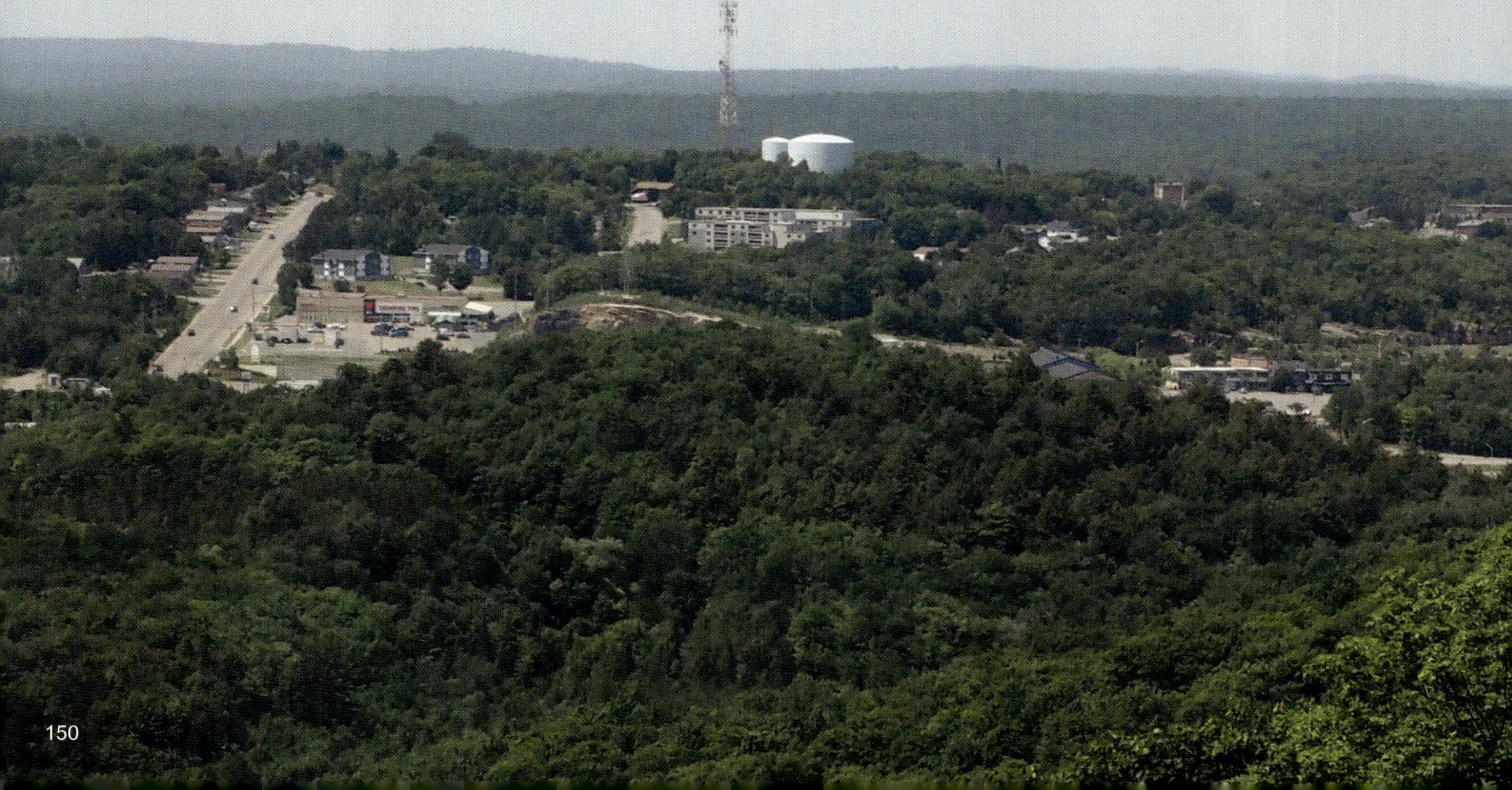

focused collaboration to envision a new future. A key aspect of that vision was reinventing Elliot Lake, incorporated as a city in 1991, as a retirement community. The context derived from the surplus housing stock at affordable prices and accessibility to the sufficiently distanced Trans-Canada Highway – an escape route from the rat race to a slower-paced life in beautiful surroundings. Also, given the remote location, retirees have a reliable income but do not need jobs, and after selling their expensive southern Ontario houses for a cheaper Elliot Lake home, they have disposable income to spare. Set up as a landlord and facilitated by the city, the not-for-profit corporation Elliot Lake Retirement Living began managing the mining companies' former housing and promoting the area as a retirement destination. Meanwhile, other initiatives focused on economic diversification based on local natural resources and assets.

Reinventing the community went hand in hand with environmental renewal. The unregulated lake deposition of 170 million tonnes of radioactive and acid-generating tailings covering a combined area of nearly 1,300 hectares had degraded large parts of the Serpent River catchment. Denison Mines and Rio Algom combined forces in a 10-year reclamation programme, including the historic mines dormant since the 1950s; all were reclaimed to the same leading standards. They relocated tailings, demolished buildings, landscaped and planted, and upgraded water treatment facilities. A million tonnes of tailings were deposited underground, but the most cost-effective option was submerging them in water after treating them with barium chloride to minimise dissolved radium and with lime to control acidity. A small area of legacy tailings became the Sheriff Creek Wildlife Sanctuary.

Despite the impressive endeavour, an environmental legacy of radioactivity and acid drainage remains, requiring environmental monitoring and management in perpetuity. Managing tailings facilities' water levels and flows are critical to keeping the wastes in a

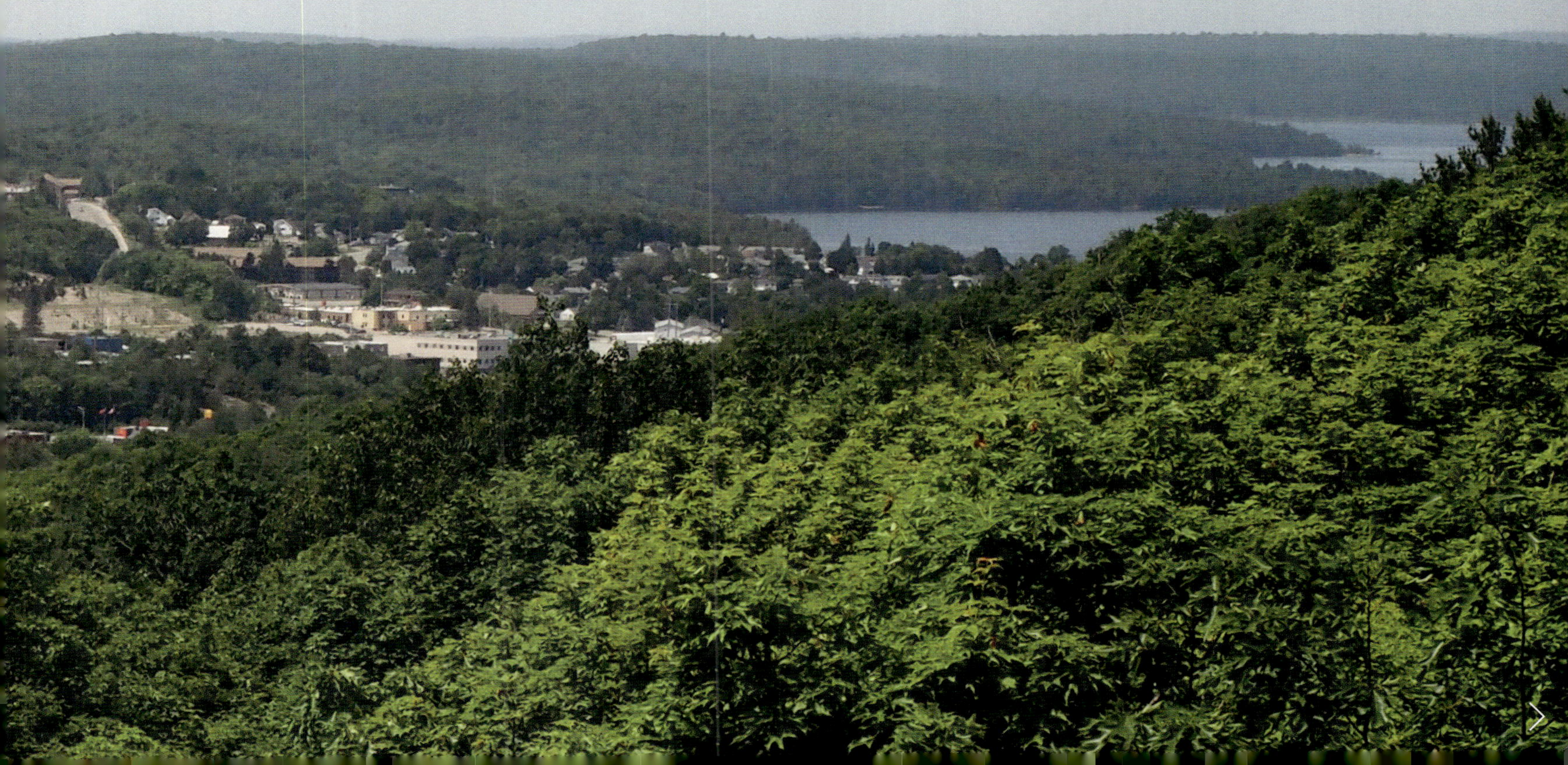

stable, non-acid state, which means, unfortunately, controlling beaver numbers. On visiting the restricted access sites, apart from the water control dams and treatment plants, and other than the occasional seep of orange water or an ochre-coloured stream, much of the land seems peaceful, green and bursting with life. Monitoring results indicate a recovering ecosystem that is, in some cases, like the Sheriff Creek Wildlife Sanctuary, flourishing.

The purpose-built mining settlement has successfully reinvented itself as a retirement community. It's an accessible, affordable, peaceful and beautiful location now recognised for its waterfront living. It's also a four-season visitor destination popular with outdoor enthusiasts of canoeing, fishing, hiking, golf, winter sports and mountain biking. The post-mining environmental management activities are generally accepted – otherwise, no one would have relocated here. However, greater access to company landholdings, given the interest in outdoor pursuits, is desired.

It's a community of independent 'doers' characterised by an ethos of volunteering; for example, the Elliot Lake XC Ski and Bike Club is a not-for-profit volunteer group encouraging the uptake of outdoor recreation and developing forest trails. The guy who set up and drives the group is also a local shop owner and fireman.

The growing trail network encourages exploration of Elliot Lake's post-mining hinterland, large tracts of which remain under company ownership. The Voyageur Hiking Trail, which crosses BHP's landholdings, is part of the Trans Canada Trail network. Many promising diversions, though, end at a padlocked gate with ample signage warning no entry and copious use of the words 'uranium', 'nuclear', 'low-level radioactive material' and 'bears', yet people feel compelled to explore beyond the gates. The XC Ski and Bike Club, with others, is working on a stewardship agreement between the council and BHP regarding access and responsibilities on restricted lands – there was an original closure goal to allow public access. Safety is the main concern with the companies liable for what happens on their land, but the inclination is to allow more public access, which the city will manage and be insured for.

Elliot Lake may be one of Ontario's most popular retirement destinations, but since Covid, younger people, attracted by the lifestyle and the prospect of remote working, have moved in, and house prices are increasing, yet remain affordable (for the moment). Its 11,000+ population is slowly climbing, and demand for housing is high, so new residential developments are planned – the first since before mining ended. Demand for visitor accommodation is growing too; there's a trailer park and campsites, lakeside lodges and a recently built hotel, with a second planned, capitalising on the proximity of the Trans-Canada Highway. Other key aspects of the economy are advanced manufacturing, some forestry, and government services.

Elliot Lake's popular Fire Tower Lookout offers panoramic views over verdant forests and sparkling lakes towards distant Lake Huron. It all looks pristine. The city does not feel or look like the erstwhile Uranium Capital of the World; rehabilitation and regeneration have been so successful that it's easy for a visitor to pass through unsuspecting. The community has moved on, but it deeply respects its mining foundations: prominently located overlooking a tranquil Horne Lake, the Miners' Memorial Park narrates the poignant story of the place and – unexpectedly – evokes a powerful connection to what went before.

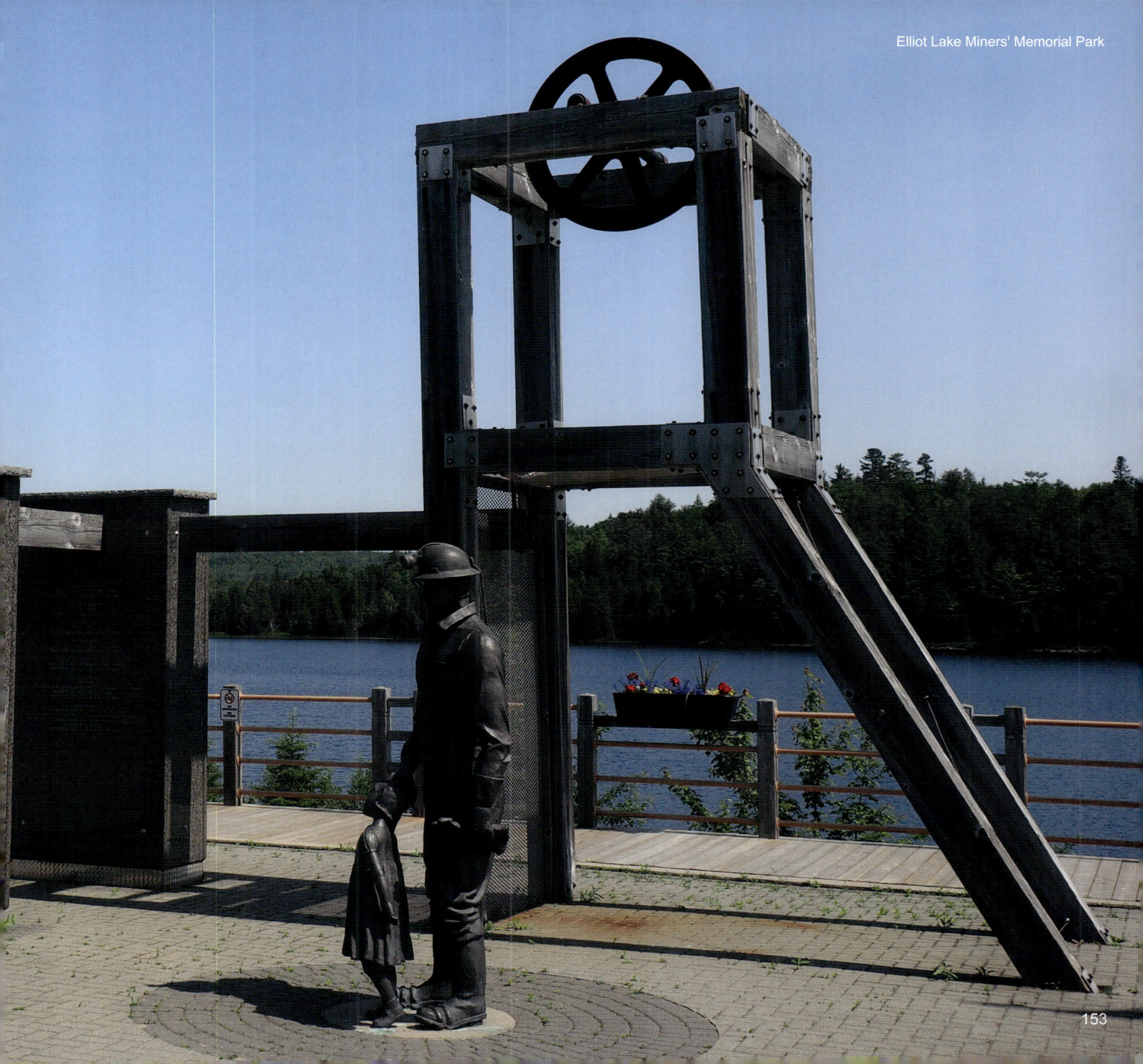

Elliot Lake Miners' Memorial Park

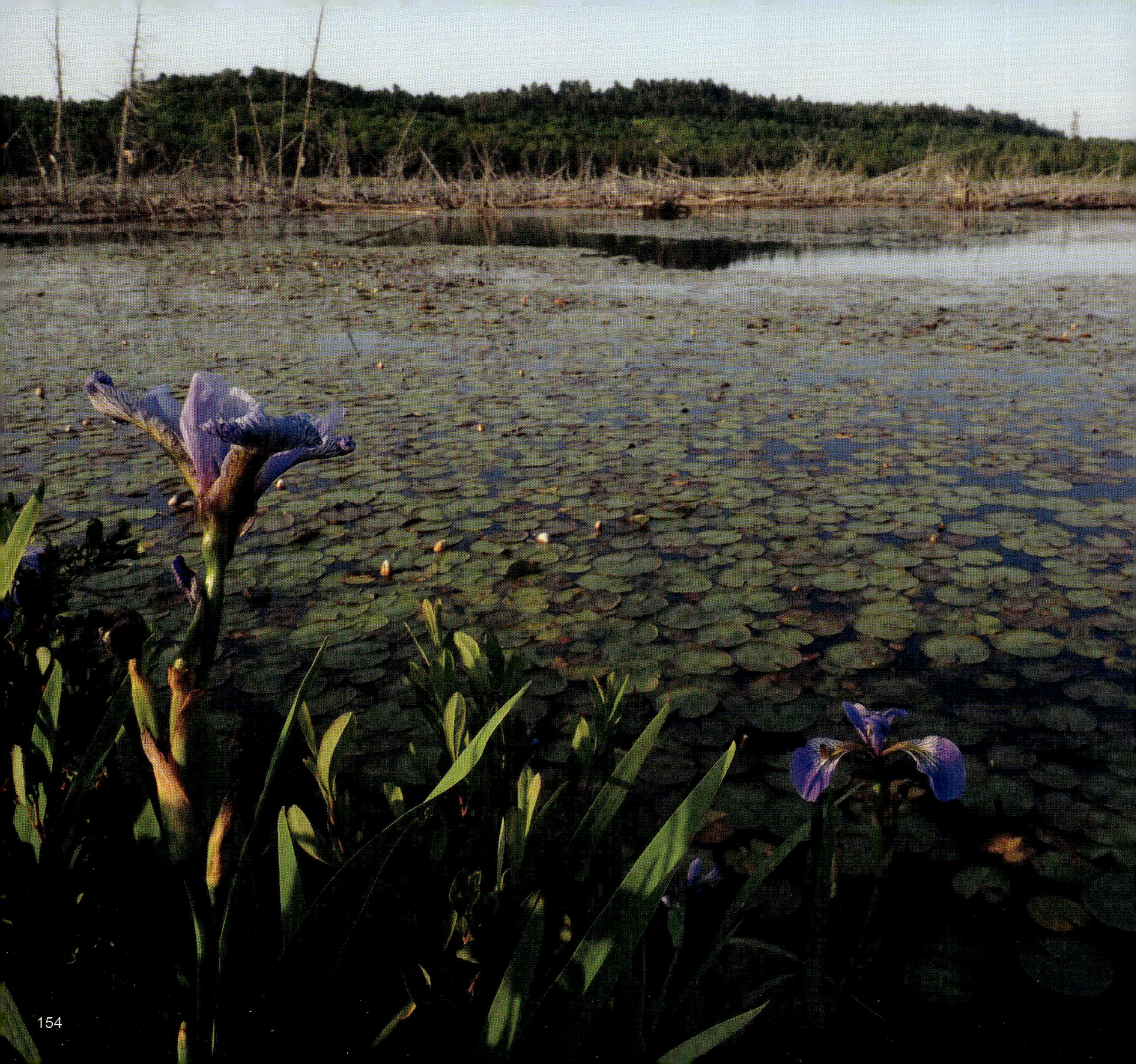

71

Sheriff Creek Wildlife Sanctuary

Canada

The Rio Algom-branded sign states that Elliot Lake's Sheriff Creek Wildlife Sanctuary is a 'licensed Radioactive Waste Facility' – between 1958 and 1964 about 150,000 tonnes of uranium ore tailings were deposited here. The site was remediated in the 1970s leaving nature to do the rest. The sign also reassuringly explains that 'Persons walking in this area are exposed to low levels of radiation. For example, walking in this sanctuary for 200 hours per year may expose an individual to about 2% of the radiation received from other natural causes during the year, such as cosmic radiation and the food we eat.' But without the sign, who would know the industrial past of this natural haven?

In 1978, as part of mine reactivation, a legacy spill of radioactive acid-generating tailings was repurposed as a ball field and recreational area in collaboration with Guelph University. The area was redeveloped as a bird sanctuary in the mid-1990s in a shared stewardship arrangement with the Penokean Hills Field Naturalists (PHFN) with support from the City of Elliot Lake. BHP (formerly Rio Algom) still manages the 407-hectare site in collaboration with PHFN.

Canada's original uranium tailings revegetation research was conducted on Elliot Lake mines by the federal government's CANMET labs and the University of Waterloo, with the support of Rio Algom. The combination of a water cover, organic matter and biodiversity reduces oxygen infiltration into the underlying tailings, minimising acid generation from the sulphide minerals within.

Three trails start from the site, which are financially supported by BHP and created and kept by the PHFN. The company also supports major physical improvements such as the construction of the entrance causeway, car park, wildlife viewing hides (aka blinds), bridges and signage, and the city has provided staff time and grants for labour to install some of the trail infrastructure.

The sanctuary packs a rich tapestry of habitats into its 182 hectares: reclaimed wetland areas, marsh-cattail and water lily wetlands, grass- and shrub-land, bog-tamarack and sphagnum wetland and transitional deciduous and coniferous forest. The Sheriff Creek Berm maintains water levels, releasing small amounts to trickle through a dense, mosquito-infested cattail and bulrush reedbed for a final polish of any remaining contaminants before entering the rest of the Serpent River catchment.

A midsummer dusk visit is a memorable audio-visual event as the sanctuary's diverse ecology performs a spectacular flourish before settling down for the night. Bleached, skeletal trees stand erect over mirror-like waters as a floating carpet of lily pads shimmers bronze in the dwindling sunlight – the stage-setting for a cacophony of animal sounds and movements. Myriad birds serenade the sunset while invisible frogs and toads croak and chirp their hearts out and whirring dragonflies hunt among irritating clouds of whining mosquitoes. And where should one's eyes settle next as agile swallows seek aerial insects; a sunset skein of geese glides in to roost; beavers motor smoothly across the glassy water, and an American bittern believes itself invisible amongst the reeds?

It's a beautiful, natural spectacle that is easily accessible and popular. The three-way partnership received official recognition in 2010 with the Tom Peters Memorial Mine Reclamation Award given by the Canadian Land Reclamation Association and in 2022, BHP celebrated the 25th anniversary of the sanctuary.

Another sign, this time on the roadside by Sheriff Creek – a yellow diamond with a bold silhouette of a terrapin warns, 'Turtle Crossing'; a terrapin duly crosses the road here en route to its night-time sanctuary.

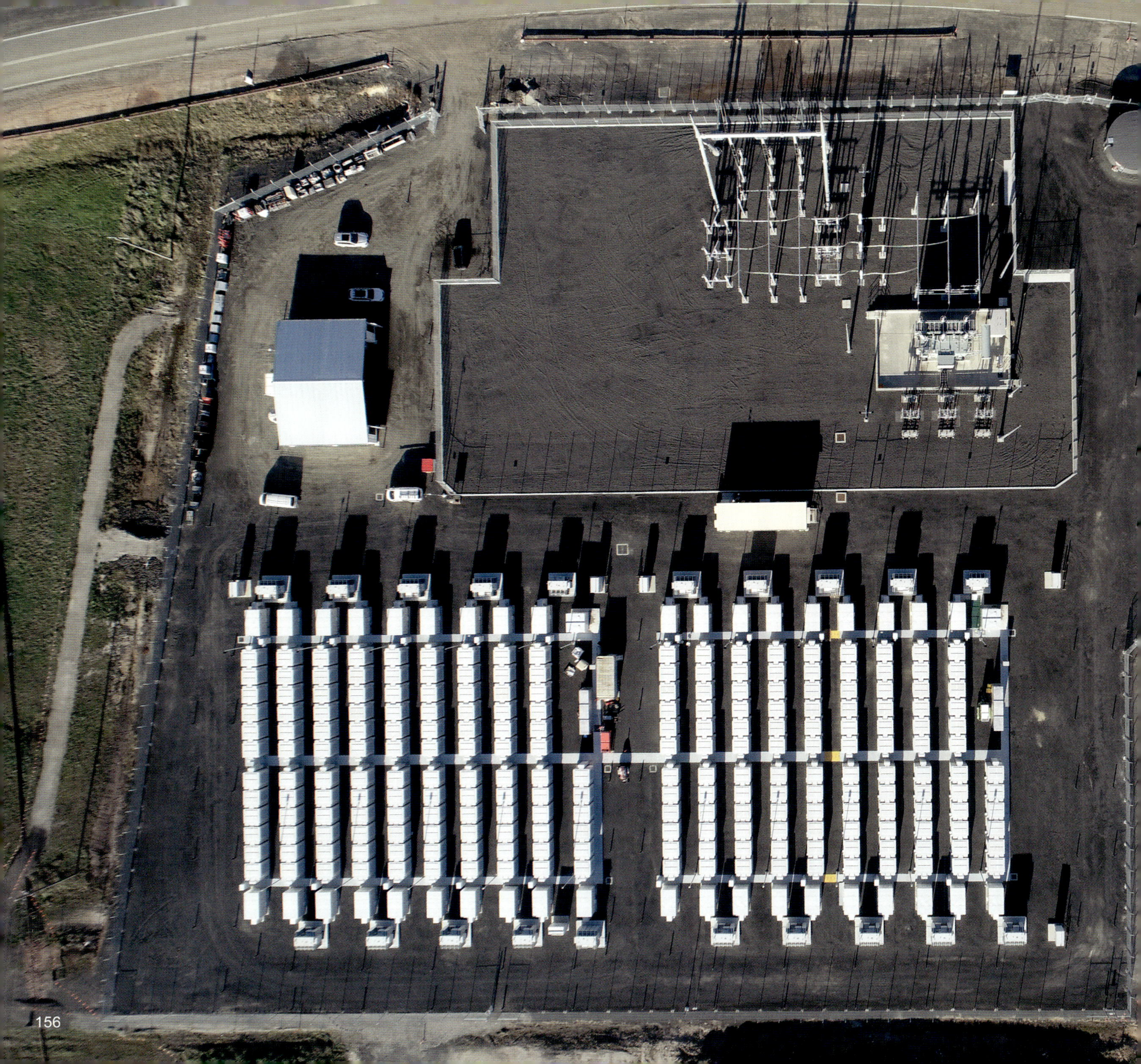

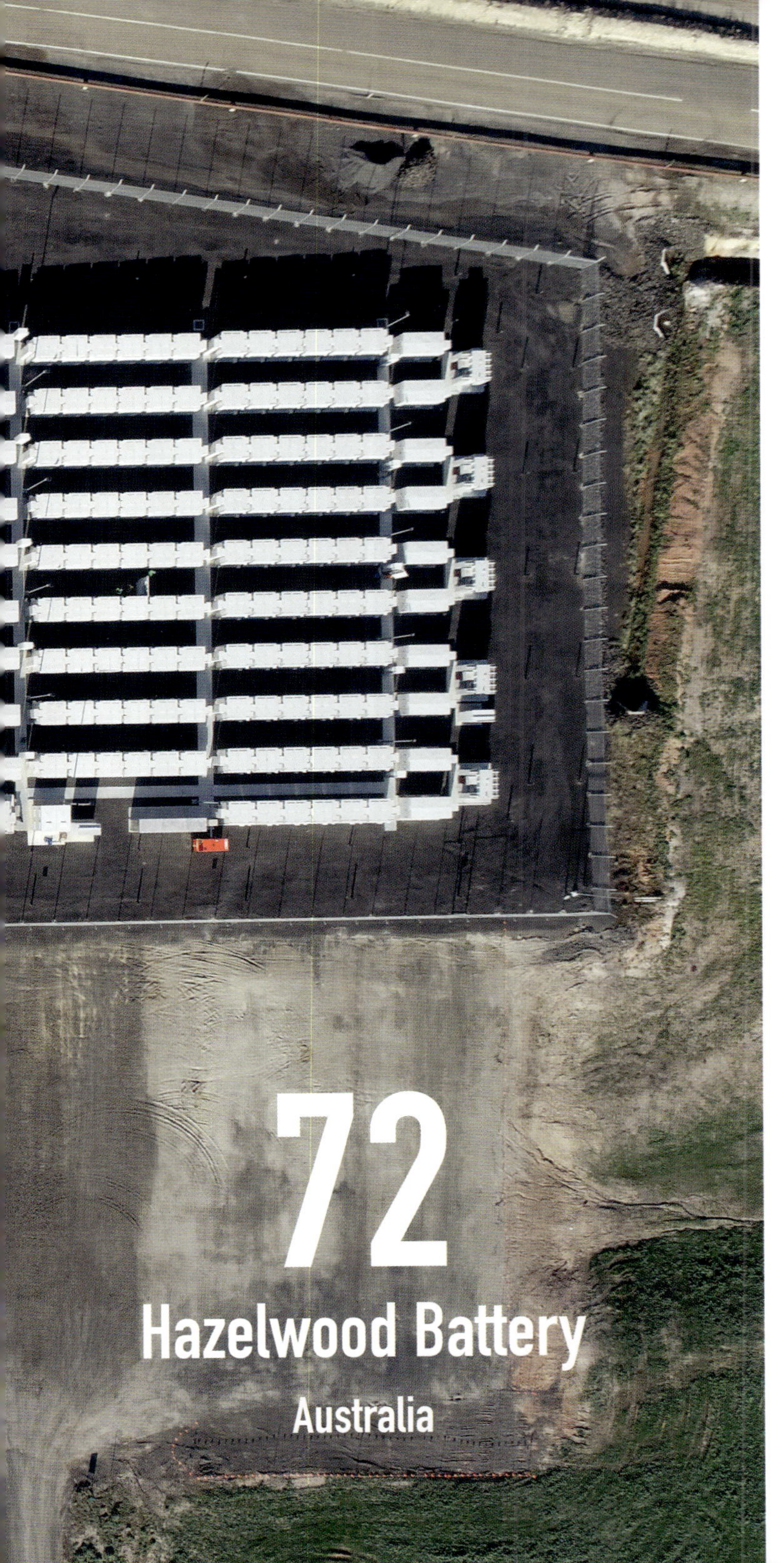

A broad, blue sky seems magnified by the pit lake filling the expansive void of the decommissioned Hazelwood brown coal mine in Australia's Latrobe Valley. For 50 years, the mine fed utility company ENGIE's neighbouring Hazelwood Power Station until its closure in 2017.

By the end of 2022 the entire power block, including its regiment of towering chimneys, had been demolished and recycled; its concrete and masonry crushed and compacted to create a vast area as flat as a chessboard, prepared for reuse.

A clue as to that use is provided by an incongruous, geometric arrangement of 342 white shed-sized metal boxes, gleaming in the sun and located close to the edge of the pit just a short distance from the footprint of the former power plant. This is the 150MW/150MWh Hazelwood Battery Energy Storage System (HBESS), which opened in June 2023 as Australia's first former coal mine and power station site to host a utility-scale battery.

Capitalising on Hazelwood's existing 1,600MW of otherwise dormant transmission infrastructure and grid connectivity, it makes a powerful statement about repurposing high-carbon generation infrastructure for a low-carbon renewable energy facility and signposts a new direction for the Latrobe Valley's traditional brown coal mining and power industries.

One function of the battery system is to flatten out the peaks and troughs of renewable energy generation – sending out extra power during periods of peak demand and storing power when demand is low, thus improving the stability of Victoria's grid. By storing the equivalent of an hour's generation from the domestic rooftop solar panels of 30,000 homes, it improves grid resilience, flexibility and reliability as wind and solar continue to carve out an ever-greater share of the state's electricity supply. It has a 20-year lifespan but can be upgraded over time as technologies advance.

On commissioning, the HBESS was Australia's largest privately funded battery system – developed, built and run as a partnership between ENGIE, Eku Energy and Fluence. ENGIE has plans to scale up the renewable energy infrastructure on the site in the future towards its ultimate target of reaching net zero by 2045.

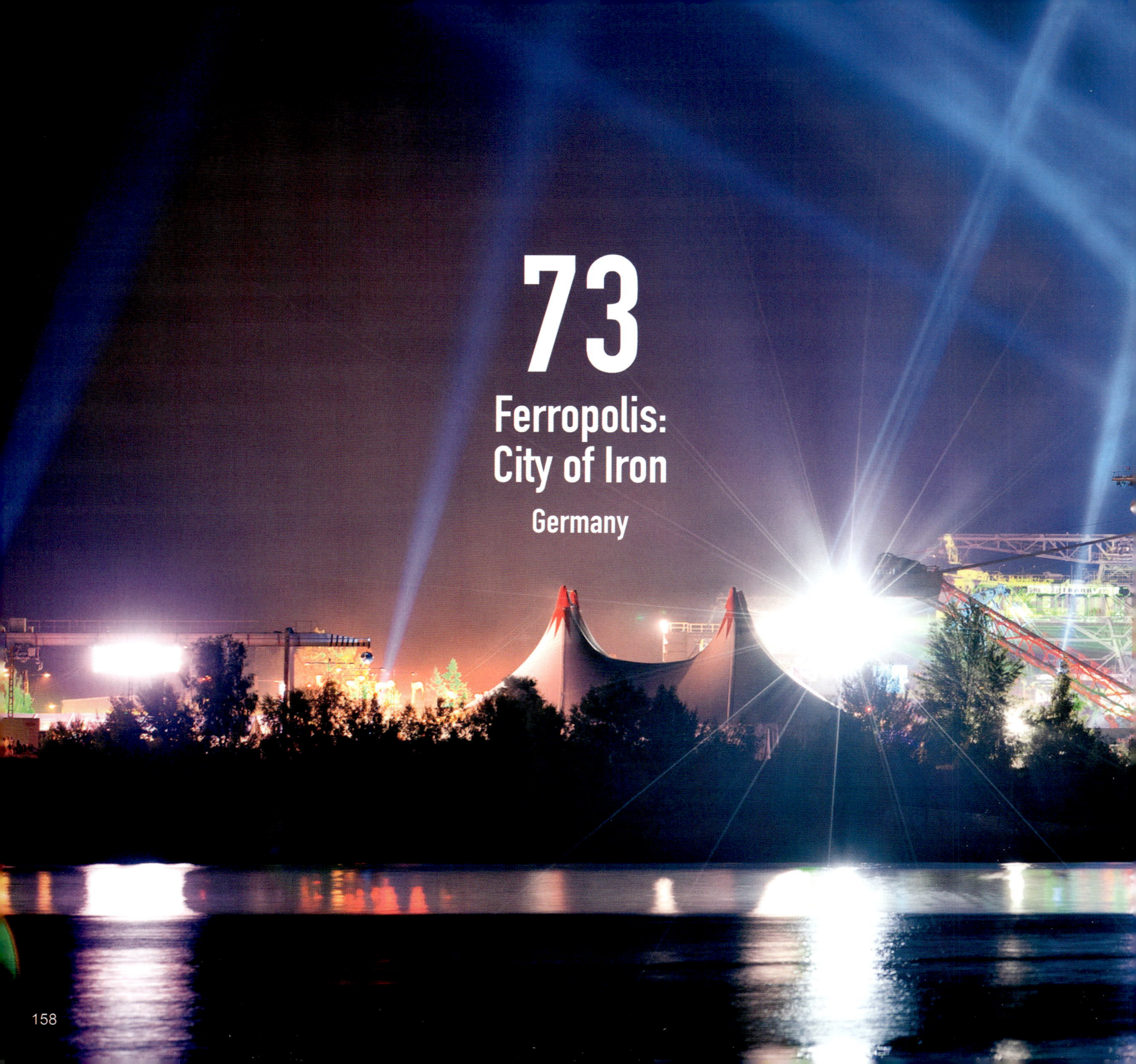
73
Ferropolis:
City of Iron
Germany

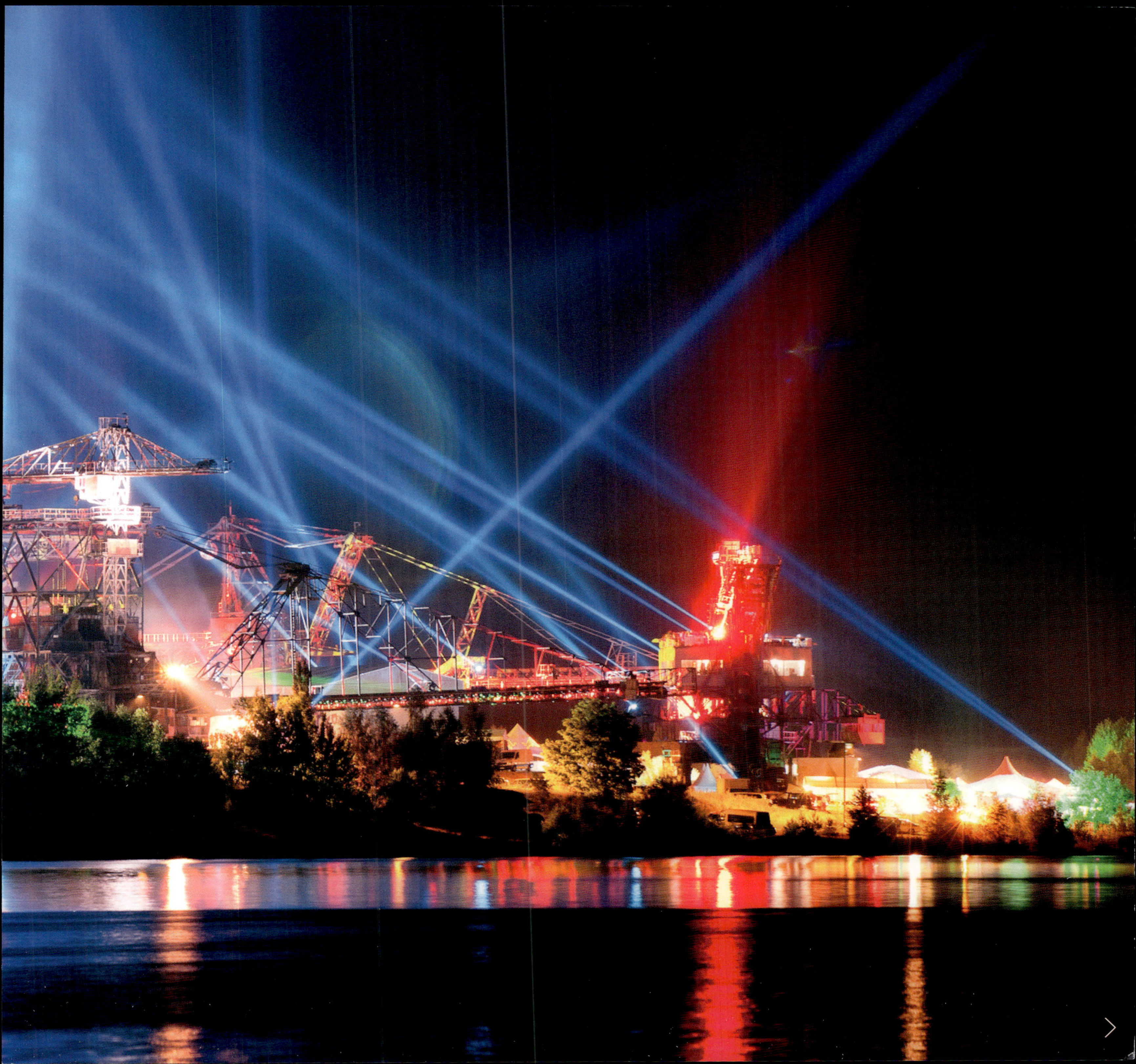

Like skeletons of immense prehistoric beasts, the hulking excavators named Medusa, Mad Max, Big Wheel, Mosquito and Gemini loom over the surrounding landscape. Located at the end of a narrow peninsula surrounded by the lake formed from a flooded opencast coal mine, Ferropolis – City of Iron spectacularly entwines past, present and future, creating an isolated and strange world.

Ferropolis is an open-air museum, concert venue and industrial monument that attracts around 300,000 people every year. The machines provide a dramatic backdrop to its purpose-built amphitheatre which also serves as a 25,000-capacity concert venue. A range of musical tastes are covered, but heavy metal rules (obviously); headliners have included the appropriately named Metallica!

Some residents of the now sleepy former mining village of Gräfenhainichen were initially disgruntled at the concerts' noise and visual effects, but once they realised each one brought EUR 150,000 into the village economy, they became more amenable. Neighbours also benefit from a free annual open day and free tickets for events, including the annual Miners' Day festival.

Over 340 million cubic metres of material was extracted from the opencast Golpa-Nord mine from the early 1960s until 1991, directly affecting nearly 2,000 hectares of land and causing the destruction of the village of Gremmin, the namesake of the new lake.

The Ferropolis concept emerged from a design workshop with students from Bauhaus Dessau after conversations between miners and students in 1991 in which the students persuaded the miners not to demolish the defunct mining

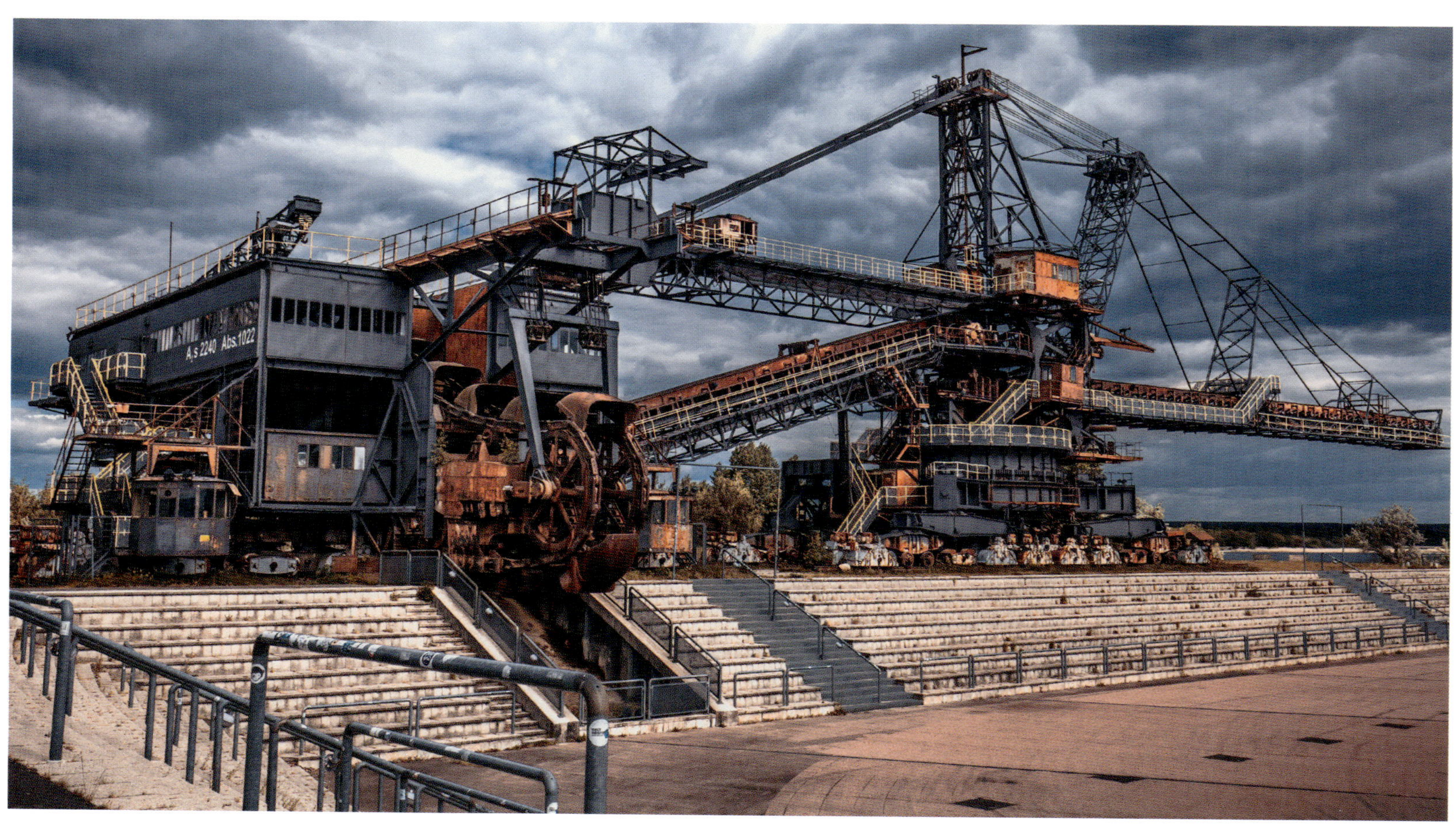

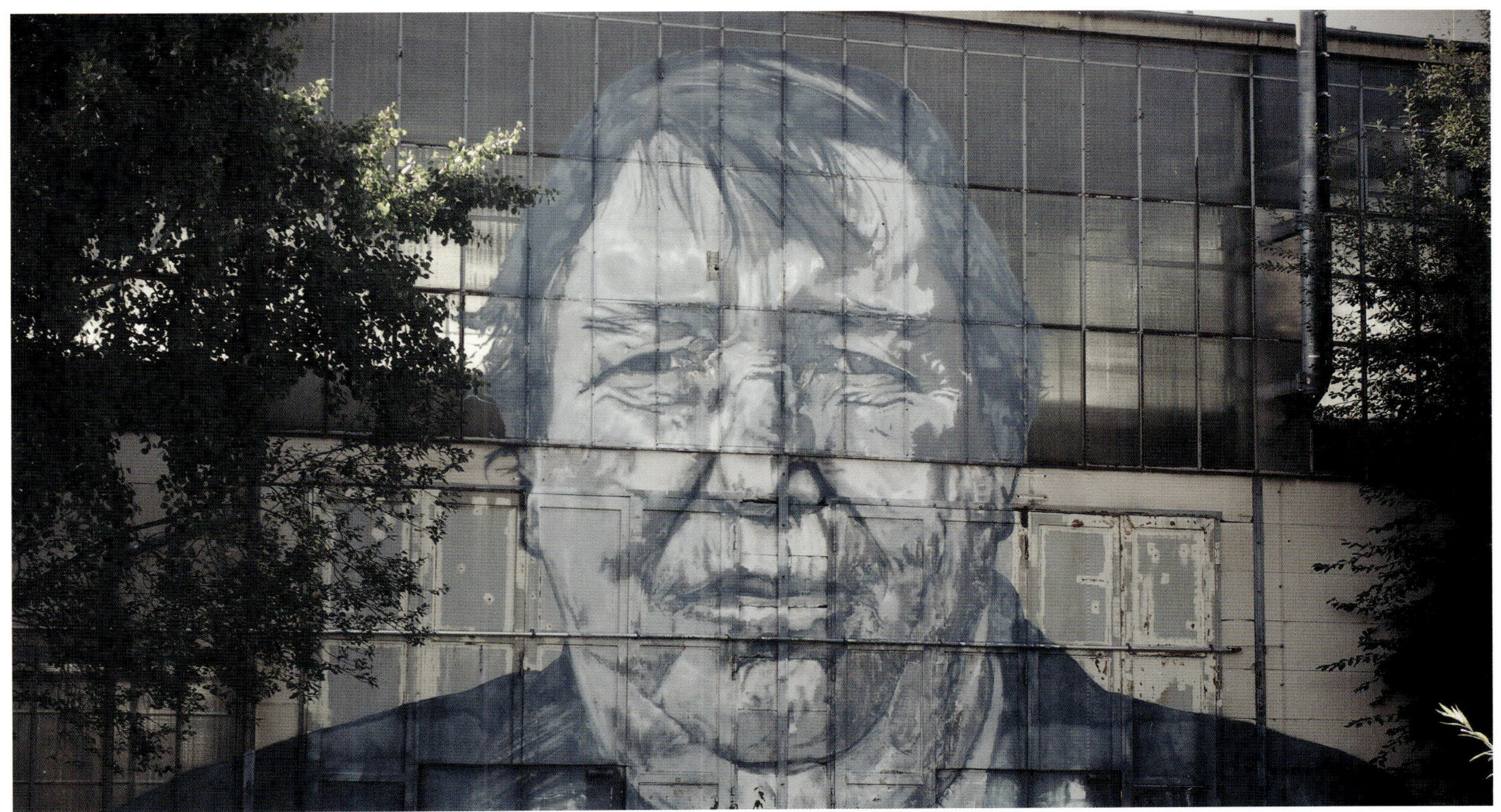

machinery. Ferropolis – City of Iron was founded in 1995 and was included in Hanover's EXPO 2000. Flooding of the open pit began in 2000, and the site and its machines have been evolving ever since. It has received national awards as a top tourism destination and is certified for its accessibility.

Former miners tell stories of their machines to visitors and help develop the museum. They are also remembered through eight huge portraits painted on the sides of mine buildings that include one of the mine's few female excavator operators, Monika Miertsch.

Visitors are encouraged to clamber aboard the machines and explore their intricate engineering while marvelling at their scale. In 2022, a lift and platform were installed to enable wheelchair users to share this experience. From the machines, impressive views unfold across the surrounding lake and forests towards Gräfenhainichen.

Ferropolis has initiated a growing number of other events that include an annual triathlon and, most recently, a campervan festival.

To ensure its long-term sustainability, Ferropolis has been restructured into a museum charity able to access a wider range of funding, and a commercial event management company, whose profits support the charity. Ferropolis is also exploring tourism opportunities with the nearby Zschornewitz power plant, which used to receive the brown coal dug from the Golpa-Nord mine. Having closed in 1992, it is one of the oldest power plants in Germany and is now considered an industrial monument.

Previous pages: View across Lake Gremmin of the MELT! music festival. **Opposite:** Two of Ferropolis's heavy earth excavators. **Above:** Portrait of Monika Miertsch, one of the female excavator operators.

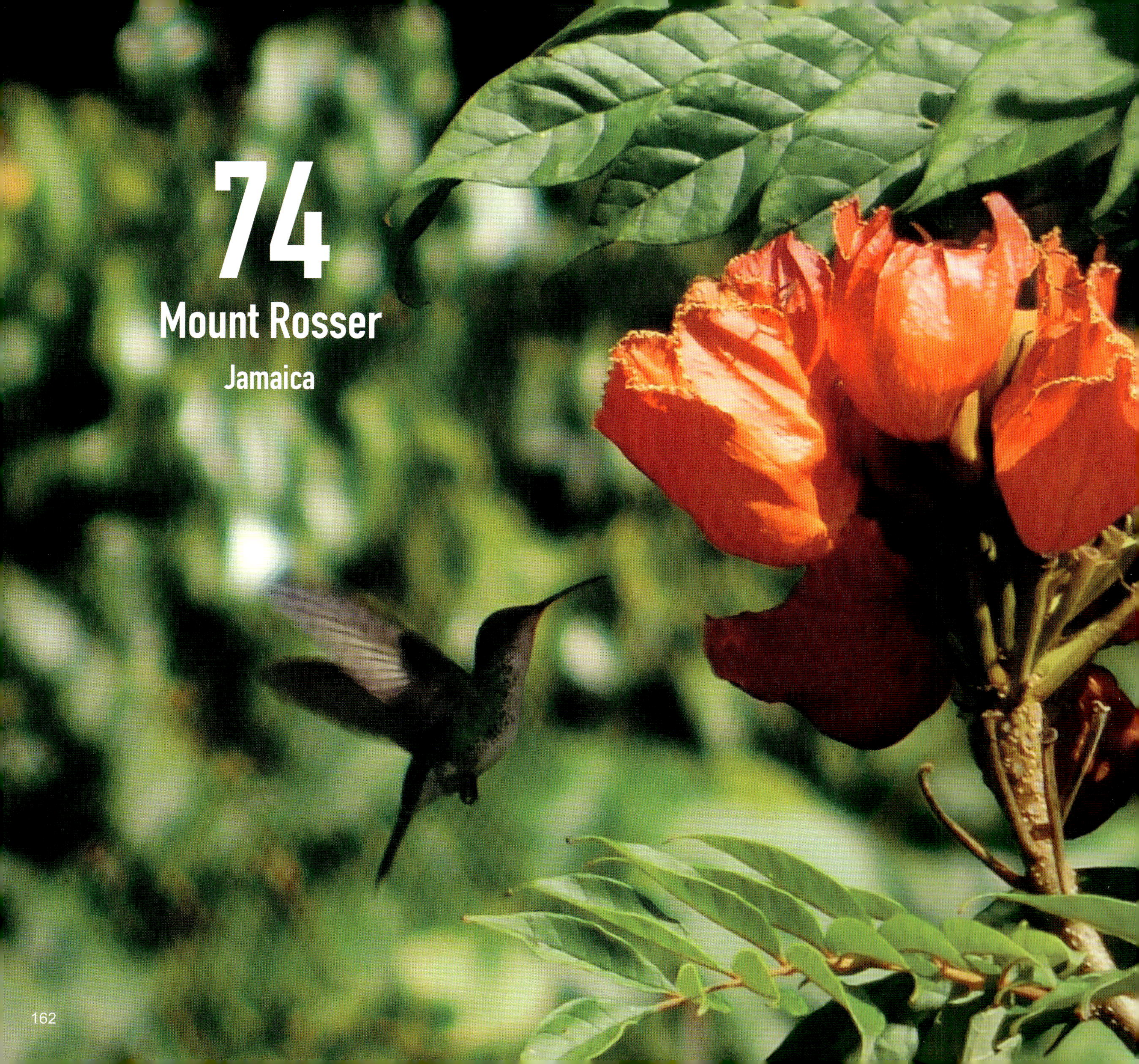
74
Mount Rosser
Jamaica

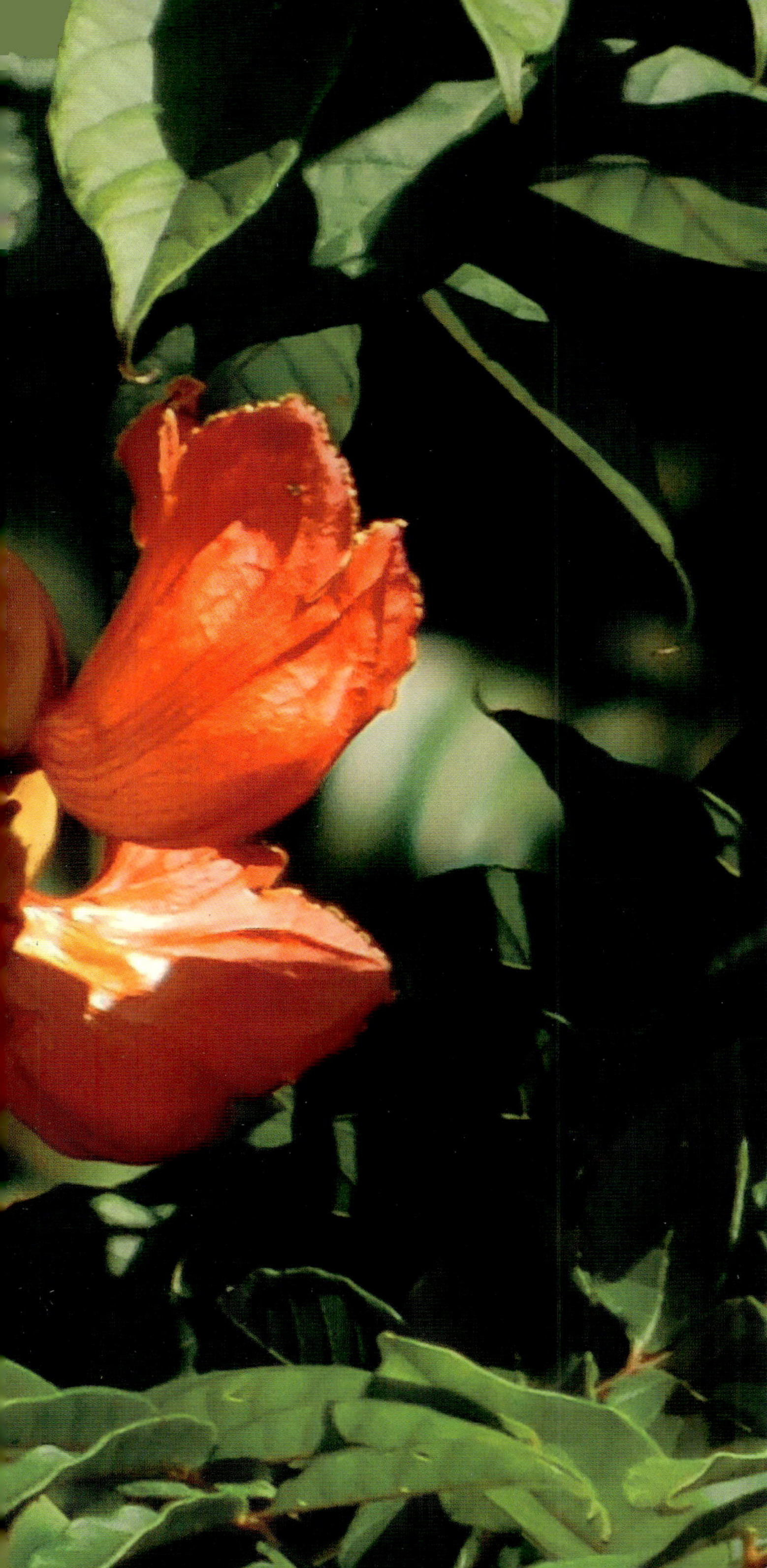

The Mount Rosser Bauxite Residue Disposal Area (BRDA) in central Jamaica was used for the disposal of bauxite residue or 'red mud' from the Ewarton alumina refinery between 1959 and 1991. Red mud is a mineral waste, not a soil, and its high water content and high pH make repurposing sites like these very challenging.

Remediation of the site's 37 hectares began in 2004 after preliminary studies and consultations between the Mount Rosser team, regulators and other government agencies. The project goal was for a self-sustaining vegetation cover, which subsequently had the benefit of enhancing biodiversity. However, red mud is a far from ideal substrate for supporting a viable new ecology, even in Jamaica's tropical climate. Also, a lack of locally available topsoil stymied the obvious approach of capping the BDRA to create a fertile substrate that would encourage plant growth.

Instead, an artificial soil substrate was made by combining both physical and chemical treatments and allowing time for natural processes to develop. Rio Tinto acquired the site in 2007 and began by draining and re-engineering the site so it didn't collect surface water. The pH was then lowered using gypsum, which also improved the soil structure; then, chicken manure was added to provide essential nitrogen, phosphate and organic matter before tolerant native plant species were introduced in 2016. Slowly the soil organic matter built up.

Within five years, some trees were over five metres tall; plant species had colonised the site from the surrounding hills. As the first industrial waste site in Jamaica to be formally closed, Mount Rosser has been transformed into a hotspot for avian biodiversity, with 42 species of bird observed, including 10 species unique to Jamaica.

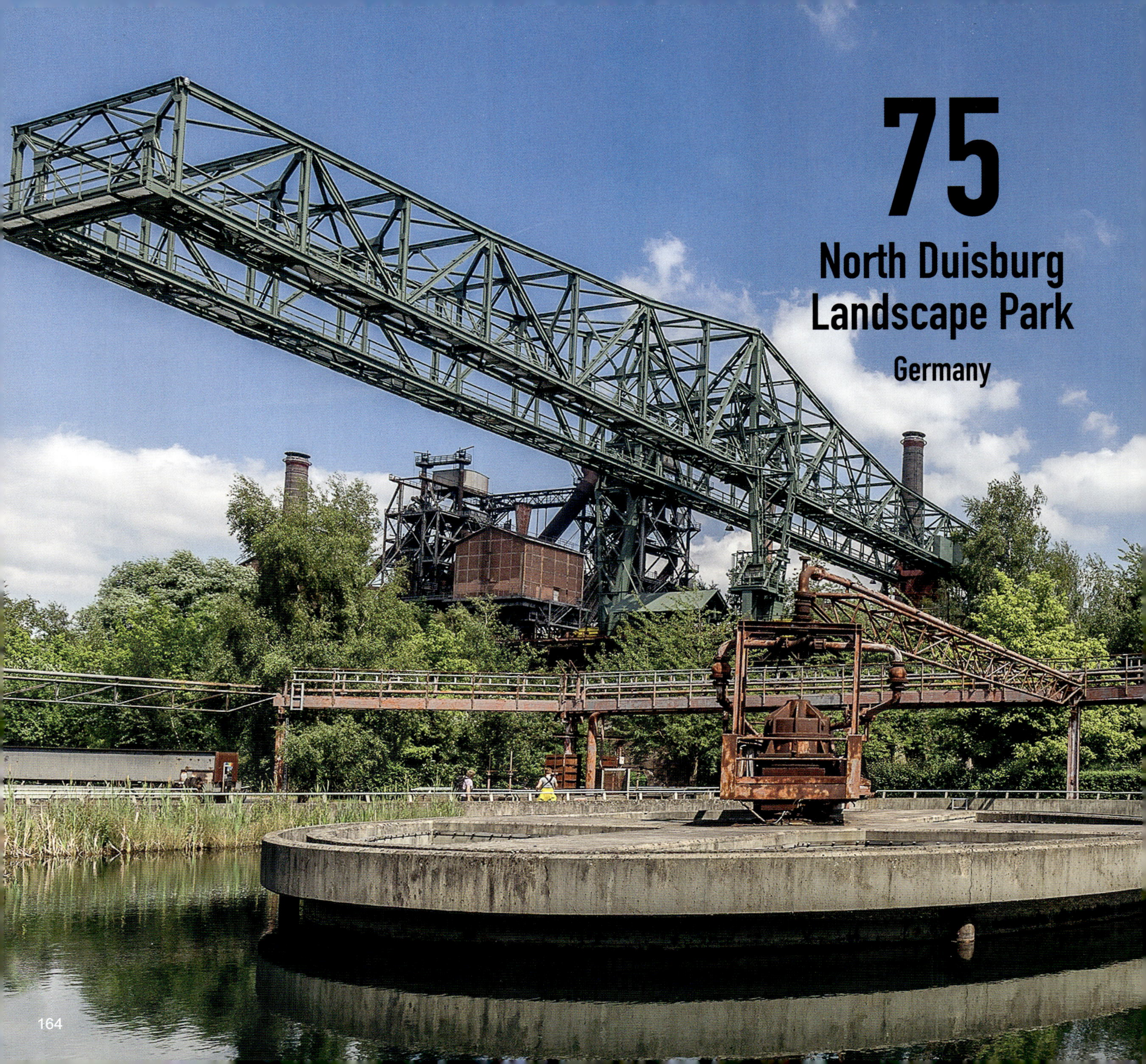
75
North Duisburg
Landscape Park
Germany

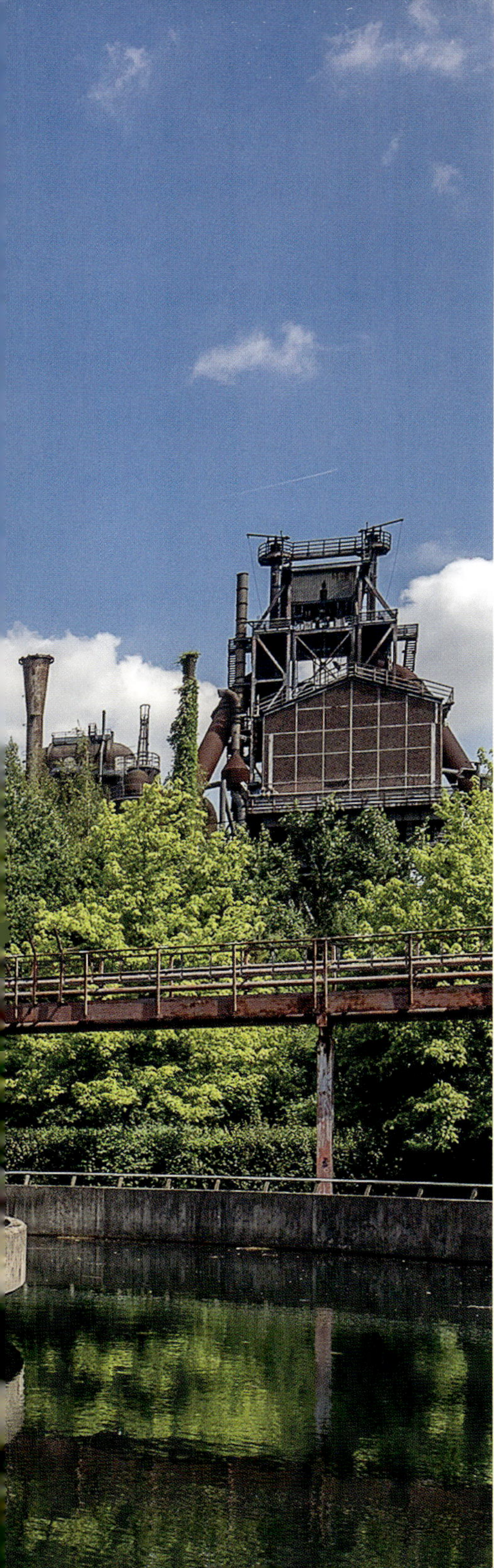

For 82 years, the Duisburg-Meiderich's Thyssen Ironworks fed German industry. Its bewildering complexity of towering furnaces, chimneys and tanks, twined tightly together by a contortion of pipework, ladders and gantries, stood tall as the tapestry into which local people's lives were woven. Unexpectedly, the plant closed in 1985 as Blast Furnace 5 produced its final molten metal; its flame extinguished for good.

The main post-closure challenge was determining what to do with its towering structures and buildings and 200 hectares of polluted wasteland. Tensions arose between local politicians who favoured demolition and local people and experts who saw potential in the site's profound cultural significance. They formed a group to resist the demolition plans and commissioned studies that proved the site's high heritage value, which led to the creation of the German Society for Industrial Culture and the Nordpark Duisburg Interest Group. Subsequent research concluded that demolition would be more costly than preservation, so in 1992, Duisburg City Council prioritised preserving the site.

At that time, the famed heavy industry of the Ruhr region was in rapid decline. The North Rhine-Westphalia state government was considering new approaches for structural change in the region and Meiderich became totemic in this rethinking of regional renewal. Alongside economic regeneration, a major consideration was to increase green spaces and access to them across the Ruhr, of which the germinal North Duisburg Landscape Park became a major component. Along with other large complex sites of the Ruhr, it became the focus of the 10-year International Building Exhibition (IBA) Emscher Park between 1990 and 1999.

The original site concept for a hybrid landscape of industry, nature and culture was conceived in 1989. The design competition was won by landscape architect Professor Peter Latz. He planned a new community asset within a literal framework of industrial archaeology and architecture. The first parts of the new landscape park opened to the public in 1994.

Today, the former contaminated wasteland of rusting, hulking structures symbolises the balance between preservation and redevelopment and is renowned for its unique ecological expression. Attracting a million visitors annually, it offers a range of recreational and sporting activities and hosts 250 events a year, including concerts, theatre, summer and Christmas markets and an open-air cinema. Many of its industrial buildings have been repurposed for hosting events or are occupied by a range of businesses and other organisations. Night-time light shows invite a different, spectacular perspective. On-site youth hostel accommodation is also available in a repurposed administration building.

Integrated with the natural open spaces are maintained gardens, orchards and vegetables, herbaceous perennials and wildflowers. Elsewhere, naturalistic plantings grow amongst industrial hardware. Water plays an important natural and structural role, connecting the site's drainage through naturalised water features that connect with the – once destroyed – Emscher River, which flows for three kilometres through the site.

The wider landscape encourages exploration with play areas, cycle trails and footpaths on former railway lines. The last remnant of the farm that once supplied food to the Thyssen company is now a farm school – the Ingenhammshof – teaching urban children about farm life and food production and home to bee colonies and rare local livestock breeds.

After an apprehensive climb through the tangle of metalwork to the 70-metre-high viewing platform on top of Blast Furnace 5, even in the rainy gloom, the power of nature to heal our industrial depredations is instantly apparent. Wild foliage weaves through the metallic latticework of the site's structures, while disordered greens and autumnal tints imperceptibly encroach over their rusty tones and drab concrete. The panorama extends over the lower Rhine and across the western Ruhr: the land is of low relief but for the occasional distant hillock of industrial waste; the continuous green blanket of trees is pierced by the slender chimneys and angular buildings of other industrial complexes.

The park and its hardy biodiversity were instrumental in establishing an influential new ecological paradigm – industrial nature. In urban areas, post-industrial brownfield sites are often the most biodiverse areas, containing many species eradicated from or squeezed out of the surrounding highly managed urban landscape, finding refuge in the brownfield's relative calm. Here, nature leads the process of natural repair while people develop the space and its culture. It's a hybrid ecology of native plants and animals, exotic garden escapees and others accidentally imported with the iron ore. Over 700 plant species occur on-site, compared with a complement of 2,000 for the whole state, alongside several bat species, natterjack toads, 100 species of beetle and over 45 species of bird, including the nightingale. Some areas have been left to nature without intervention, forming an industrial 'wilderness' – an undisturbed sanctuary for sensitive species, off-limits to people. The park has had a major influence on the way other post-industrial sites have been restored across Germany and beyond. Indeed, the Western Ruhr Region Biological Station even has an outpost here.

The park receives public funds from the city, region and state to preserve and develop the site. But the cost should be weighed against the enormous price originally anticipated for demolishing, decontaminating and regenerating the site. It has become an international icon for a sensitive and positive way to restyle and reuse post-industrial structures and land. This status has been recognised in numerous international awards and in the site's role as an anchor point on the Ruhr's 400-kilometre Industrial Heritage Trail and on the European Route of Industrial Heritage, which extends across 51 countries.

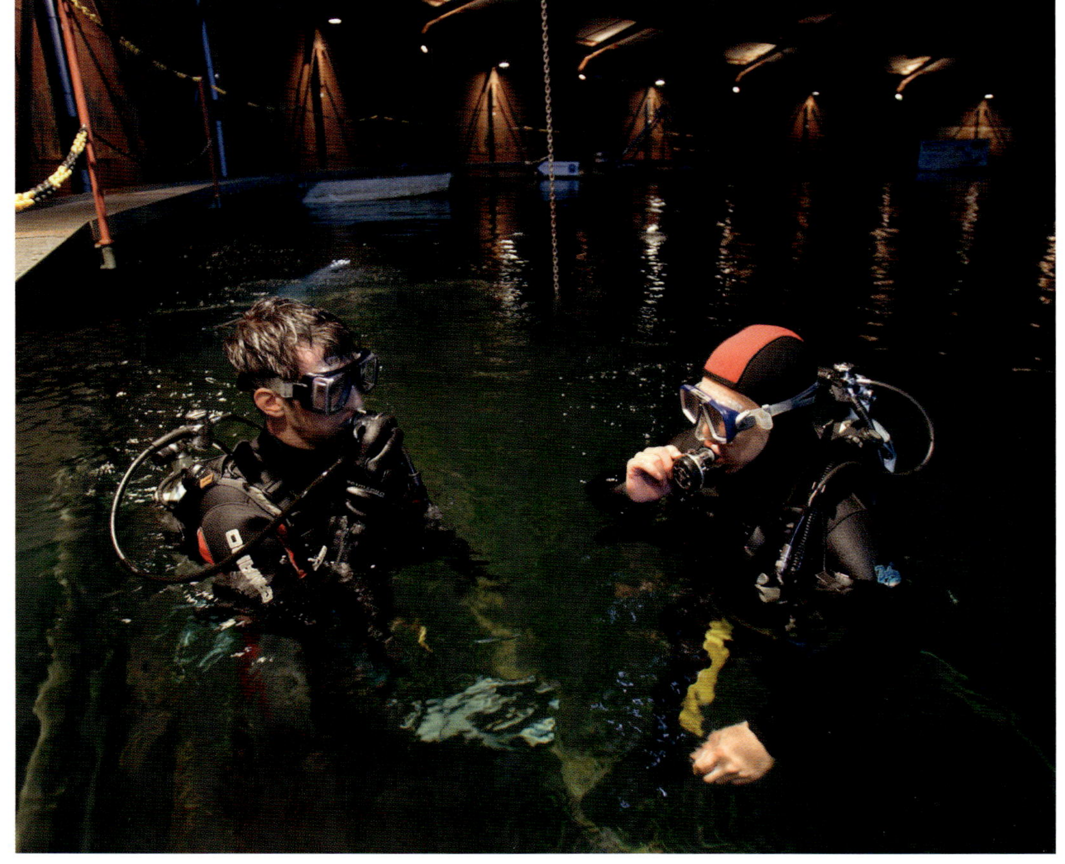

Jonathan Park lighting installation (right)

A different visual experience occurs after dark. British artist Jonathan Park's installation of coloured lights illuminates the contorted architecture with vibrant primary colours. Each colour represents a different element of the iron-making process; green is for gas, blue represents water, and red is for fire and heat. Visible for kilometres around, the lightscape has added another dimension to the region's cultural backdrop.

Diving in the gasometer (below left)

It's incongruous to see people wandering around a former ironworks in wetsuits heaving compressed air cylinders, but the park is home to the largest indoor diving pool in Europe, which is easily missed at first sight. Constructed in a repurposed 1920s' gasometer originally built to store blast furnace gas, the pool's 21 million litres of rainwater cover an artificial reef, a boat wreck, an aeroplane fuselage and more. The facility is run by the Nordpark Diving Club, which offers diving courses and training for emergency services and hosts special events like underwater weddings!

Climbing, via ferrata and high ropes activities

The site offers a variety of gravity-defying activities. The German Alpine Association's Climbing Garden (right) takes advantage of the heights and angles of the coke and iron ore bunkers' concrete walls and towers as cliff faces, while the renowned 530-metre-long Via Ferrata Monte Thysso takes intrepid visitors on an alarming journey across sheer concrete faces. Power-Ruhrgebiet runs high-wire parkour adventures that explore the former casting hall and a disused blast furnace. The climactic 50-metre height offers breathtaking views across one of the world's great urban-industrial landscapes.

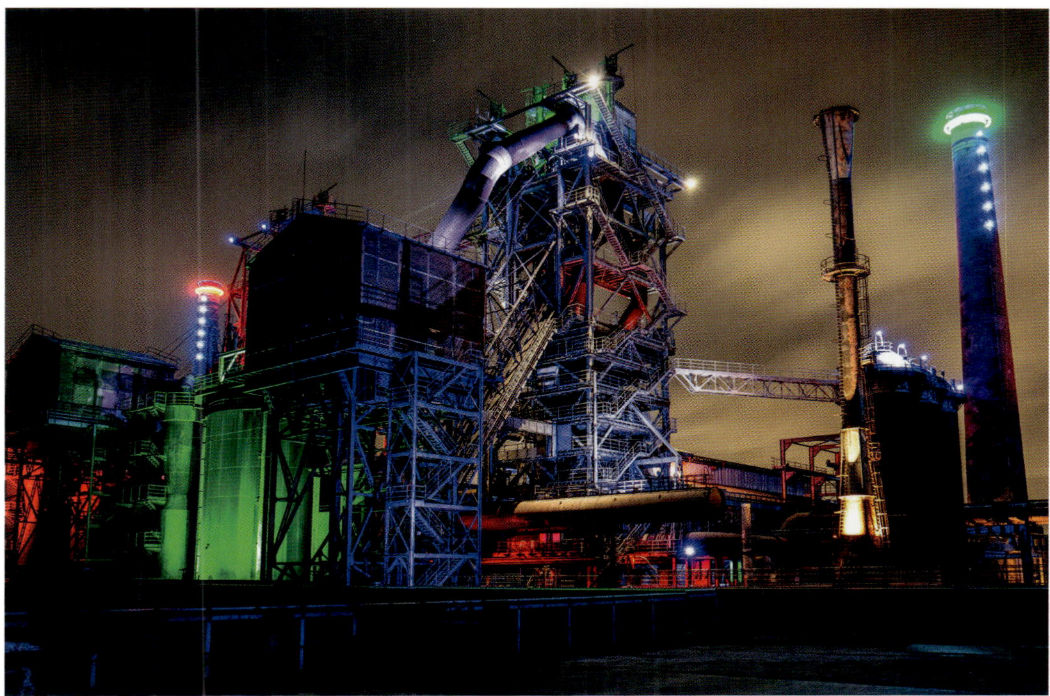

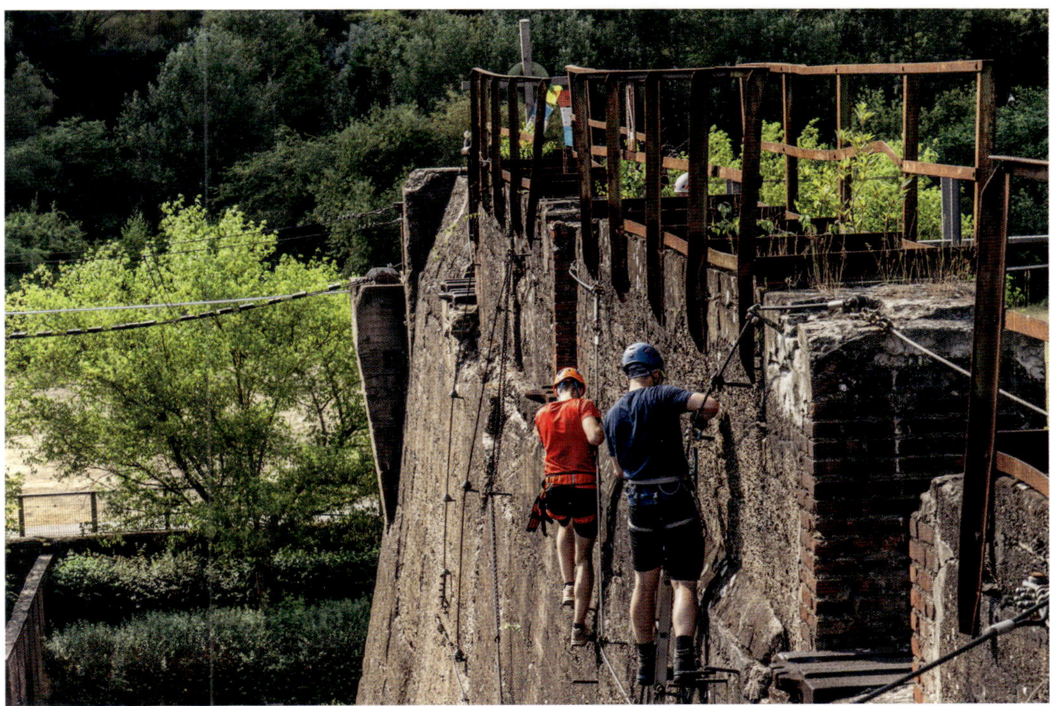

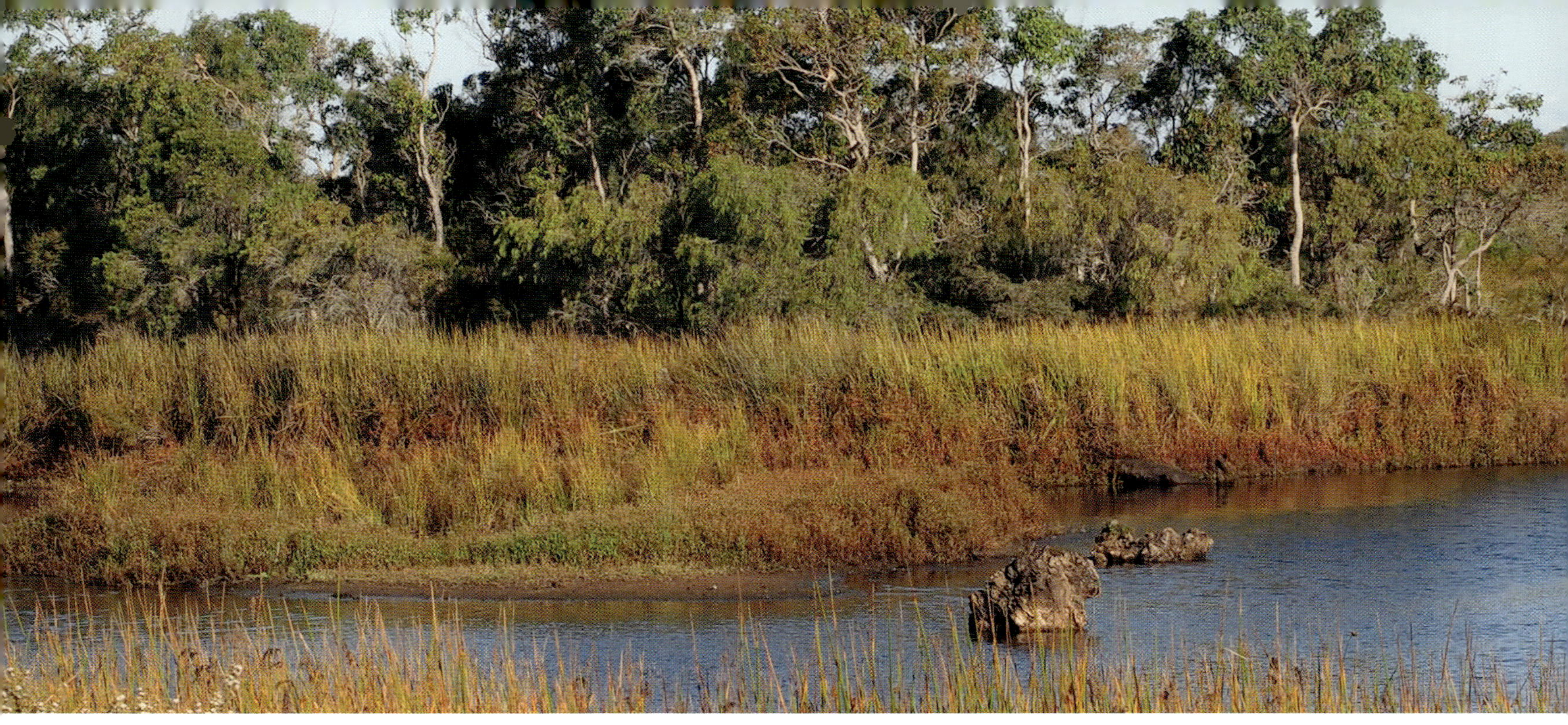

76

Beenup Consultative Group

Australia

Meaningful participation by local communities and other stakeholders is central to today's good practice mine closure planning. Before such social engagement was commonplace, BHP convened the Beenup Consultative Group (BCG) in 1989 in advance of developing their Beenup titanium minerals mine in Western Australia (WA) as a communications conduit with the Augusta-Margaret River community.

Beenup's location on the doorstep of the Scott National Park and its serene Blackwood River, all within an internationally recognised biodiversity hotspot, imparts a peaceful ambience of rural beauty. It is close to coastal tourism and surrounded by beef, dairy and sheep farming, viticulture, forestry and fishing. It is a popular retirement area with an engaged, influential and informed community. Until BHP arrived, there had never been any thought of large-scale mining.

The BCG comprised a cross-section of society representing local government, the local community, a conservation group, neighbouring landowners, farmers and fishers, the business sector – and BHP. The community was split with roughly half against the mine, owing to environmental concerns, disturbance and a lack of empowerment or influence, and half for, due to the economic opportunities it could bring to the area.

The mine began production in 1997 with a 25-year mine life. Unexpectedly, BHP announced Beenup's closure just two years later because – despite its mineral riches – it was technically too

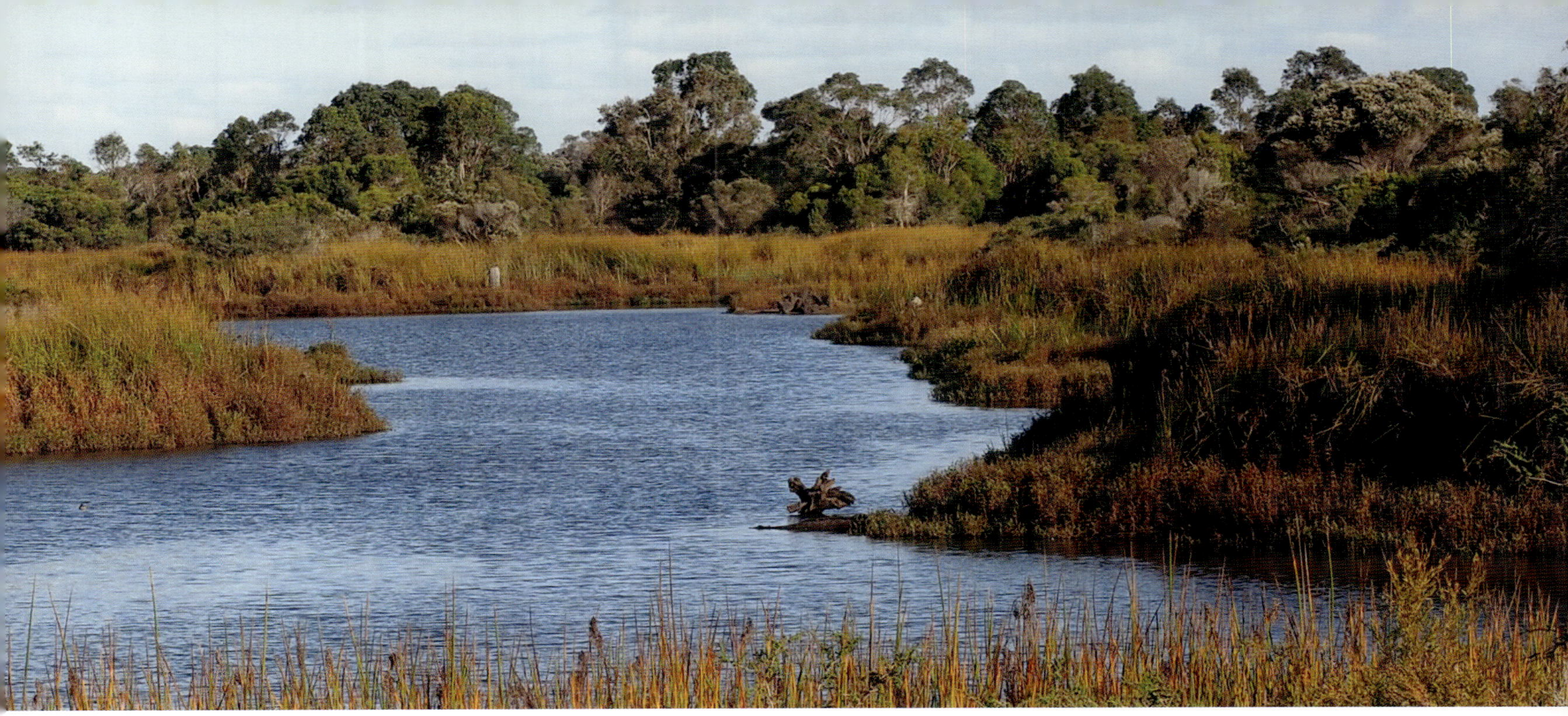

challenging to mine. The sand's high clay content stymied efficient tailings deposition, and the soil's pyrite content acidified the dredge pond water, which had to be treated, and precluded the planned option of dry tailings storage. However, over 300 hectares of land had been damaged in this time, including a 55-metre-deep dredge pond over two kilometres long and a 40-hectare dam.

BHP acknowledge that in the lead-up to closure, engagement with the BCG had become informative rather than consultative, making many suspicious that information was being withheld and distrustful of the company's motives. The community was not happy about the mine's failure. Misgivings grew that BHP would be unable to pull off a high-quality mine rehabilitation in a time when there

was little written government closure guidance, but the BCG now came into its own.

Recognising that trust needed to be rebuilt, BHP offered up all the site's environmental data to the BCG to encourage participation. With the BCG's input, BHP determined the closure goal should focus on managing water quality, which would be best achieved by maintaining a water cover. The agreed final plan was for landscaping to reflect the local geomorphology with natural shapes to waterbodies and self-sustaining habitats of native vegetation, permanent and seasonal wetlands and pastures. As implementation began, the BCG requested an independent peer review of the plans and was heartened by its positive findings, and BHP agreed

to accept the review's recommendations including some progressive ideas contributed by the BCG.

The AUD 44 million project was technically complicated, but an environmentally focused design had become central to the plan. It involved pumping millions of cubic metres of pyrite-rich tailings back into the deep hole where a wetland cover would prevent acid generation. The material placed above water was mixed with three times the amount of lime sand than calculations showed was necessary to ensure geochemical stability. Millions of litres of excess water needed to be removed from the site, so rather than discharge into the river, a neighbouring farmer used it for irrigation. The site's wetlands now effectively store run-off and filter out nutrients transported from upstream farming areas, and control water movement off site. The resulting self-sustaining ecosystem links with the neighbouring national park; it is not a walk-away solution, however, and site access is currently restricted.

BHP finally achieved regulatory sign-off for Beenup against their rehabilitation completion criteria in 2018, although the land has not been relinquished. In 2003, the BCG received the Golden Gecko Award for Environmental Excellence from the WA government, and there is an ongoing internal discussion about whether the company should keep the site as a reputational asset.

And, in an industry first, Beenup was recently used for a pilot study of natural capital accounting for a mine to evaluate 'nature-positive' outcomes. Natural capital accounts help to describe ecosystem changes and how they affect well-being and economies. Alongside the expected biodiversity improvements, the study found unforeseen benefits, such as carbon storage and improvements in the quality of water flowing through as its wetlands help clean farmland run-off improving downstream water quality. The study's illustrative figures suggest that the restored ecosystem had an estimated natural capital value of AUD 30-40 million.

Today Beenup is used by birdwatchers, and by regulators and academics as a model for mine closure. It's almost 25 years since rehabilitation started, and although it is still a work in progress, the site is now richer in nature than before mining began. It hosts 15 ecological communities and 251 plant species, including four designated rare plants, aided by experts from Kings Park and Botanic Garden in Perth. It's also a birding hotspot with over 100 bird species recorded.

Sentry-like kangaroos watch your car as you enter Beenup. In the late afternoon warmth of another glorious WA day, as flocks of birds wheel overhead, the sinking sun saturates the natural greens, browns and blues of this once degraded land. Large, in-situ photos of the old workings jolt you back, but as you lift your gaze your view is of a place that looks and feels like a wild, natural space.

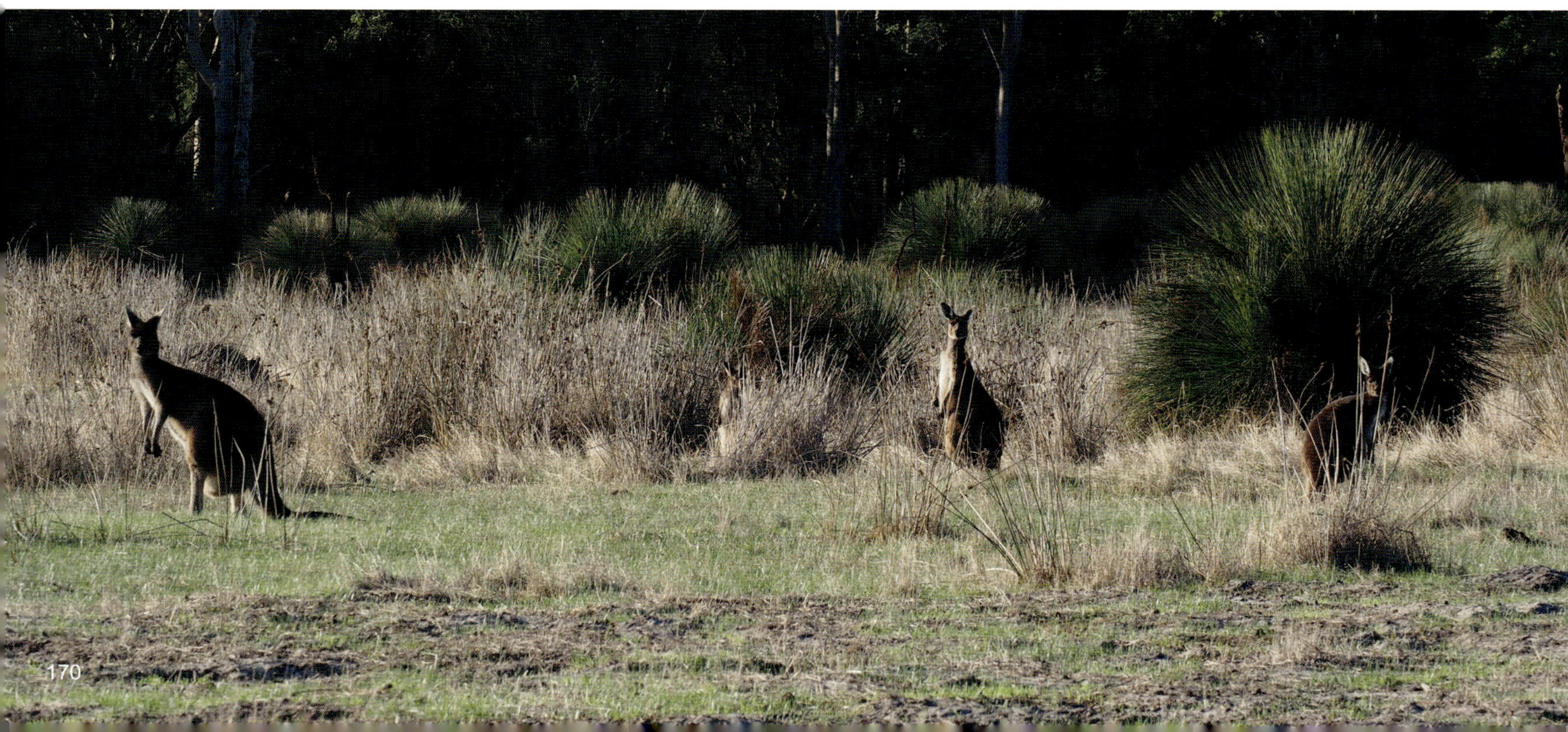

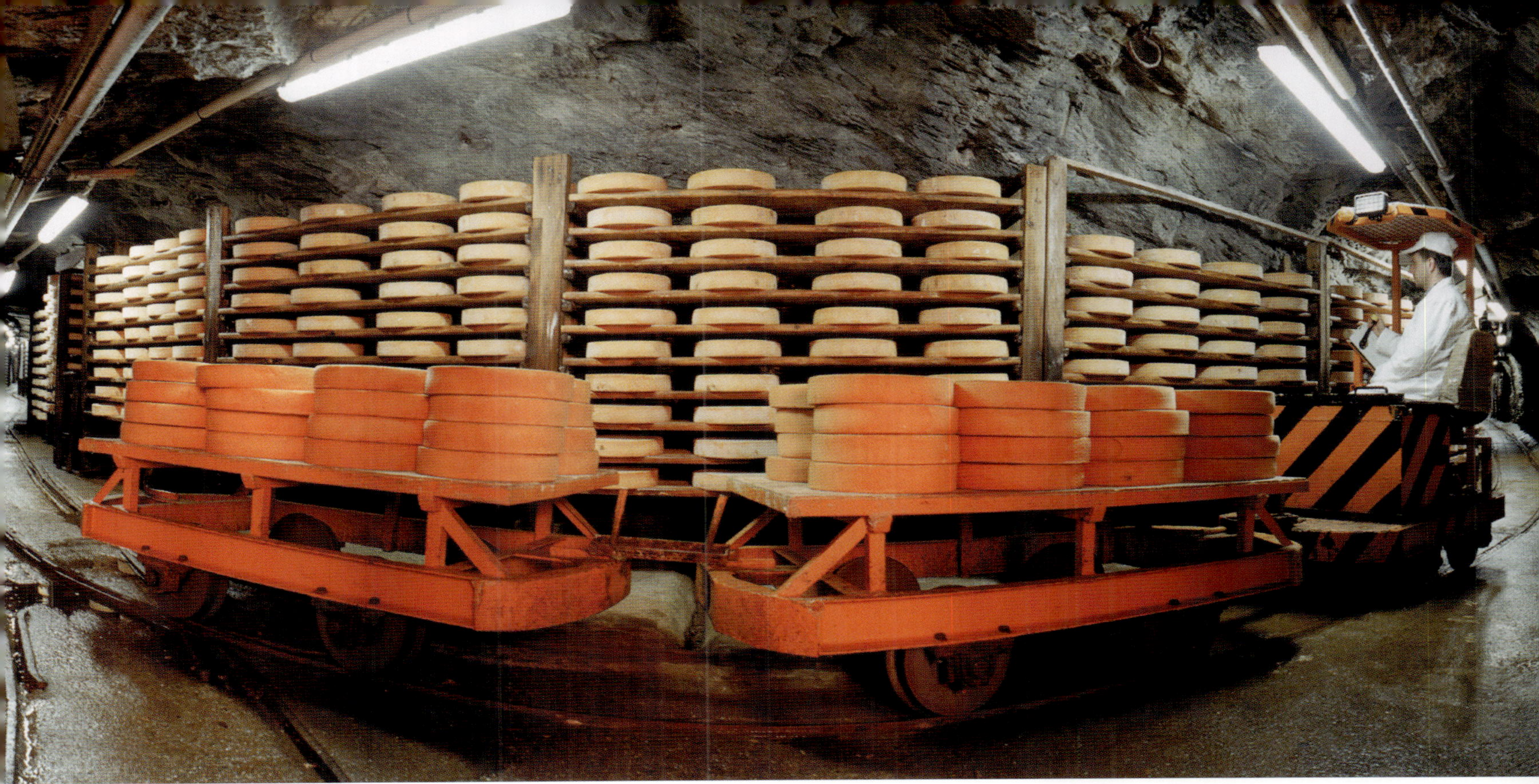

77
Fontina Cheese Mine
Italy

High in Italy's north-western corner, within sight of Mont Blanc, the Aosta Valley's alpine pastures have produced its legendary Fontina Val d'Aosta cheese for centuries. The area's dramatic mountain peaks and glaciers and peaceful meadows attract tourists too, but iron and copper mining were the economic mainstay for generations.

Fontina cheese, with its soft, elastic texture, subtly sweet nutty flavour and rich orange-brown rind, is the valley's main claim to fame these days.

The 200 member organisations of the Milk and Fontina Producers' Co-operative make 250,000 truckles of the delicious cheese annually, mainly for the home market (about 90%), turning over EUR 20 million in the process.

The cheeses mature for about three months at the ideal temperatures and humidities found underground in repurposed caves, originally hewn out by the military during World War II. One warehouse in the Valpelline side valley is different; an adit of the Preslong copper mine, abandoned in the mid-forties, houses up to 60,000 maturing truckles, their intense earthy, woody aroma filling the underworld void. The original tram tracks that once transported copper ore are now used to move cheese in and out of the old mine.

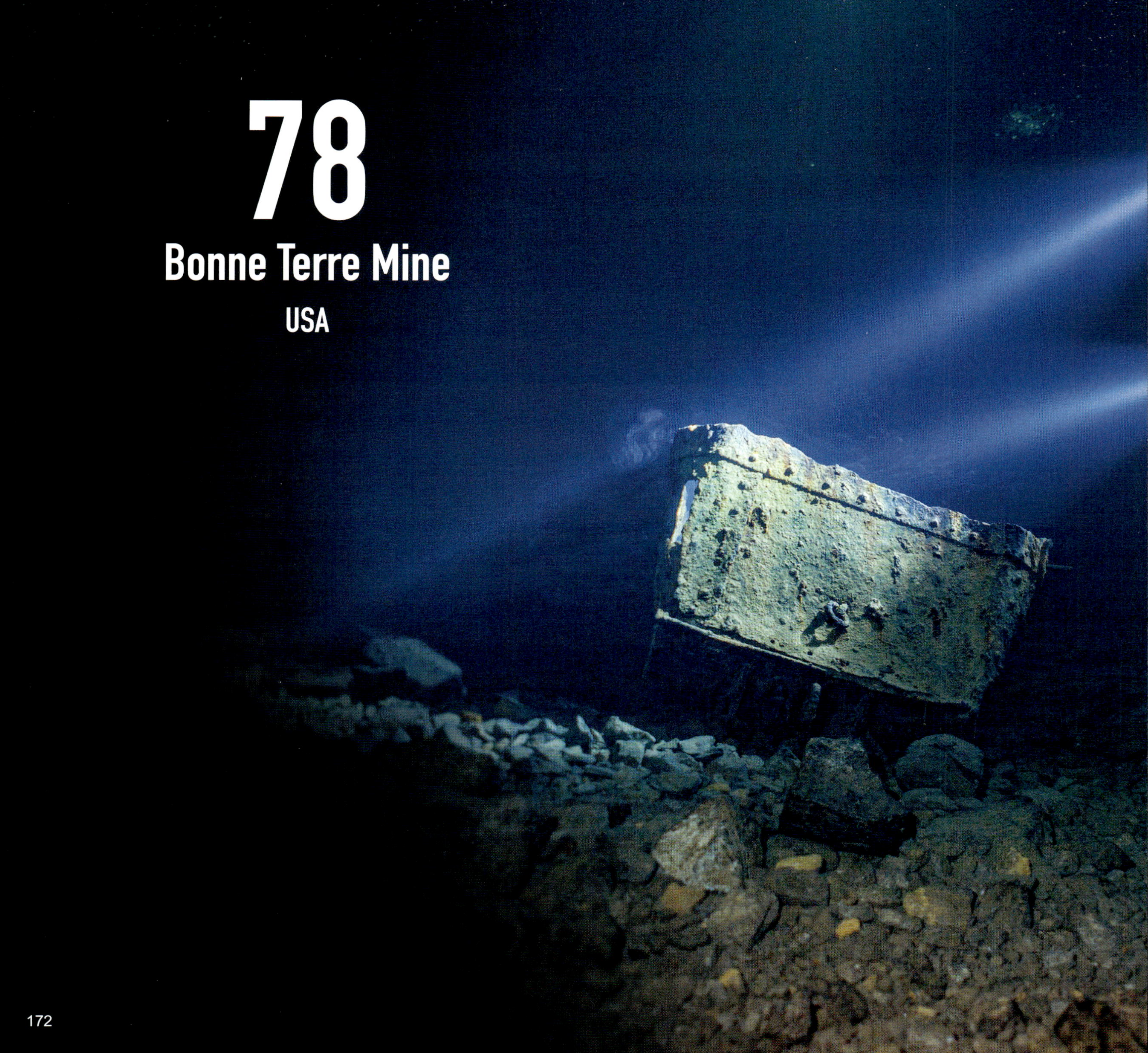

78
Bonne Terre Mine
USA

Bonne Terre means 'good earth', but the French who named the area weren't thinking about farming but war. Bonne Terre was rich in lead, which their soldiers used for shot. The mine proper was established after the American Civil War, and the town was created by clearing the surrounding wilderness.

At its peak, it was the biggest lead ore producer in the world. Over 35 million tonnes of ore was excavated in its lifetime, and by the time it ran out in the late 1960s, you could have fitted a 20-storey building in its largest cavity.

Left to flood, it's now one of the world's largest subterranean lakes and within its depths are five levels of workings, including offices, machine shops and the remnants of a railway, including abandoned mine carts.

The mine was bought in the 1970s by Douglas and Catherine Goergens, who saw the opportunity in the mine's limpid depths. Now, around 25 diving trails start at the stadium-lit Billion Gallon Lake.

The mine water is naturally clear, which has made it attractive to divers and filmmakers alike. Some of *The Abyss* was shot here, and Jacques Cousteau filmed a documentary in its depths. He planned to be there for half a day but loved it so much that he stayed for almost a week. Cousteau advised the Goergens not to fill it with fish and instead make the most of its clear water and mining heritage. Despite Cousteau's warning, they introduced two small-mouth bass, named Bonnie and Terry, the water's sole full-time inhabitants.

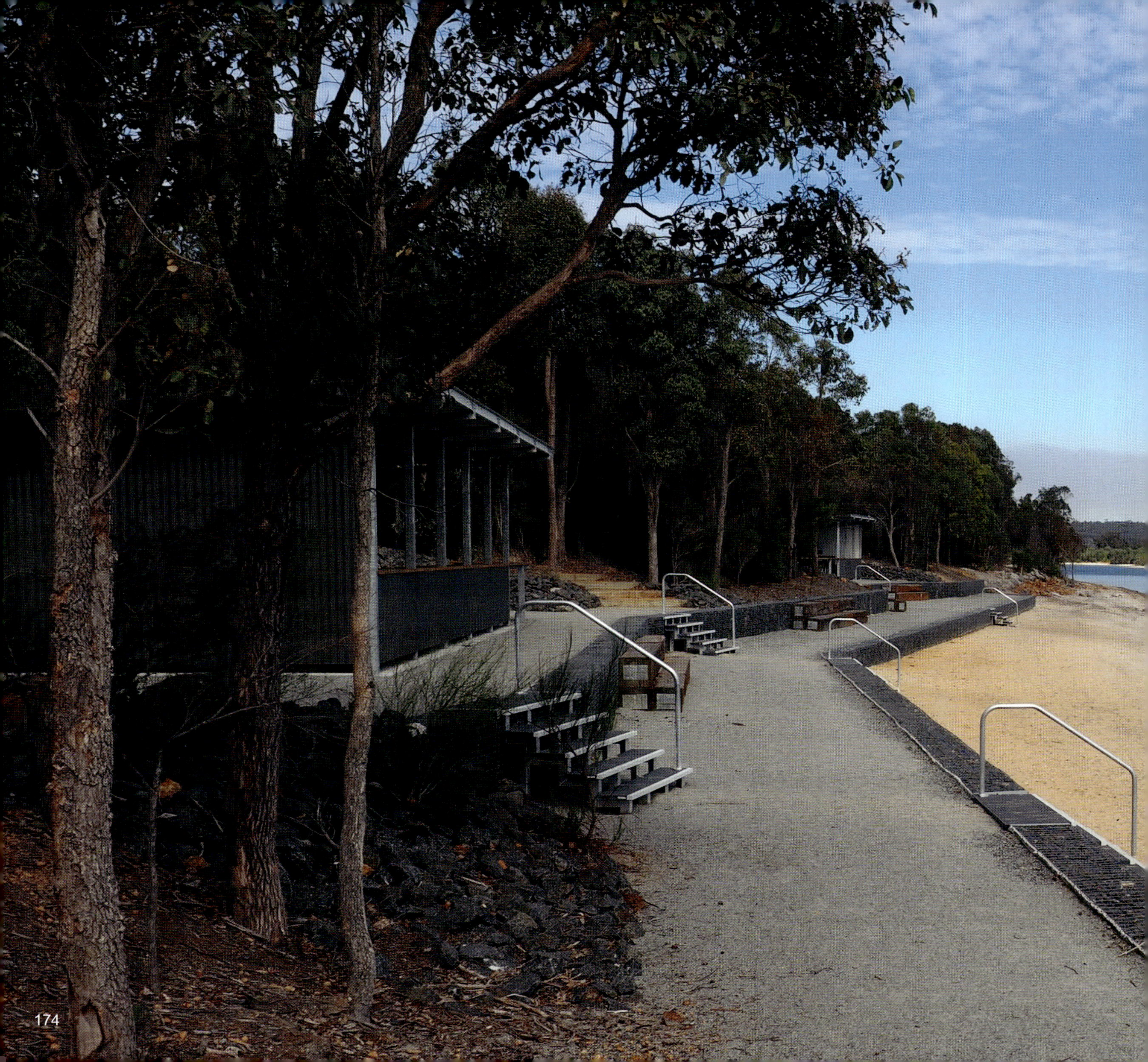

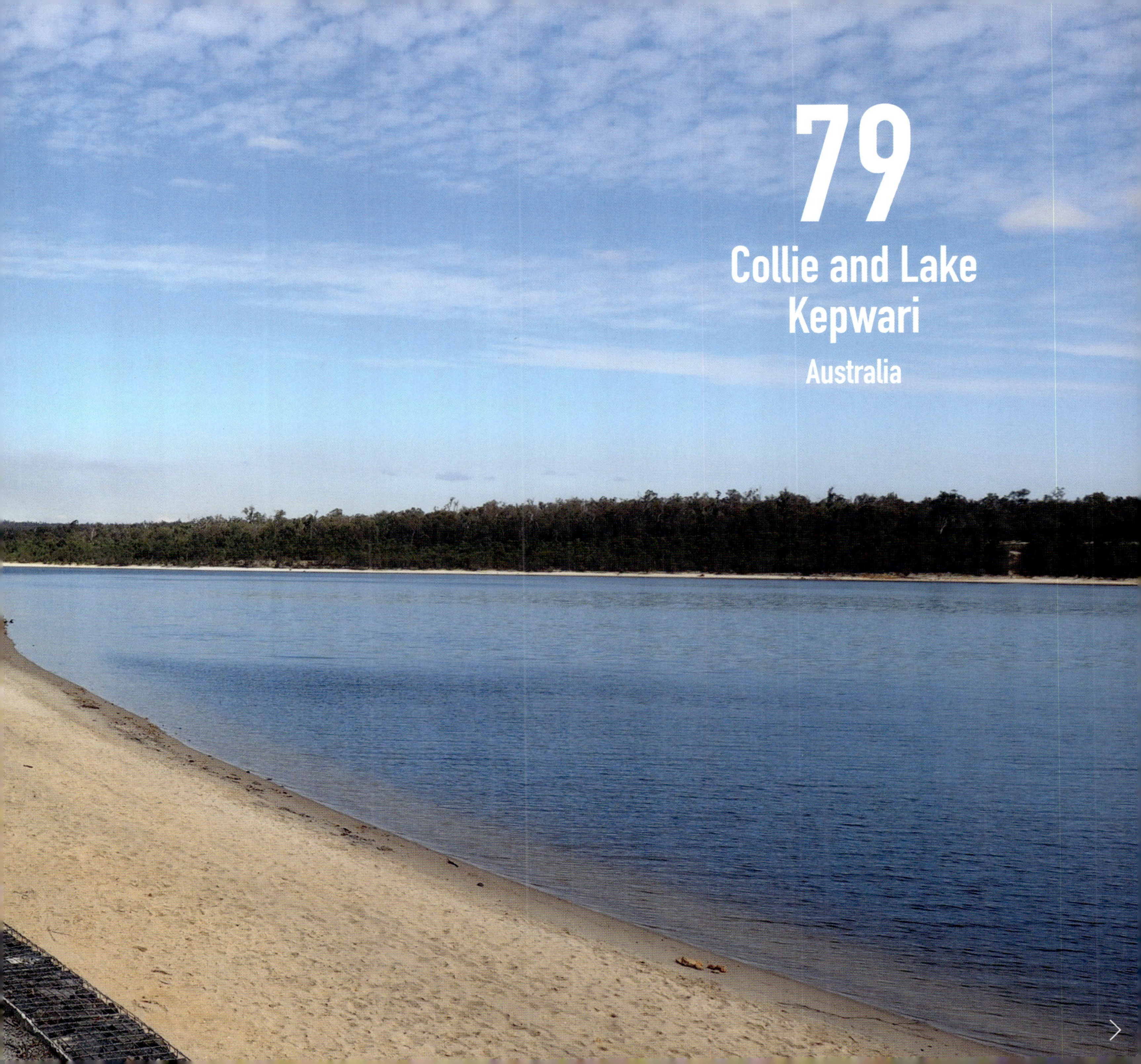

79
Collie and Lake Kepwari
Australia

For 125 years, the town of Collie has been home to Western Australia's coal industry, providing most of WA's electricity from its coal-fired power stations. But, even in the 1990s, when coal mining and power generation were thriving, there was a growing call to diversify away from the town's main industry. Local people and politicians saw the area's natural beauty as the foundation for a new socio-economic direction.

An easy drive from Perth and the coastal resort of Bunbury, its picturesque countryside features the Wellington National Park and its well-loved, forested Collie River valley. Mick Murray, a charismatic former Collie mineworker and activist, began pushing this agenda. Simultaneously, the W05B opencast coal mine, 15 minutes from Collie, ceased mining in 1996 after 50 years of operations. Murray and the community spearheaded the concept of the site as a recreational lake to catalyse Collie's transformation. This was instrumental in Murray's election to the state government in 2001; in 2017 he was promoted to Minister for Sport and Recreation, giving him a major platform for driving the Lake Kepwari vision (*Kepwari* is a local Aboriginal Noongar word meaning 'playing in water'). Importantly, the cabinet also got behind diversifying Collie's economy.

The regulator originally wanted the mining void to be backfilled with mine waste, but, following debate, the stakeholders agreed on a closure vision that went beyond a basic safe and stable landform closure approach and instead followed the community's preference for a recreational lake that would stimulate new business opportunities. Relying solely on rainfall and groundwater sources to fill the void, as described in the approved 1997 closure plan, would have taken over 100 years, with poor water quality from sedimentation and acid drainage limiting the lake's usefulness. Rapidly filling the pit would improve the lake's water quality and speed up access, so a new plan was devised to convert the closure strategy's unconnected lake with its resulting poor water quality to one that was safe for recreational use, with a sustainable ecology and aquatic life fit for human consumption. But this raised many challenges.

The south branch of the Collie River had been redirected around the mine's perimeter during operations. The plan was to divert its seasonal high flows into the void to create the lake, but four years of trials showed that the lack of flow through the lake led to poor water quality. In 2011, however, an unexpected river flood into the lake caused it to overflow and showed a possible solution; the flow-through it created improved water quality with corresponding ecological benefits. Discussions ensued between the company and regulator until, in 2018, approval for a permanent flow-through was granted and implemented the following year. The diversion channel was backfilled and restored. The rehabilitation work was carried out by Griffin Coal and Premier Coal.

Access to the site was restricted, but community pressure, led by Murray, grew for public access, which also oiled the bureaucratic handover process. Lake Kepwari was formally relinquished into public ownership in 2020, including 220 hectares, of which 120 were restored land, and was officially opened the same year by the WA premier and two ministers, one of whom was Murray. It had taken a quarter century of research to generate the evidence to allow the creation of what it became on relinquishment, but to turn it into a recreational hotspot required government funds and expertise. Grants were given to create car parks, toilets, camping and picnic areas, swimming beaches, a boat ramp and jetty, new entrance roads and a bridge over the river; millions more were spent on enhancing the site's walking trails. Aquatic life has been successfully reintroduced, and the shimmering blue lake, ringed by white sand beaches and a deep green forest, is now a popular watersports and camping destination. Adventure tourism is also being encouraged with investments into extensive mountain bike trails through the area's jarrah forests. Culturally, the site has now been incorporated into the Collie River Waugal Aboriginal Heritage Site. The money was well spent, with 51,000 visitors recorded in 2022 with knock-on benefits for the local economy.

The lake's formal opening coincided with the launch of Collie's Just Transition Plan – a five-year state government staged plan guided by a multi-stakeholder working group, with industry support, to drive the town's socio-economic transformation. 'Just Transition' is an international concept for greening the economy in a fair and inclusive way

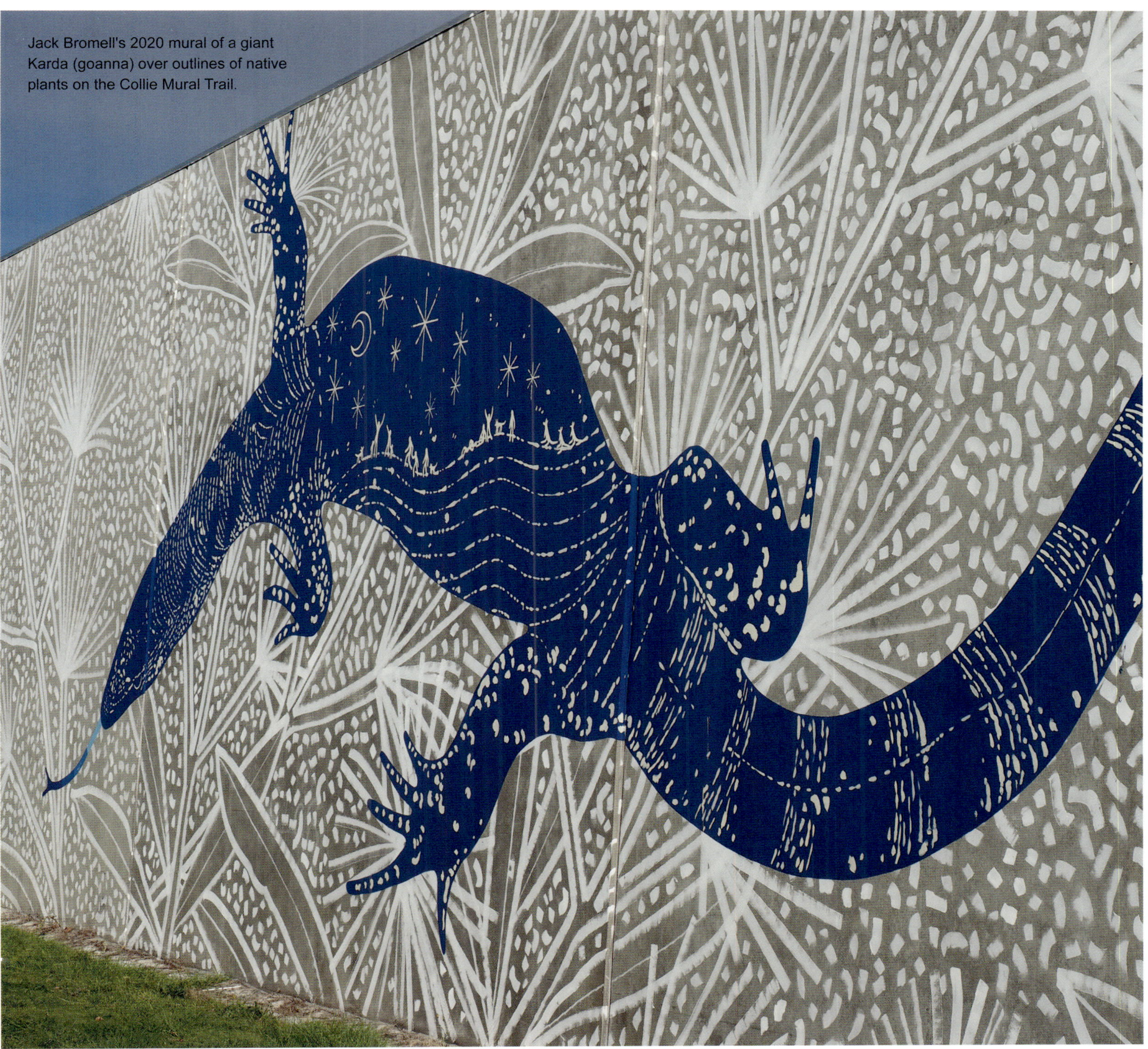

Jack Bromell's 2020 mural of a giant
Karda (goanna) over outlines of native
plants on the Collie Mural Trail.

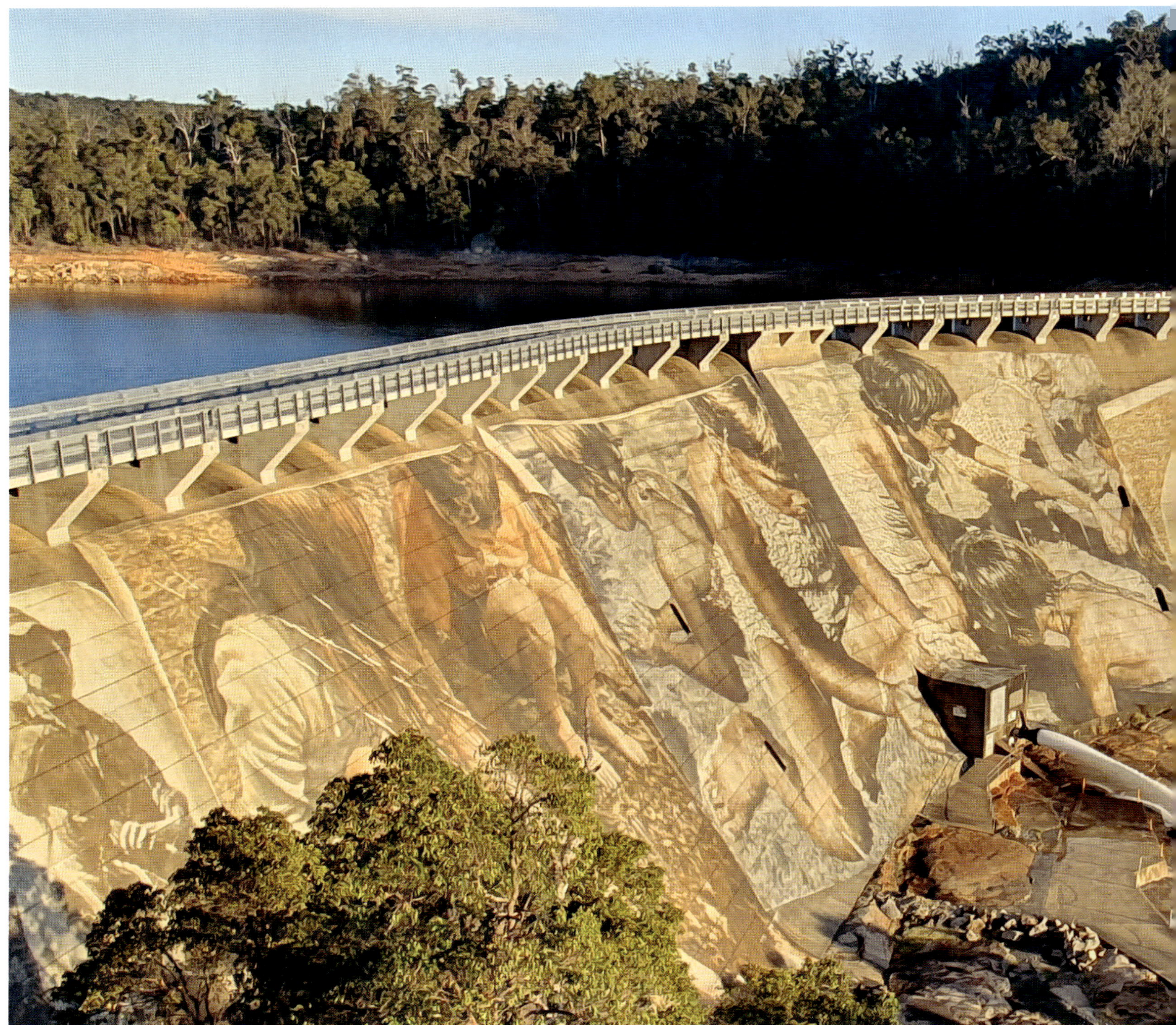

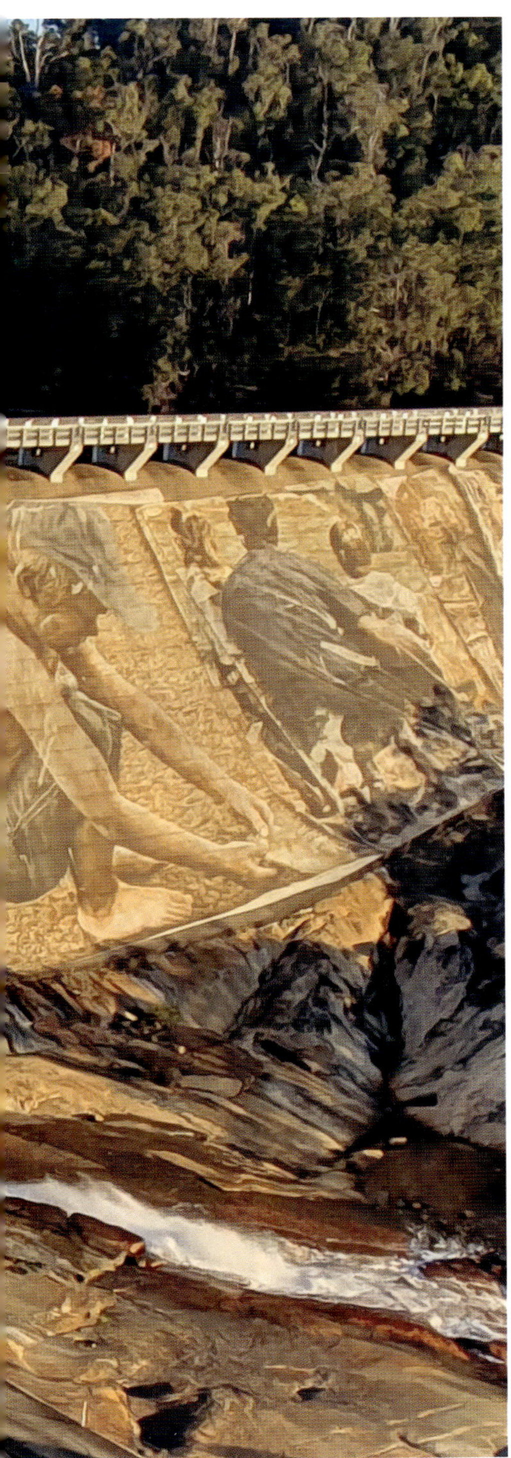

by creating decent work opportunities and leaving no one behind. The Collie plan is forecast to be worth over AUD 600 million to adapt the economy and society to the retirement of the power stations by 2030 and create hundreds of high-quality blue-collar jobs.

Collie is an unemployment black spot, but there are many exciting new initiatives underway, such as the construction of a magnesium smelter, plans for a 2,000MWh grid-connected battery, billions of dollars' worth of renewable energy investment in the region, and a medicinal cannabis farm which will employ up to 80 workers. A further AUD 300 million is dedicated to power station decommissioning, which will support many jobs for a while. Millions have also been invested in the Collie Jobs and Skills Centre to train and guide the town's workforce in new directions.

A WesTrac/CAT Technology Training Centre was opened in 2020 on former power station land to provide training in the maintenance of house-sized, autonomous CAT-manufactured mining trucks. Their use in WA's mines is growing rapidly and each one requires three workers to operate and maintain it. CAT selected Collie through industry contacts and state government incentives. This is a new global industry and the WesTrac centre is one of only two of its kind in the world (the other is in Arizona). Hundreds of new technicians have already been trained, with many from outside WA and even beyond Australia, with associated indirect spend in the community.

The most visible sign of the town's emerging new direction is the Collie Mural Trail – a stunning outdoor art gallery of murals connecting the town centre to the Wellington Dam. Funded by the WA government and the shire council, the 40 murals depict local stories of the area's Aboriginal people, coal mining and more. The murals encourage meandering exploration of the community and out into the Collie River Valley, with Artist Guido van Helten's 8,000-square-metre 'Reflections' mega-mural painted on the dam – a powerful evocation of place. The publicity around these vibrant murals in a coal town challenged external perceptions but also encouraged locals to view their place through new eyes – a crucial step in the journey of change.

There is a tourism 'buzz' in Collie – the strategy is, so far, working with a 73% increase in visitors between 2019 and 2021. The long lead time needed to plan Collie's economic regeneration while the coal economy was still in play was critical. Also essential was the leadership of Mick Murray and others – one of those 'unusual suspects' from outside conventional circles who was able to motivate change. The standout achievement of Lake Kepwari's creation shows that as well as the rarity of its relinquishment (defined as the return of ownership, residual liabilities and responsibility to the corresponding jurisdiction or original owner, or transferral to a third party), community motivation and political leadership are essential, as is the importance of being able to capitalise on fortuitous events, like the flood that filled the pit and changed a strategy.

Opposite: Guido van Helten's 'Reflections' mural on Collie's Wellington Dam.

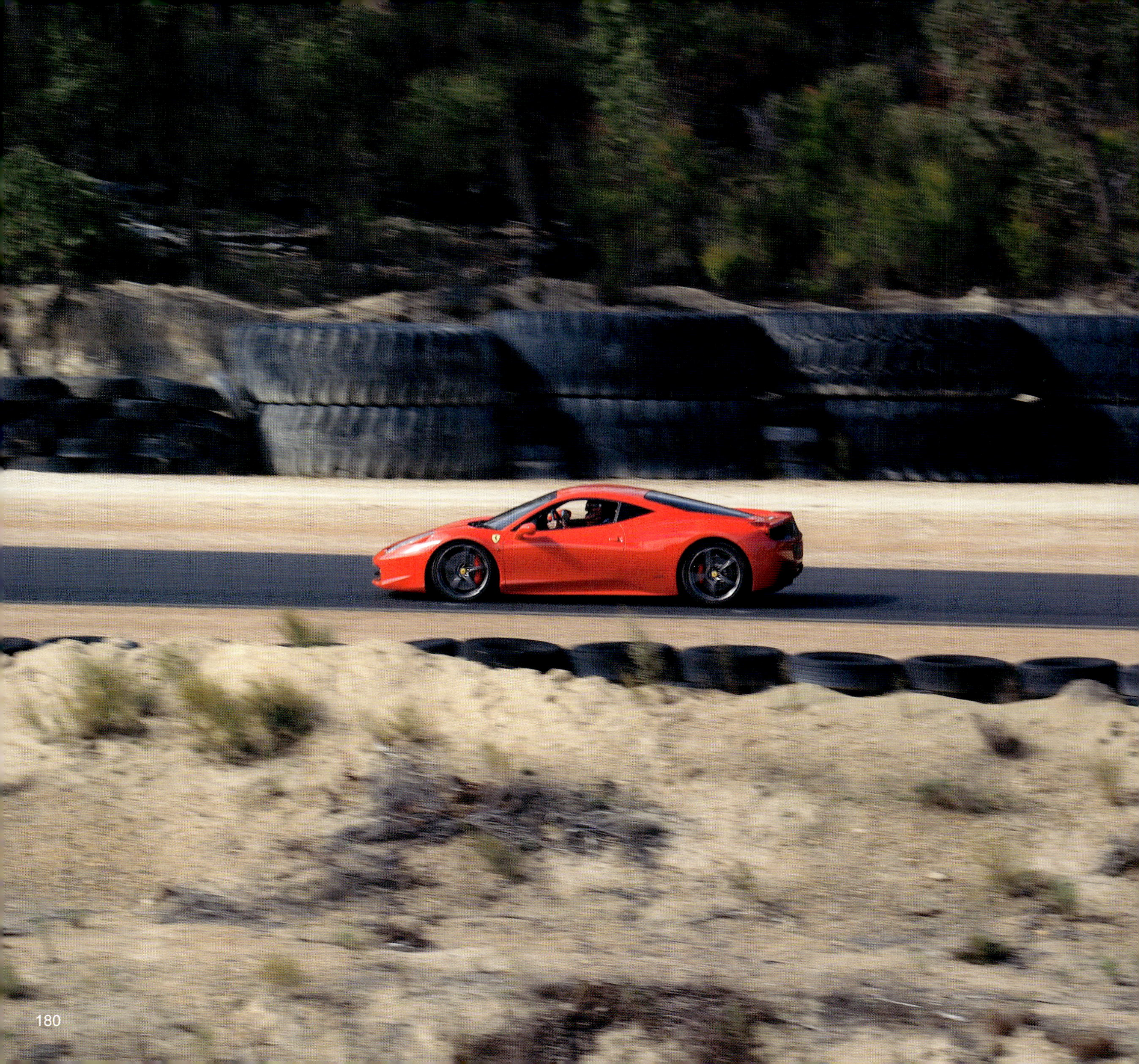

80
Collie Motorplex

Australia

It's an unexpected sight after the short drive from the coal town of Collie: 10 gleaming Ferraris parked in rows under an azure Western Australian sky. Taking turns to scream around a smooth, black ribbon of track that loops through the jarrah forest, they speed past recycled mining truck-tyre crash barriers, brilliant against the dark backdrop.

Collie Motorplex is easily accessible from WA's main population centres and is the most popular regional motorsports track in WA. Built on the site of the former Western Collieries No. 2 underground coal mine, which was decommissioned in 1998, it's an early example of the drive to diversify the local economy from coal mining and power generation.

The 349-hectare site lease was acquired by Motoring South West Inc., supported by the South West Development Commission, to develop into a high-speed motorsports playground. The idea had long been discussed amongst local motorsports enthusiasts, and the former mine manager had a hand in the site's transformation.

Collie Motorplex was officially opened in 1999 and immediately became a popular venue for blue light services driver training, club-level motorsports, four-wheel-drive enthusiasts and youth and corporate driving events. In the early 2000s, it was already attracting crowds from beyond the locality to watch and compete and spend in the local area.

In 2003 the state government granted AUD 1.3 million to improve the facility, enabling Collie Motorplex to put on licensed car and motorbike races. In 2006 it held the inaugural Historic Race and Regularity event for historic vehicles competing over two days for the Collie Coalfields 500. The first official race of the Collie Motorplex State Championship Track was started by the South-West Minister.

Further upgrades since mean the facility now possesses a 1.6-kilometre sealed track, with a longer 2.55-kilometre loop that includes 13 turns, making it WA's longest bitumen race circuit. Alongside the two loops, the site offers a host of other adrenalin-fuelled motoring features, such as the drift stadium, hill climb and emergency vehicle training roads. Improved spectator facilities include a central hill where 80% of the track is visible. There are also training rooms and clubrooms, a medical centre, on-site catering facilities and an ablutions block, as well as bunkrooms, some of which are in repurposed mine buildings, and camping facilities. The track has been certified by both Motorsport Australia and Motorcycling Australia, enabling it to host national events.

Collie Motorplex is run as a not-for-profit organisation overseen by a management board of representatives from the Shire, the Chamber of Commerce, the South West Development Commission and Premier Coal. Other members include people with an interest in motorsports or representing the community. The site is run by two full-time staff and two part-time staff plus a handful of casual-hours catering staff. Ownership of the site is a three-way lease arrangement between the Department of Biodiversity, Conservation and Attractions, Premier Coal and Collie Motorplex.

Socially, Collie Motorplex is well-known for supporting grassroots motorsports. It is fully booked every weekend, with mid-week bookings becoming more popular with private practice and corporate days. Most track users hail from out of town and generally stay, eat and shop in Collie. In addition, the Wellington Dam, Collie River Valley and Lake Kepwari are on offer close by, helping the town diversify away from mining and burning coal.

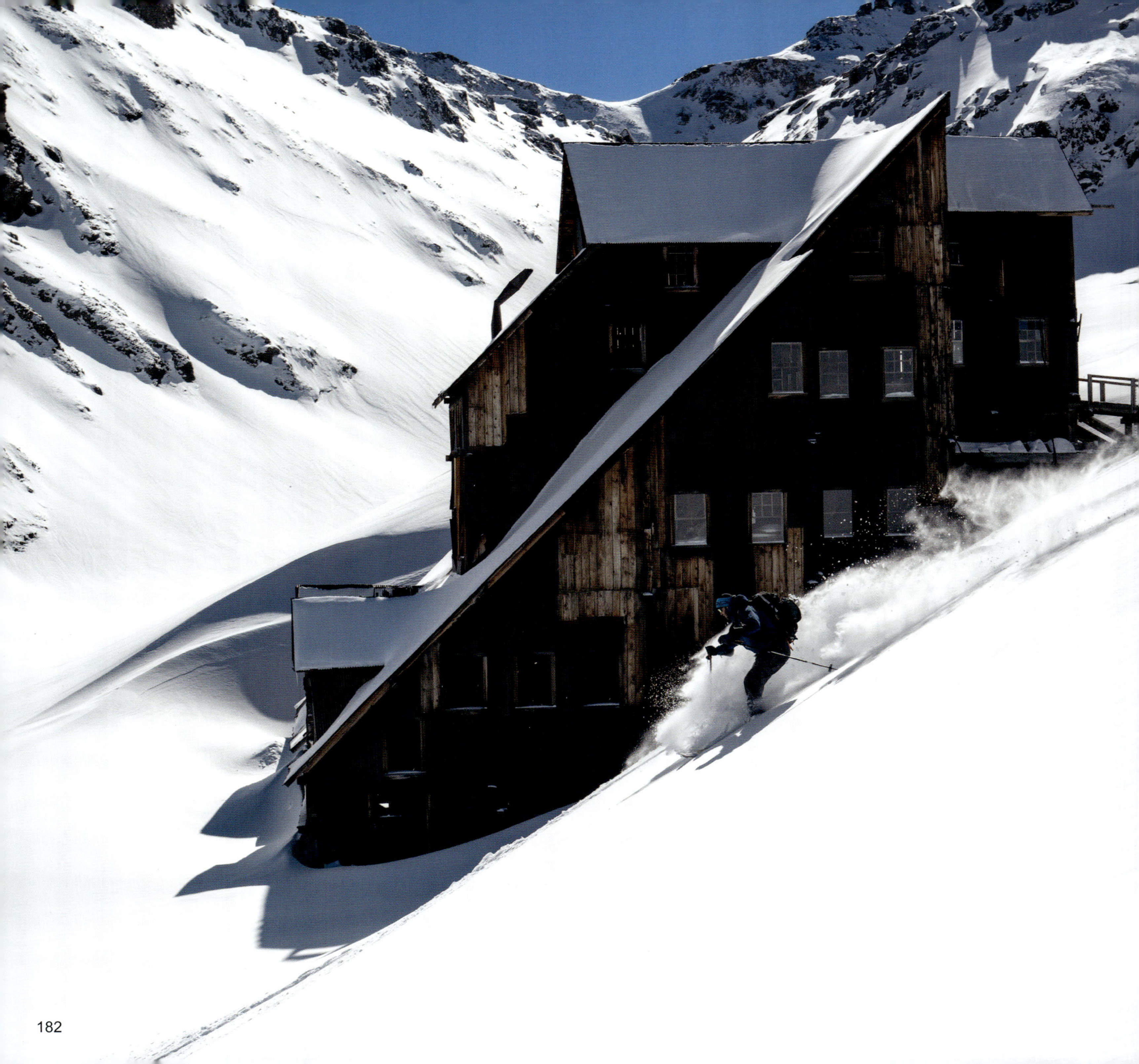

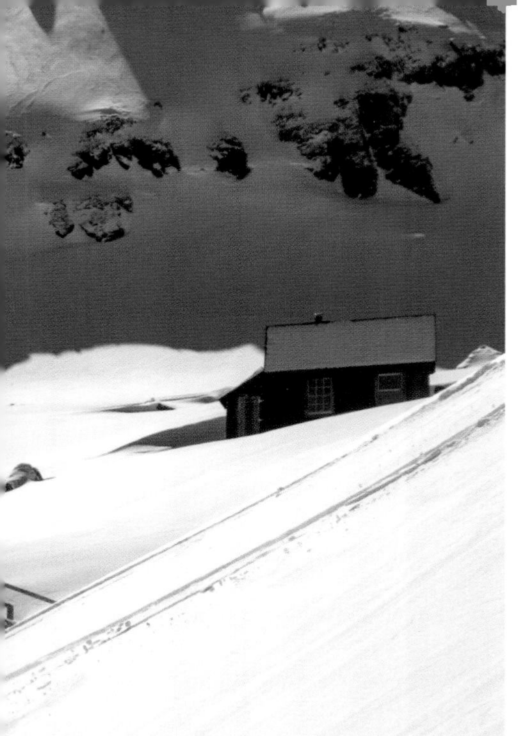

81
Idarado Mine Reclamation

USA

Dramatically located in southwestern Colorado's San Juan Mountain range above the small ex-mining town of Telluride, the many underground metal mines in the Telluride and Red Mountain mining districts – working and abandoned – were acquired by Newmont Corporation and consolidated to form the Idarado Mining Company in 1939. It was soon being worked to supply gold, silver, copper, lead and zinc for the Allied war effort. Since the mid-1800s, mining had been the main economic engine that drove the development of the region and, specifically, the towns of Telluride and Ouray, but falling mineral prices forced Idarado to close in 1978.

The environmental impacts continued after mining. Acid drainage flowed from mine portals into mountain streams already polluted with zinc washing in from processing wastes, while the air around Telluride contained contaminated dust blown from an adjacent tailings pile. Dilapidated, unsafe – but historic – buildings and myriad industrial remnants became impromptu monuments to those

who toiled in such difficult places. Despite these impacts, the mountain landscape was enriched by mining heritage, its defunct structures contributing to rather than diminishing the sense of place.

In 1983, the State of Colorado sued Idarado/Newmont for historic water pollution. A decade of bitter legal battles ensued, accompanied by parallel environmental studies, legal wrangling and negotiation. Eventually, in 1992, Idarado/Newmont and the State agreed on a pragmatic truce by which Idarado would apply current environmental standards to clean up over a century of environmental impacts, including those created before the company formed and many the company never owned.

Implemented through the ground-breaking Idarado Remedial Action Plan, the five-year clean-up began the following year, stabilising and revegetating tailings piles and addressing poor water quality in the Red Mountain Creek and San Miguel River. For the time, many of these actions pushed the technical envelope, and many of the techniques employed influence remediation practices in use today.

Thirty years later, acid drainage and water quality issues remain in some areas, but the collaboration is in this for the long haul, trialling new technologies and adapting to changing water dynamics in the high country's underground workings. What began as a bitter dispute evolved into a respectful collaboration. Local citizens' groups from Telluride and Ouray and local and county governments were critical to this successful outcome, understandably demanding environmental improvements and strongly behind the preservation of the area's mining heritage and its unique sense of place.

With most of the remediation completed, in 2000 Idarado began to return most of its land back to public ownership. Together with the non-profit Trust for Public Land, the US Forest Service and the local Red Mountain Task Force, they established the Red Mountain Project to preserve mining structures and historic landscapes which were being threatened by residential development. Congress granted USD 14 million to the project, which led to the transfer of over 3,600 hectares of historic mining lands to the Uncompahgre National Forest, including 2,225 hectares of Idarado's land sold by the company to the Trust for Public Land. Idarado also donated a further portion of land and funded a Forest Service-managed interpretative exhibit overlooking the heart of the Red Mountain Mining District.

Other initiatives were developed. The 2020 Idarado Houses Restoration project, in partnership with the Trust for Land Restoration and the Ouray Country Historical Society, demonstrates the long-term commitment to cultural heritage. The two-year project restored and preserved century-old miners' houses as a focal point of historical interest along the remote mountain road of the San Juan Skyway Scenic and Historic Byway.

The sublime Bridal Veil Falls Powerhouse – perched precariously on the edge of the highest unbroken waterfall in Colorado – supplied hydroelectricity to the mines for fifty years from 1907. After legal battles over water rights and easements, Telluride and Idarado worked together to update the powerhouse, now listed on the National Register of Historic Places, to produce electricity once again and use its water supply infrastructure to provide clean water to the community.

The powerhouse once also breathed life into the famed Lewis Mill during its short operational life between 1910 and 1913. The mill stood idle and decaying for decades in the harsh conditions found at 4,000 metres. Its cultural significance outweighs its short years of active duty, however, and it was placed on the list of Colorado's Most Endangered Places. The five-storey structure became the focus of stabilisation work, and it is now on both the state and national registers of historic places. Idarado contributed to the stabilisation work and passed ownership of the mill to San Miguel County.

The Idarado Legacy Project created 37 houses in the upper San Miguel Valley to help fund community recreation projects and donated land for Telluride's Pandora Water Treatment Plant, using the Pandora Mill water infrastructure to supply drinking water. It also gave access to the Telluride Via Ferrata and Bridal Veil Trail, both important to local tourism.

Today, jeep tours take visitors through the historic mining areas, and golfers, backcountry hikers, mountain bikers, ice and rock climbers, base-jumpers, paragliders and fishermen frequent the landscape. Since the 1970s, skiing has become a large part of the economy. Culturally, Telluride has changed almost beyond recognition. Now over 50 years old, the Telluride Film Festival has grown to become one of the country's most popular – and laid-back – film events, frequented by Oscar hopefuls every year.

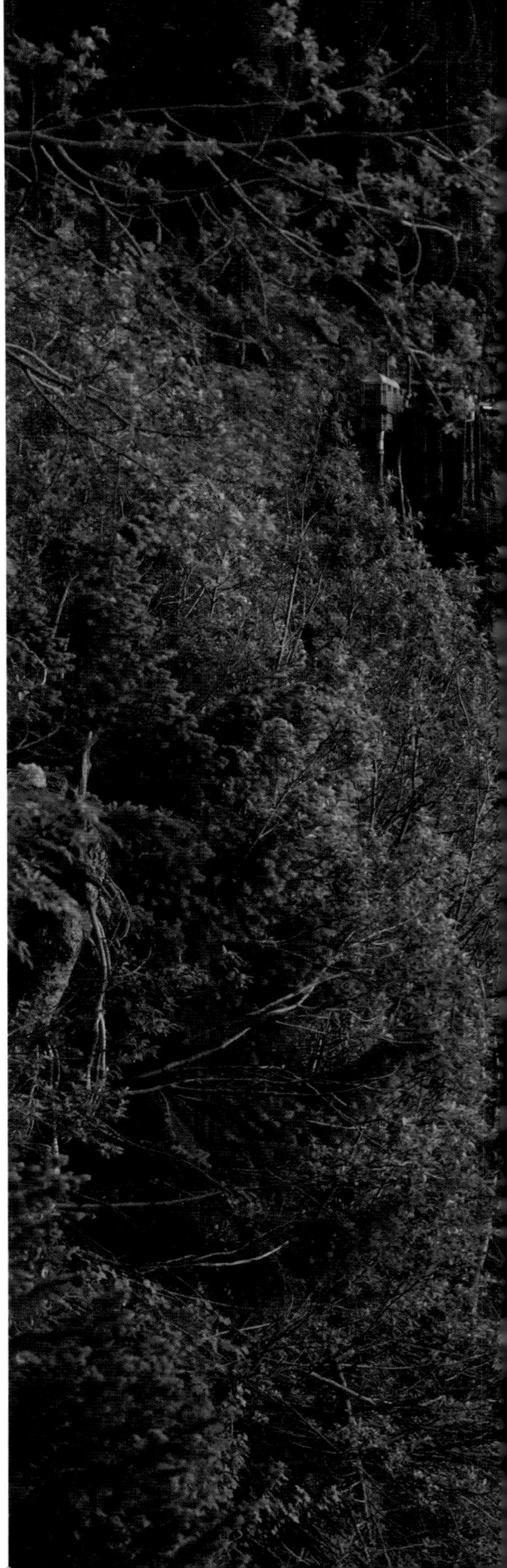

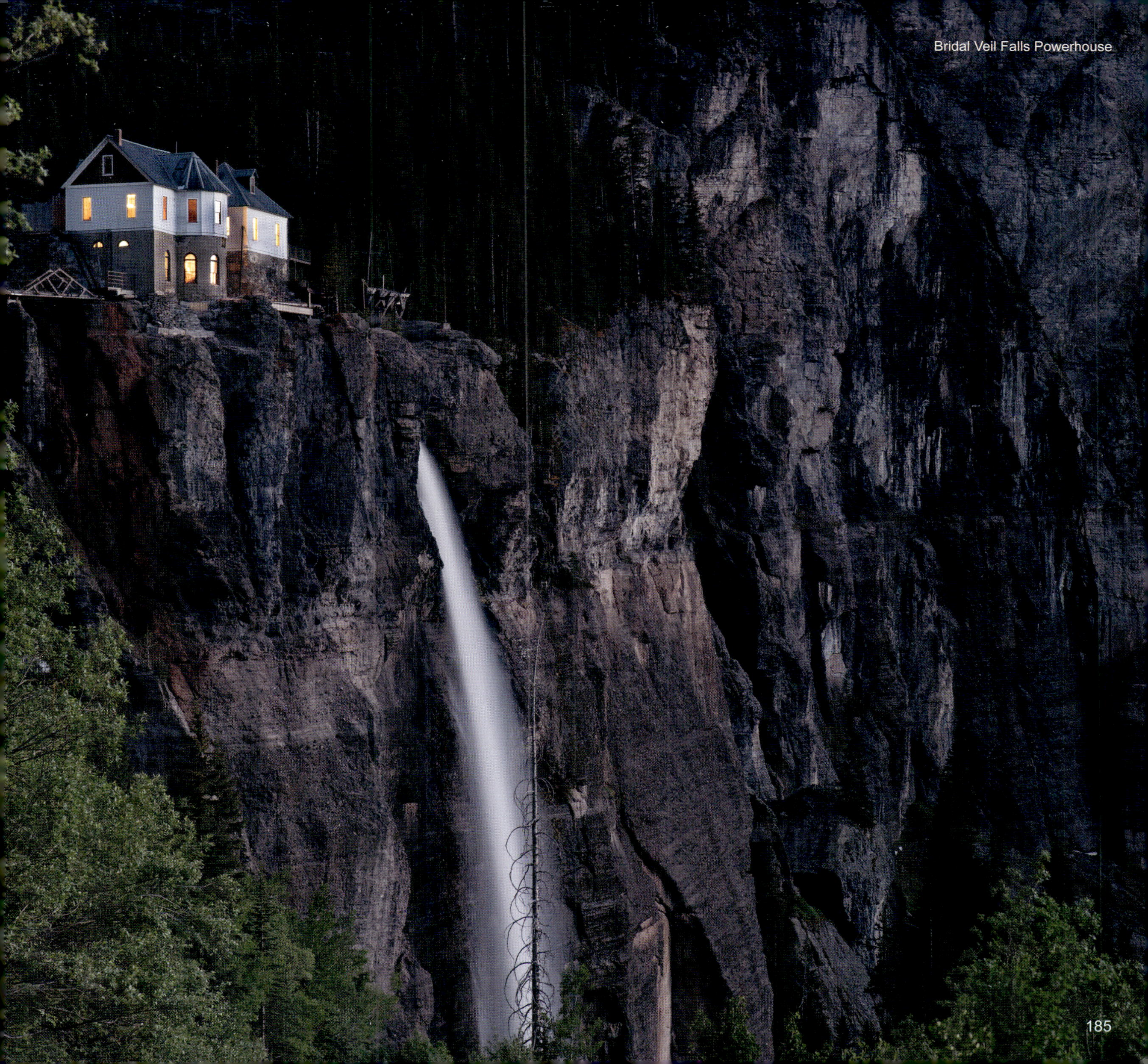
Bridal Veil Falls Powerhouse

82
Mettiki Fish Hatchery
USA

Secreted away amongst the Appalachian hills and forests of Western Maryland's Garrett County coal country is a white, plastic-skinned structure filled with circular blue tanks of bubbling freshwater teeming with young rainbow trout. They are growing in treated mine water at the Mettiki Hatchery, from where they will eventually populate Maryland's rivers for recreational fishing – a big business.

Mettiki Coal operates a large mining complex near Oakland and must treat its mine water before discharge. The water is alkaline with low concentrations of potentially toxic elements but with elevated levels of dissolved iron and manganese. It's a sensitive water environment due to its proximity to the eastern continental divide and the headwaters of the Potomac River.

The fish hatchery idea was hatched in the 1990s by a retired Mettiki mine environmental manager and Maryland's Department of Natural Resources. Trials began, initially in nets, then later in tanks. Everyone held their breath until the concept was proved and it became operational.

A disease outbreak caused a rethink and, in 2013, Mettiki Coal and Fishing and Boating Services of Maryland's Department of Natural Resources (DNR) renewed a co-operative agreement and upgraded the facility. The company funded and constructed the new hatchery, while Fishing and Boating Services equipped it with the paraphernalia needed to raise rainbow trout bio-securely and operates and manages the site. It is funded from the sale of DNR 'Trout Stamps', or permits.

Mettiki hatchery was completed in 2014 and uses treated water in a gravity-driven flow-through system, which drastically reduces pumping costs. Nurturing trout takes large amounts of cold water and the minewater ranges between 7 and 15°C are optimal. It is treated by lime dosing, which also raises calcium concentrations to levels ideal for raising trout. Before entering the hatchery, the water is degassed to remove excess nitrogen and carbon dioxide, then re-oxygenated before disinfection with ultraviolet light.

The hatchery receives two deliveries of around 150,000 rainbow trout eggs each year from the federal hatcheries programme. They hatch after two to three weeks. The fry are then moved into two 800-litre tanks to grow to fingerling size (about 15 grammes and the length of a human finger) at about six months. Then, between 80,000 and 90,000 fingerlings are transferred to the nearby Bear Creek Hatchery for their spring river stocking season. The remaining fingerlings are transferred to twelve 3,000-litre tanks to reach 'stockable' juvenile size (about 230 grammes). Each tank holds about 6,000 fish. About 37,000 trout are grown out for nine to ten months. They are then used to stock rivers across an area of 155km^2 for Maryland's 'Put-and-Take' recreational fishing programme.

There used to be several mine water fish hatcheries in Appalachia, but Mettiki is the last one. It is expensive to treat water and raise trout, but the operating coal complex makes it financially viable. Another reason for its success has been the collaborative and positive relationship between the miner and the regulator.

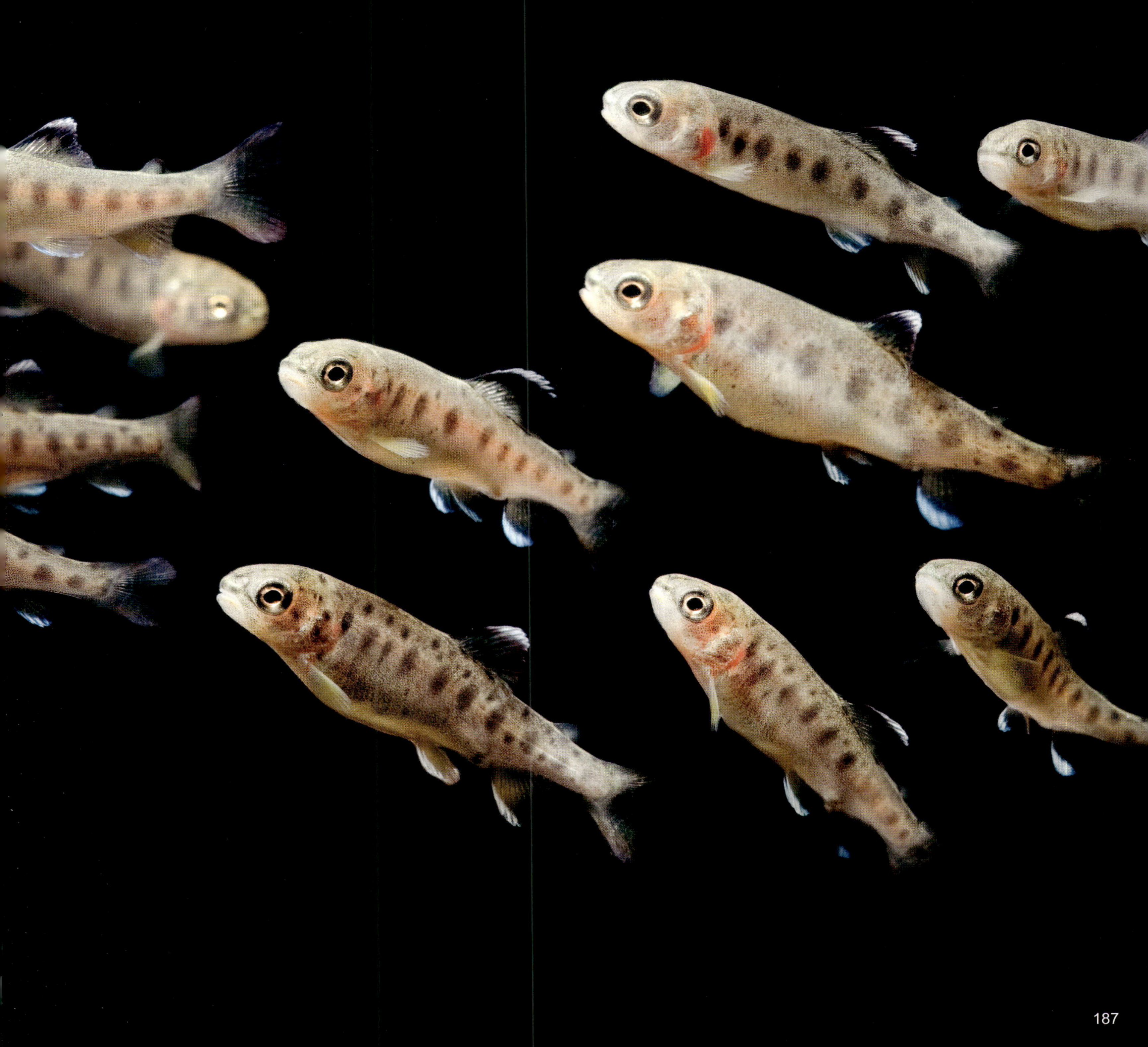

83
Sovereign Hill Living Museum
Australia

Mining museums commonly aim to preserve community origin stories and a sense of place and to support jobs in – often declining – post-mining societies. Less common are those that are economically successful while acting as a cultural glue and supporting hundreds of livelihoods. Sovereign Hill's living museum in Ballarat, conceived by local enthusiasts in the 1960s to preserve the town's historic buildings and to tell its golden origin story, celebrated its 50th anniversary in 2020. It relates the wild story of Ballarat's founding decade, beginning with the 1851 discovery of gold and the subsequent gold rush – one of the world's biggest. Ballarat was the site of the 1854 Eureka Stockade rebellion against the British administration, considered by many as the birth event of Australian democracy.

Sovereign Hill's founders intended more than just an assemblage of old machinery and buildings – they wanted historical accuracy and to purvey 'the atmosphere of the gold fields'. They set up the non-profit Sovereign Hill Museums Association and the museum was 'born in 1970 with a rebellious spirit and an innovative vision' at a time of growing national interest in settler history.

The 25-hectare museum complex on Ballarat's Sovereign Hill is closely linked to old gold diggings. On entry, a meandering route passes first through an early gold rush, grubby tent township aside a stream where you can pan for real gold. Then, it runs past the weathered wooden structures of the recreated Ballarat Red Hill Mine, where Cornish miners discovered the world's second-largest gold nugget. Known as the Welcome Nugget, it weighed close to 70 kilogrammes.

Up the hill, more substantial wooden buildings with corrugated iron roofs denote the community's progression as you approach the Gold Smelting Works where you can experience real gold being poured. Towering over these is the Sovereign Quartz Mine's headframe, next to old 'mullock' (waste rock) heaps. Here, you can take an underground tour or stay on the surface and marvel at its steam-powered machinery. 'Trapped' is a recent exhibit that relates the story of the 1882 New Australasian gold mine disaster in which 22 miners died. A right turn onto Main Street brings finer wooden buildings with ornate frontages and large shop windows, a pub, a hotel, a theatre, a bowling alley, a bank, blacksmiths and more. Behind these is a district of small houses with vegetable and decorative gardens and outdoor 'dunnies' surrounded by rough picket fences.

The collection of more than 60 recreated and relocated historic buildings feel authentic rather than pastiche, thanks to their exquisite attention to detail. They form the set for a multi-sensory experience with sounds of musket firing, excited children, horse-drawn carriages, music, and apothecary perfumes wafting into the street. In a touch of theatre, the sensory experiences are animated by actors in period costume with an infectious enthusiasm for the place and the tales they tell and a willingness to answer questions and pose for photographs. Schoolchildren and their teachers in 1850s' attire also enliven the site as they crocodile along the streets on organised school visits. To avoid taking authenticity too far, the experience had to be sanitised to avoid discouraging visitors with the reality of clagging thick mud, human and animal effluent and piles of rotting sheep offal from the 1,000 or so sheep butchered each day for meat!

AURA, a popular sound and light spectacular, attracts crowds after dark, projecting stories of the Wadawurrung creation, the Eureka Stockade and gritty tales of gold onto buildings as an innovative moving theatre across the site. There are also seasonal events, such as the Winter Wonderlights festival.

Sovereign Hill has received numerous national and state tourism awards over the decades. The much-loved visitor destination attracts several hundred thousand visitors per year, including thousands of schoolchildren on immersive learning experiences connected to the curriculum.

It employs over 350 workers plus over 200 volunteers, and generates hundreds of millions of dollars for the state. Key to its success has been the proximity of Melbourne, just 90 minutes' drive away, enabling day trips from Victoria's capital.

In its 50th anniversary year, Sovereign Hill launched its 20-year master plan,

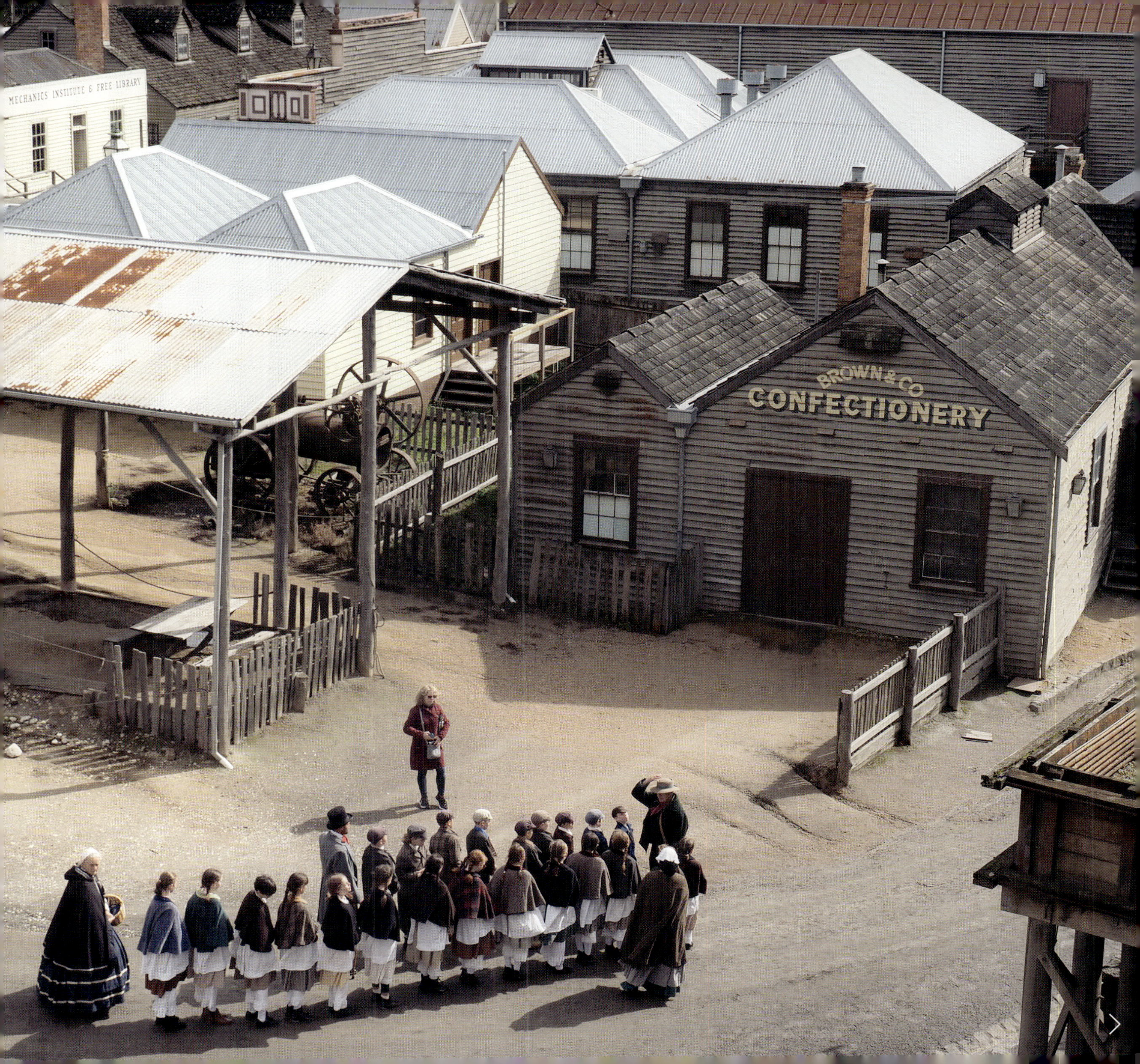

MECHANICS INSTITUTE & FREE LIBRARY
BROWN & CO
CONFECTIONERY

supported by the Federal Government's National Tourism Icons Program, with consequent impact on the state economy and employment. The museum recently launched two new centres for deep learning: the Australian Centre for Rare Arts and Forgotten Trades (CRAFT), dedicated to preserving and protecting the skills of the past; and the Australian Centre for Gold Rush Collections, with its archives of 150,000 gold rush artefacts.

Crucially, it is also evolving stories in a dedicated new centre about the impacts of the gold rush and its aftermath on the Traditional Custodians of the land, the Wadawurrung people, and is redeveloping the Chinese Protectorate Camp to better relate the gold rush story of Chinese immigrants.

As a living museum that explores, educates and entertains, Sovereign Hill is far more than a theme park, with its curation of history and artefacts, where education and research are central. Even though it's recreational, it's learning by osmosis. It feels authentic and invites visitors to step subconsciously into the muddy shoes of an 1855 Chinese miner, or inhabit a draughty run-down shack in poverty, or work in a sweet shop in an early Victorian town.

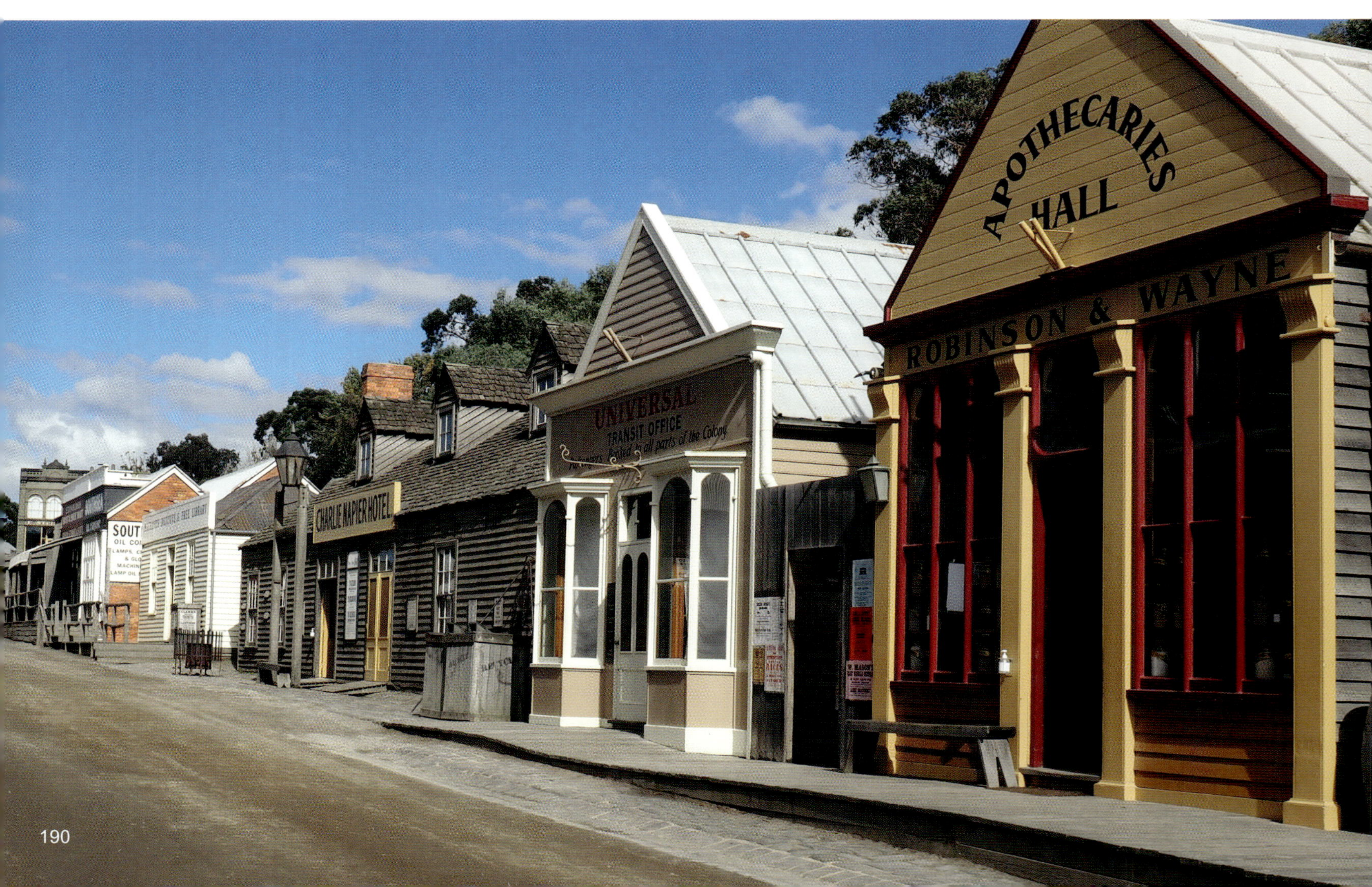

84

Quarry Karts

UK

The gaping grey void of Penrhyn Quarry in Snowdonia, north Wales, was once the largest slate quarry in the world. At its peak, it employed nearly 3,000 quarrymen (and women) and still produces a small amount of high-quality slate today.

But its main draw these days is Zip World's adrenalin-fuelled recreation centre (if that doesn't sound like your thing, you can just watch other people scream while you sip a coffee!). Zip World has leased the flooded quarry and its surroundings to deliver high-quality, high-speed experiences using some of its original infrastructure.

Gravity is the accelerator for the three-wheeler downhill carts. Thankfully, their soft, fat tyres tightly grip the loose slate surface – with only the occasional heart-wrenching skid – as you descend three kilometres at speed from the high edge of the pit, along quarry roads and through tunnels and chicanes and tight hairpins. The most challenging obstacle is the person in front forcing you into the tricky choice of the right moment to undertake!

Your velocity means there is little incentive to take your eyes off the road to gaze at the serene vistas of the surrounding Eryri National Park. You may also hear screams other than your own when you pass beneath one of Penrhyn Quarry's other main attractions – the UK's longest zipline and the world's fastest.

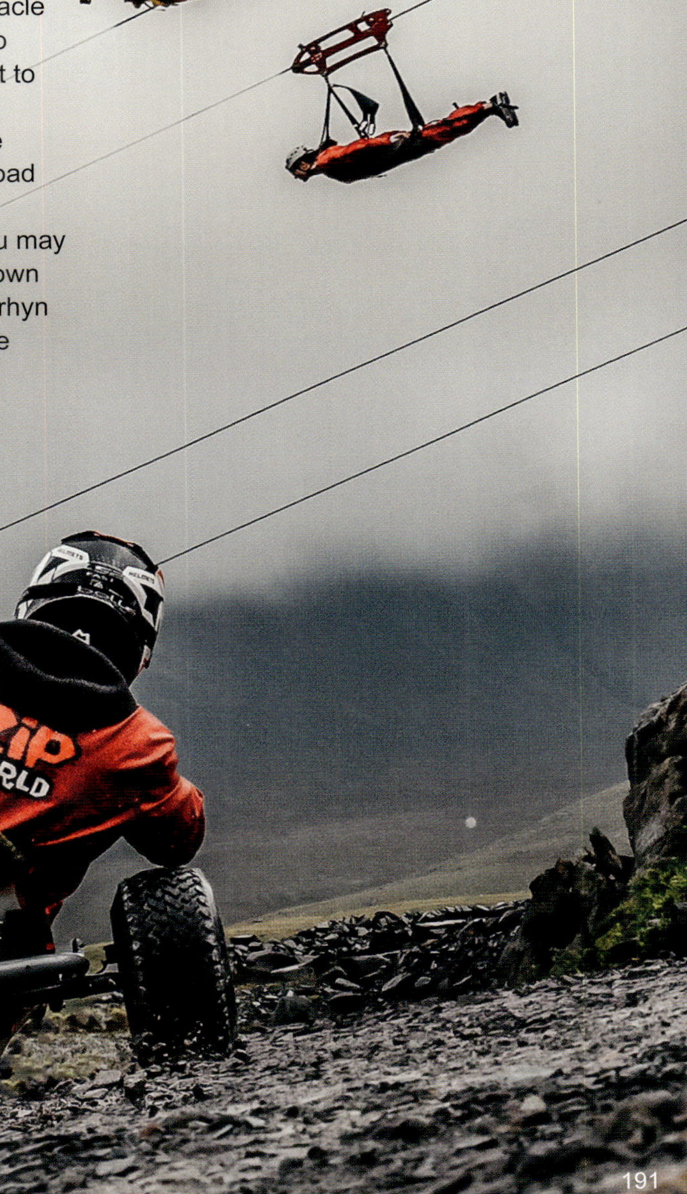

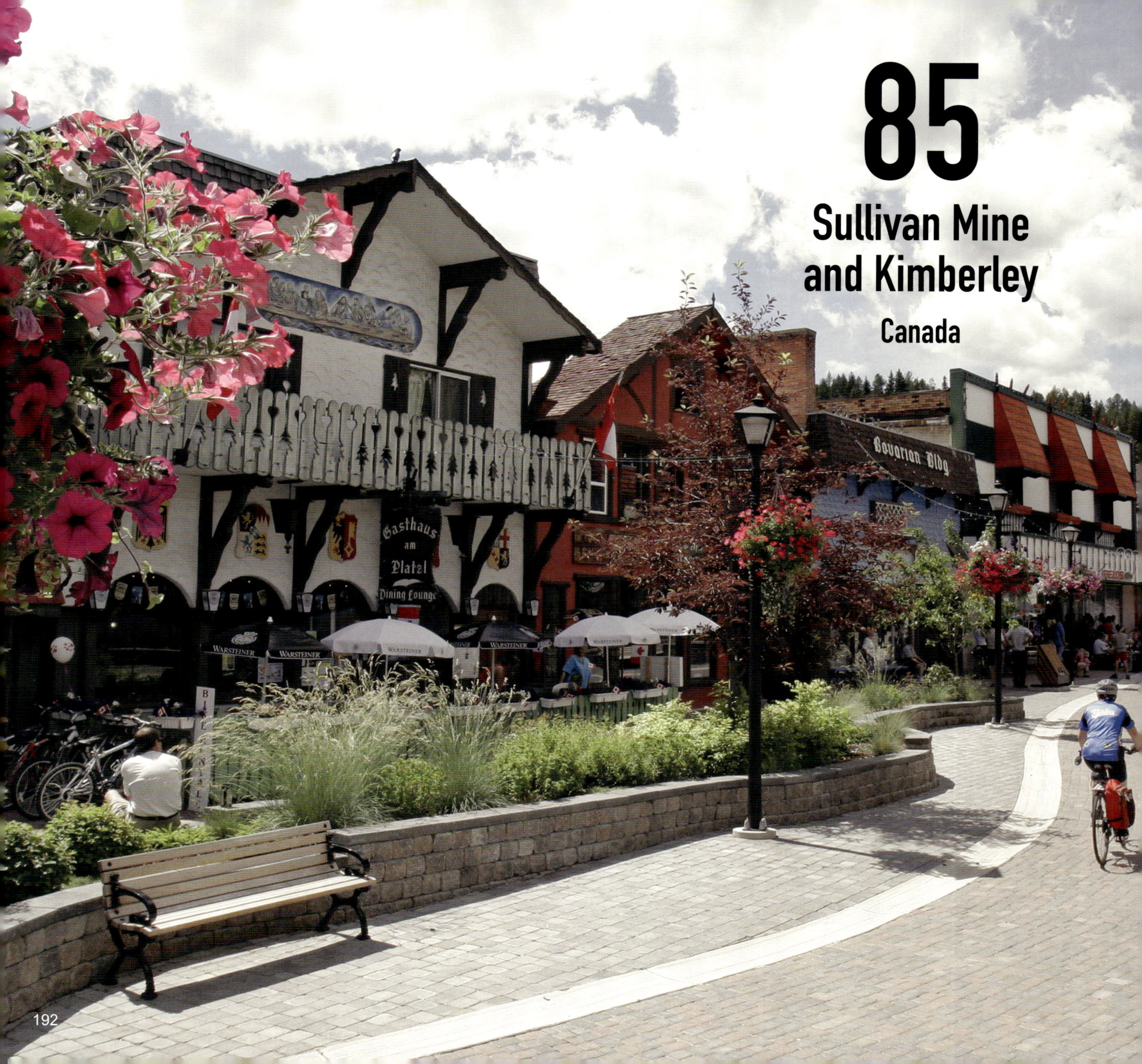
85
Sullivan Mine
and Kimberley
Canada

The former mining community of Kimberley, British Columbia, was built to serve Teck's giant Sullivan lead and zinc mine. It is Canada's highest city at well over 1,000 metres, nestled between the picturesque Purcell and Rocky Mountain chains of the East Kootenays.

In the 1970s, Kimberley's long-serving mayor, Jim Ogilvie, realised the mine would not last forever, so the town needed to diversify away from its single-industry focus. Kimberley sits conveniently on the main highway through BC's southern Rockies, but few motorists had reason to stop; so, to entice them, Ogilvie ambitiously drove the regeneration of the downtown plaza into a faux Bavarian *Platzl* (after Munich's public square). The *Platzl's* pastiche of ornate wooden facades, pretzels and beer steins was home to the largest free-standing cuckoo clock in the world! And it worked, creating the foundation of today's year-round tourism and priming the community for further regeneration when the mine closed in 2001.

Sullivan Mine was one of the world's largest underground mines. Over 90-plus years its 500 kilometres of tunnels produced 26 million tonnes of lead, zinc and silver concentrates worth more than CAD 20 billion, easily making it Kimberley's largest employer and tax contributor. Its 2001 closure was announced in 1990, after which planning began in earnest. Kimberley's strong leadership and Teck's collaborative approach were fertile ground for new ideas to address the city's future economic viability and the mine's environmental legacy. Comprising government bodies, the Ktunaxa Kinbasket Tribal Council,

the steelworkers' union, Kimberley councillors, the public, and the East Kootenay Environmental Society, the Sullivan Mine Public Liaison Committee (SMPLC) was set up as the communication pathway between mine and community during the closure planning and implementation activities.

Teck spent over CAD 100 million on Sullivan's closure activities, transforming the site to knit almost seamlessly into its surrounding mountain and valley landscape. After demolishing buildings and reprofiling the site, waste rock and tailings were progressively reclaimed to forest and rangeland. There was no soil, so 1,100 hectares were capped, then covered with glacial till (unsorted glacial sediments left behind when the ice melted) and planted with an agronomic sward mix. This treatment combination aimed to reduce erosion and minimise acid generation. Promoting wildlife was a closure aim for the tailings facility, so islands of thicker till were created for planting native trees on the tailings impoundment. Today, a 300-strong herd of elk roams across it and neighbouring land.

The sulphidic geology and its derived mineral wastes generate acid, which has been responsibly managed for decades. Sullivan's mine water has been collected and treated for over 40 years, with major improvements noted since the 1970s. Today, a collection system of 25 pumps and 30 kilometres of pipeline directs mine water to the Drainage Water Treatment Plant, which treats the equivalent of between 400 and 1,200 Olympic swimming pools of water annually. Mine water treatment will be needed in perpetuity. Upgrades and new

technologies are regularly considered, including research into alternative water treatment such as passive biological treatment systems. Teck received the 2020 BC Mine Reclamation Award for Outstanding Reclamation Achievement from the British Columbia Technical and Research Committee on Reclamation for their water management efforts.

As the community continues to develop, and to address groundwater contamination issues, Teck re-established a public liaison and community forum in 2020 to provide regular updates on site activities and closure progress and to seek input for planning Sullivan's ongoing management.

Sullivan's closure represented a local tax shortfall of CAD 2 million – a lot for a small population. Company and city collaborated to create a regeneration vision that capitalised on the natural surroundings involving a year-round resort based on skiing and golf and as a retirement destination. The non-profit Kimberley Community Development Society was set up in 1984 to implement the vision and manage and develop many of the community's new assets.

Known for their famous powder, the Purcells accumulate up to five metres of snow each winter. Skiing was primarily a local pursuit for decades until the necessity for a post-closure economy drove investment into upgrading and expanding North Star Mountain's skiing facilities to create the Kimberley Alpine Resort. Close to the Canadian Rockies International Airport, the resort now boasts 80 runs and five lifts on a geology honeycombed with mine workings.

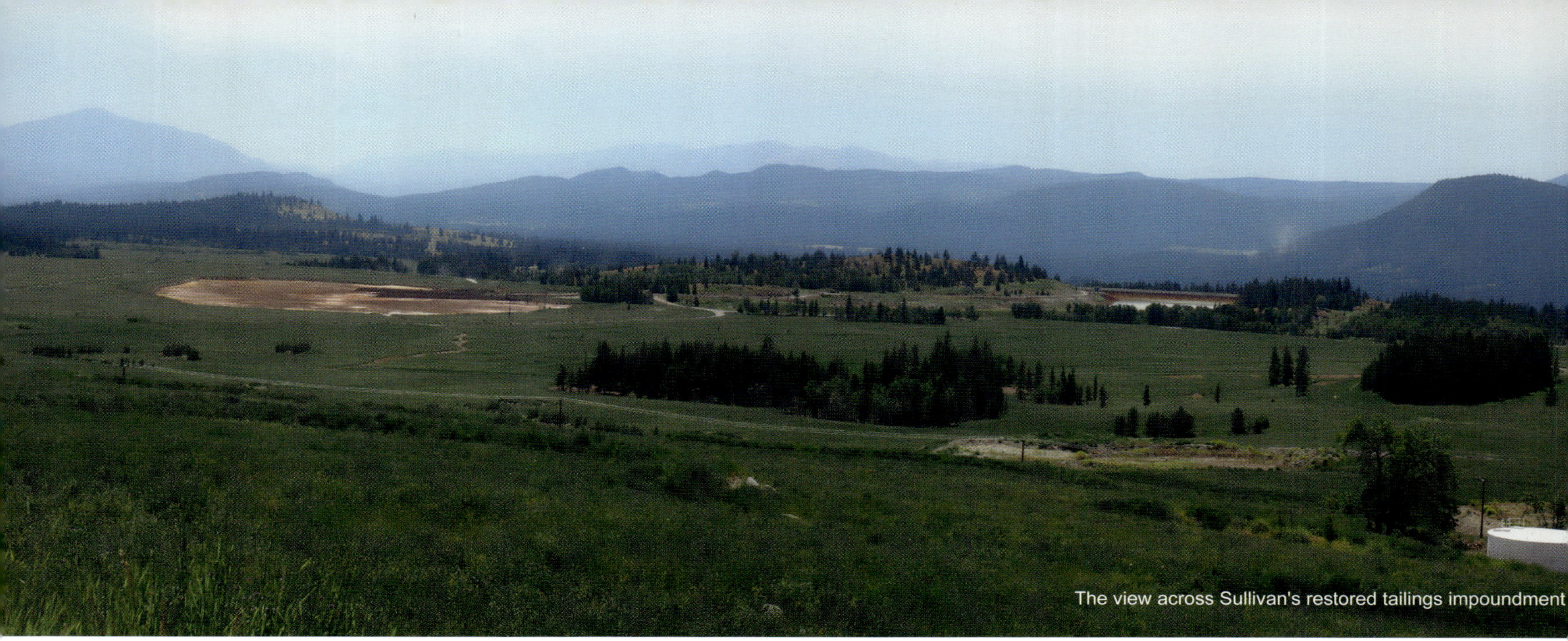

The view across Sullivan's restored tailings impoundment

Kimberley's Underground Mining Railway is a small tourist railway constructed in the mid-1990s to link the *Platzl* and the Alpine Resort. With government and Teck support this evolved into a mining history train tour as a 230-metre tunnel was driven into the mountainside to house the Sullivan Mine Interpretive Centre. Other on-site attractions include the recently restored 1924 mine powerhouse that provided compressed air to the underground operations and power to the electric trains that transported ore, workers and equipment.

For summer tourists, alongside the 'great outdoors', two championship golf courses were developed in the mid-1990s from land that the company sold to the city at below market value, while supporting the renovation of a third. They also arranged a land swap to facilitate the construction of new residential and holiday homes and provide the city with new parkland. The company also created a joint venture with a real estate developer to build holiday homes.

In a first for the province, the city, Teck and renewable energy partners collaborated to develop the 1MW SunMine solar farm where the mine's concentrator used to be. Operational since 2015, its 96 angled solar trackers slowly follow the sun across the sky, generating sufficient power for 275 homes. SunMine provided the city with a long-term revenue source before it was sold to Teck as part of the company's commitment to taking action on climate change.

It is now over 20 years since Sullivan closed. Kimberley has transformed from a mining town to a tourism and retirement community, and unlike many post-mining communities, its 8,000-plus population is growing. The roots of its post-closure success were local leadership, transparency, collaboration and active participation between the key actors. A shared vision meant the city and the company had a unified voice when attracting external investment and support.

From the hillside overlooking the restored site, the revegetated tailings impoundment has a hint of prairie – small tree islands and wood piles support and shelter predators and prey as herds of elk casually saunter across the site and sunbathe by erstwhile mine buildings. Sullivan's successful closure is recognised internationally for its environmental and community renewal, a renewal aided by a community primed for transition since the 1970s because of a visionary mayor with a post-mining dream for Kimberley.

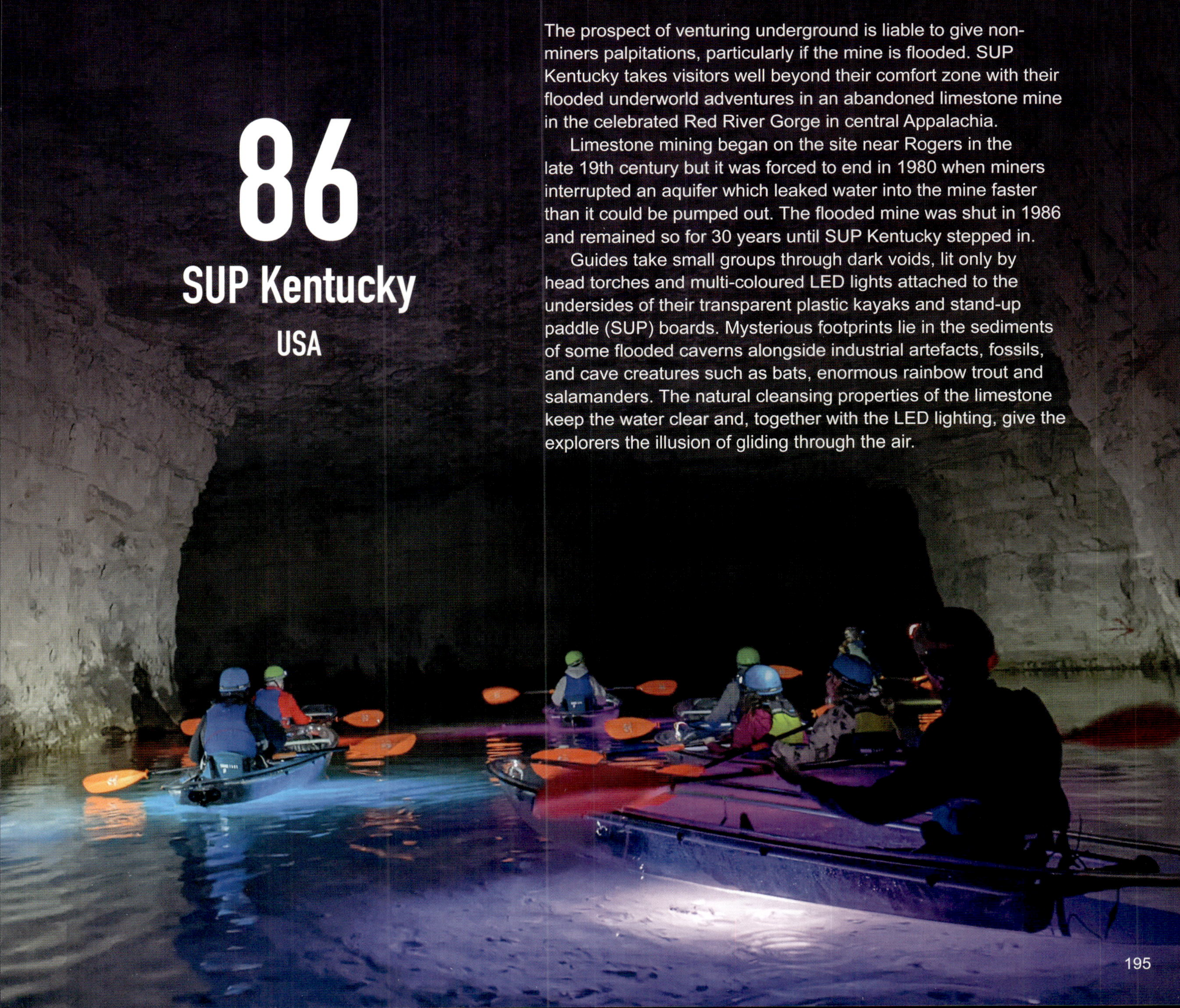

86
SUP Kentucky
USA

The prospect of venturing underground is liable to give non-miners palpitations, particularly if the mine is flooded. SUP Kentucky takes visitors well beyond their comfort zone with their flooded underworld adventures in an abandoned limestone mine in the celebrated Red River Gorge in central Appalachia.

Limestone mining began on the site near Rogers in the late 19th century but it was forced to end in 1980 when miners interrupted an aquifer which leaked water into the mine faster than it could be pumped out. The flooded mine was shut in 1986 and remained so for 30 years until SUP Kentucky stepped in.

Guides take small groups through dark voids, lit only by head torches and multi-coloured LED lights attached to the undersides of their transparent plastic kayaks and stand-up paddle (SUP) boards. Mysterious footprints lie in the sediments of some flooded caverns alongside industrial artefacts, fossils, and cave creatures such as bats, enormous rainbow trout and salamanders. The natural cleansing properties of the limestone keep the water clear and, together with the LED lighting, give the explorers the illusion of gliding through the air.

87

Wieliczka Salt Mine

Poland

Exploring Wieliczka's underworld is to travel through centuries of – often turbulent – Polish history. Surface salt production from brine began here in Neolithic times and the salt mine has produced salt continuously for over 700 years, doubling as a tourist attraction for 500. In its heyday, it underwrote much of the Polish economy; its cultural significance is profound, and it continues to generate millions of Euros for the local economy through tourism.

Wieliczka's extraordinary labyrinth of tunnels and chambers is a testament to the ingenuity and creativity of countless generations of miners. The 14th-century ruler King Casimir III the Great was instrumental in its development, and it generated a third of the royal treasury's income – the salt was referred to as 'white gold'. By the 1500s, the mine was one of Europe's largest enterprises. Tourists first arrived in the 15th century, including Renaissance genius Nicolaus Copernicus, followed later by Goethe and Pope John Paul II.

The mine's story is intricately linked to the waxing and waning of Poland's national fortunes over the centuries, not least the 20th century.

The communist government of the mid-20th century prioritised salt mining over tourism and destabilising parts of the subterranean tourism infrastructure.

Tourism became increasingly important again in the 1970s, as did awareness of Wieliczka's cultural significance and the need to preserve the mine. In 1978, it was listed as a UNESCO World Heritage Site and as a national Historic Monument in 1994. Protective efforts were intensified during the 1990s and salt mining ended in 1996.

Extending to a depth of 327 metres over nine levels, Wieliczka's 245km of underground workings encompass 2,400 excavated chambers and were accessed by 26 shafts dug during the mine's life. Today around two kilometres of tunnels and 20 chambers are open to visitors. Almost two million people visited in 2019, making it probably the world's most-visited underground mine.

The visitor finds the air clean and the temperature stable. Surprisingly, the rocky environs are a waxy, lead grey. Glistening white crystalline salt grows from thin cracks in the grey, often coalescing to form 'salt cauliflowers', or coating supporting timbers, or dangling as thin stalactites overhead. Huge chambers open up periodically, lifting your gaze from the tunnels' confines. Some are filled with many forests' worth of majestic, angular wooden structures to prevent the ground above you from collapsing.

There is a lot going on in Wieliczka's underworld. Guided tours include: the Tourist Route, the Miners' Route and the Pilgrims' Route. The first takes visitors

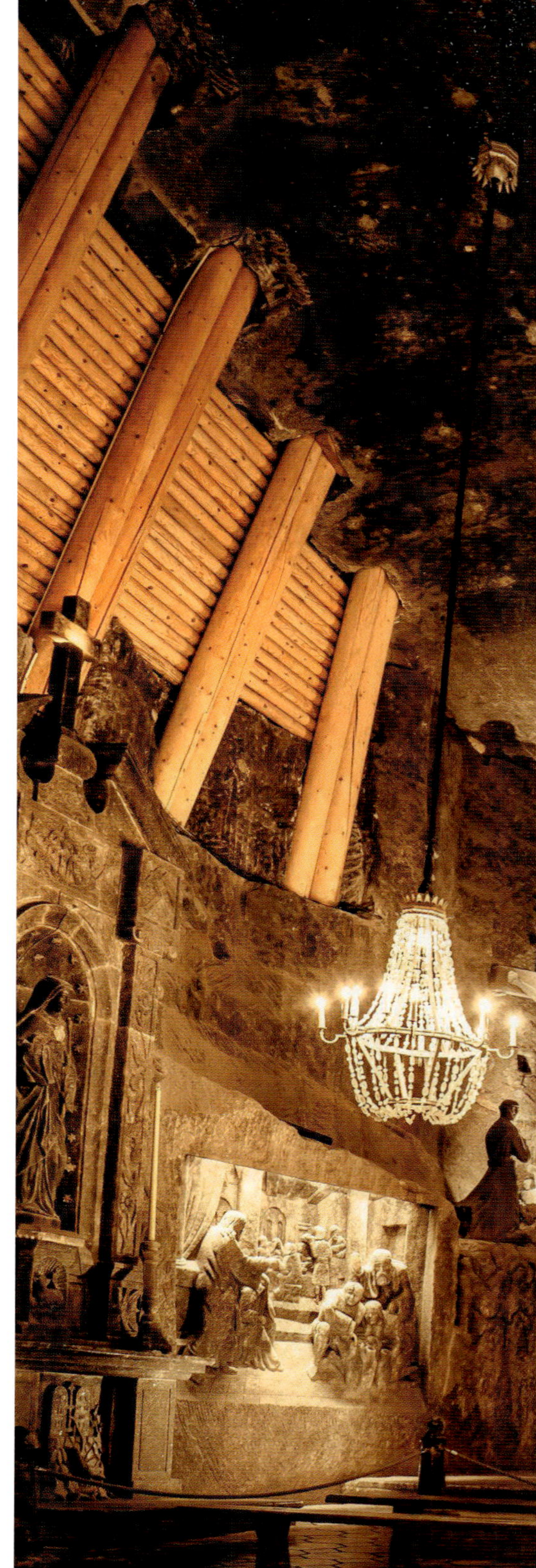

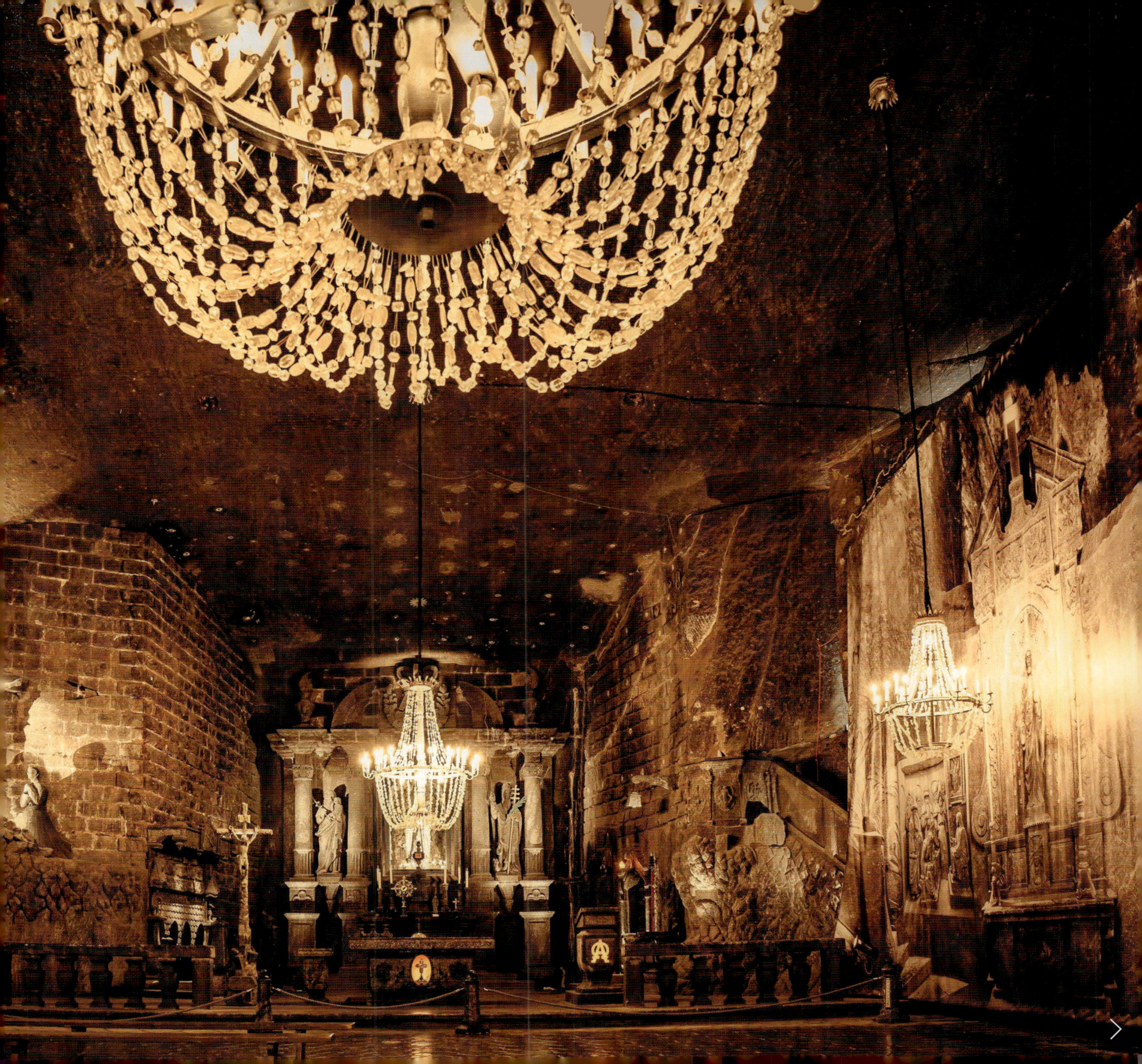

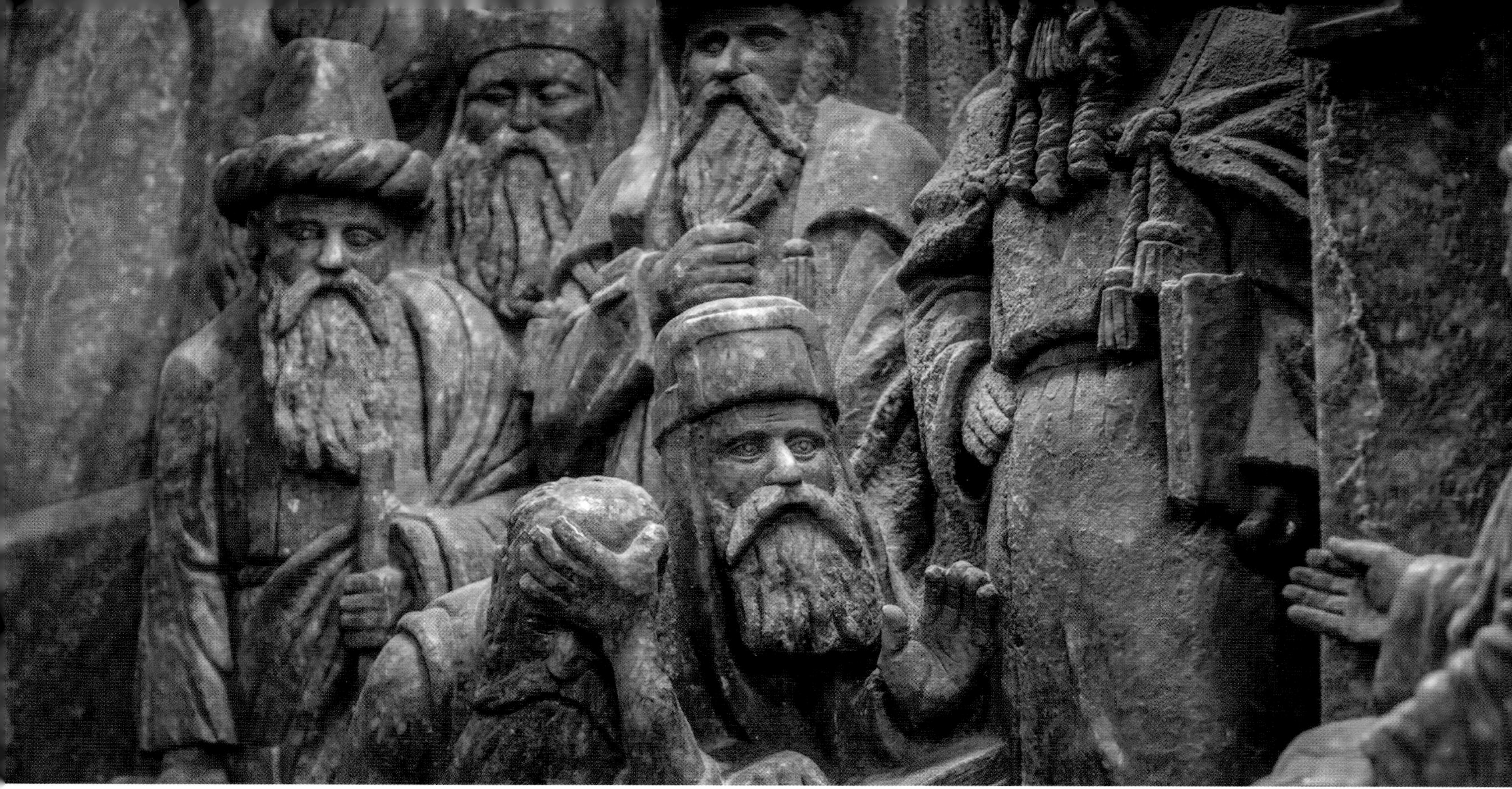

through a subterranean landscape inhabited by life-size rock salt figures and mannequins in period costume and industrial artefacts that relate how salt was mined. Other rock-salt sculptures carved in intricate detail by miners long ago document the roots of Polish culture, history, kings and queens. The scale of the chambers is breathtaking, and each is unique. They have been repurposed as chapels, event venues, function rooms, a health centre, restaurant and souvenir shops. You can even spend the night in the bunk room.

The Miners' Route offers an underground tour of 'education and adventure' where 'every visitor becomes a novice miner'. As well as narration of the place's history, geology and mining practices, visitors carry out some of the mining tasks of old.

The Pilgrims' Route joins the many underground religious places and artefacts. The old miners were deeply religious and sculpted many of the religious statues, reliefs and chapels in their own time, creating more than 40 places of worship. A highlight is the stunning St Kinga's Chapel – its grey walls, floor and ceiling are brightly illuminated by the white light of diamantine salt crystal chandeliers. Inspired by St Kinga, the patron saint of salt miners, it was hewn from the rock salt between 1896 and 1963 and is Europe's largest underground place of worship.

In the 1800s, the restorative properties of salt water and dry air were used to treat everything from a runny nose to infertility and from hysteria to 'failures resulting from excesses in love'. The treatments died along with the mine's physician, but Wieliczka re-emerged a century later as an underground sanatorium for treating respiratory conditions and some allergies. Built around a saline underground lake, today's Wieliczka Salt Mine Health Resort offers 'subterraneotherapy' treatments for respiratory problems, taking advantage of its microclimate and air quality.

On the surface, the saltworks still produces tonnes of salt per year from brine pumped from underground. Pumping out the water also prevents the dissolution of the underground workings.

The Graduation Tower, close to the shaft building from where visitors descend below ground, is an observation tower coated in dark blackthorn branches dripping with salt-laden water, which flows in channels around the site and is sprayed as a mist, marketed as a restorative and relaxing experience.

The salt mine is owned by the government. Approximately 350 employees run the visitor destination. Additionally, although salt is no longer mined, hundreds of miners still work underground behind the scenes. They ensure visitor safety, protect historic areas, manage water leaks, and stabilise old workings with timber structures and by backfilling with sand, and engineer spaces for new tourist attractions.

The salt mine is still a substantial operation that affects its neighbouring community of Wieliczka. Public funding over the past 20 years has revitalised the townscape and infrastructure, as the town capitalises on its star attraction, creating its own brand as a 'Spa Town'.

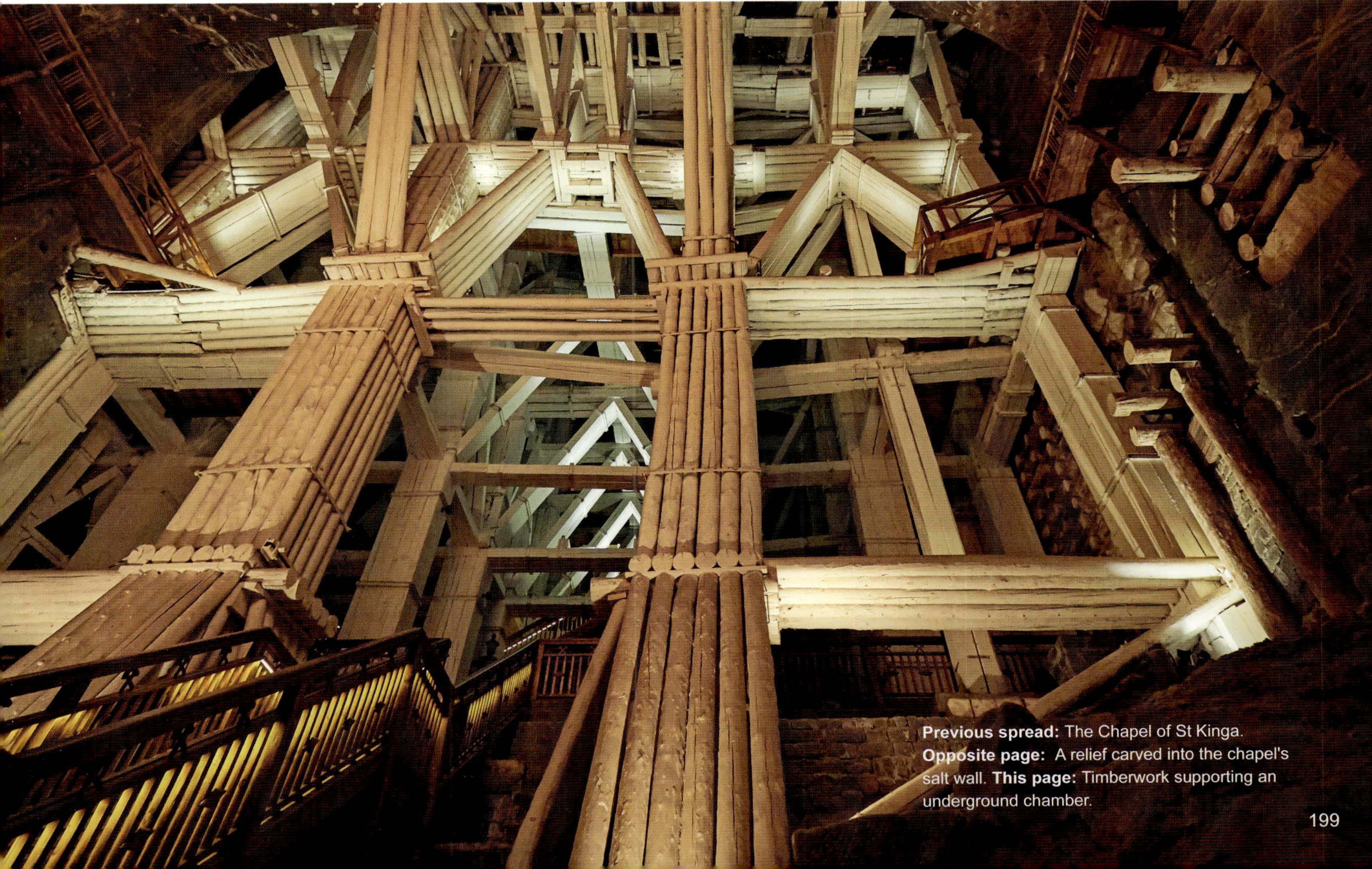

Previous spread: The Chapel of St Kinga.
Opposite page: A relief carved into the chapel's salt wall. **This page:** Timberwork supporting an underground chamber.

88
Woodlawn Eco-Precinct
Australia

Filling mining voids with waste materials is a very common after-use, but Veolia, the multinational energy, water and waste management company, has taken this to the next level with its Eco-Precinct in New South Wales (NSW), Australia.

Woodlawn Mine produced almost 14 million tonnes of copper, lead and zinc ore from its surface and underground operations. It shut down in 1998 after 20 years of operation, and the site was left derelict with significant land contamination and acid drainage. However, its 800-metre diameter, conical open pit with impermeable rock walls was ideal for repurposing as a landfill site.

Veolia conceived a new type of landfill for the site to manage Sydney's municipal waste. Traditionally, landfills used old 'dry tomb' technology to prevent water ingress, which would otherwise percolate through the waste, leaching out contaminants that would then need to be treated. Veolia's 'wet bioreactor' concept enhances leachate recirculation, aiding microbial degradation of the waste to produce methane, a major greenhouse gas, which is captured and used on-site to generate power. Although CO_2 is released from burning methane, the latter is a much more potent greenhouse gas which, without the site's technology, would have been released into the atmosphere. Each tonne of waste sends 1.33MW of electricity to the grid. Veolia persuaded the regulator to approve the bioreactor concept in 2000. (Simultaneously, a change of stewardship was needed because waste management had traditionally been carried out by government, rather than the private sector.) It became operational four years later, an early step in the transition to a circular – and greener – economy.

The 250-kilometre railway line from Sydney offered the advantage of drastically reducing truck movements to other landfill sites, significantly minimising the carbon footprint. Woodlawn handles two trains per day of containerised municipal solid waste, which is then trucked 10km to Woodlawn. Ten trucks tip about 110 containers per day – over a million tonnes per year, equivalent to about 40% of Sydney's putrescible waste. The pit now holds over 13 million tonnes, buried in a way that accounts for settlement and enhances gas production, driving a seven-fold increase in the power generated between 2008 and 2023 from 1MW to 7MW. This equates to 60,000MWh into the grid annually – enough electricity to supply 10,000 homes, rising to 30,000 homes by 2047.

Other activities were developed to take advantage of convenient power, waste heat (from the power plant) and the large landholding, leading to the development of the Woodlawn Eco-Precinct. Waste heat from the power plant supports a horticultural facility and a 3.6-tonne-per-year barramundi fish farm that supplies Canberra's restaurants.

Opened in 2017 at a cost of AUD 100 million, a mechanical-biological treatment facility (MBT) removes metals for recycling and extracts and composts organics from the waste for use in land rehabilitation.

Turned on in 2011, the 23 turbines of the Woodlawn Wind Farm produce enough electricity for 32,000 homes, and since 2019, a 2.5MW array of 2,500 panels tracks the sun across the sky. The solar farm powers on-site activities, making the MBT self-sufficient. Veolia also plans to spend AUD 600 million on on an Advanced Energy Recovery Centre at Woodlawn, providing further power for on-site activities and export to the grid.

The Eco-Precinct's footprint spreads across 6,000 hectares of countryside, including a working beef and sheep farm run as a separate business. The mine site proper is heavily affected by metal contamination and acid drainage, with potentially toxic wastes in the tailings storage facilities and contamination of some of the farmland. As part of Veolia's regulatory agreement, they are committed to rehabilitating the site over the 25-year-life of the bioreactor, including restoring the local ecology and managing habitats.

The MBT's composted organics are used to rehabilitate the old tailings facilities, while a leachate treatment facility handles the toxic acid drainage. From above, managed water bodies in various hues of green look like stained glass windows around the open pit.

These, as well as the tailings facilities and stormwater management systems, are maintained to prevent run-off from the site. Contaminated ground and surface waters are collected and pumped to leachate dams where the water is sucked into powerful evaporators that propel arching plumes of spray into the air producing rainbows as excess water evaporates.

For a rural NSW location, the Eco-Precinct provides welcome employment opportunities. About 100 people, including contractors, currently work at the Eco-Precinct, some of whom used to work in the mine. They lost their jobs without warning when the mine closed. The village of Tarago and neighbouring communities have been the beneficiaries of the Veolia Mulwaree Trust. Since 2005 the Trust has dispensed over AUD 13 million of funds generated by the Eco-Precinct's waste management work to 1,500 community projects. Without Woodlawn's landfill, the mine site might have remained a polluting eyesore, but there are still mineral riches below ground that are the focus of exploration by a separate company. Maybe one day, beneath the bioreactor and Eco-Precinct, there may be another mine, too.

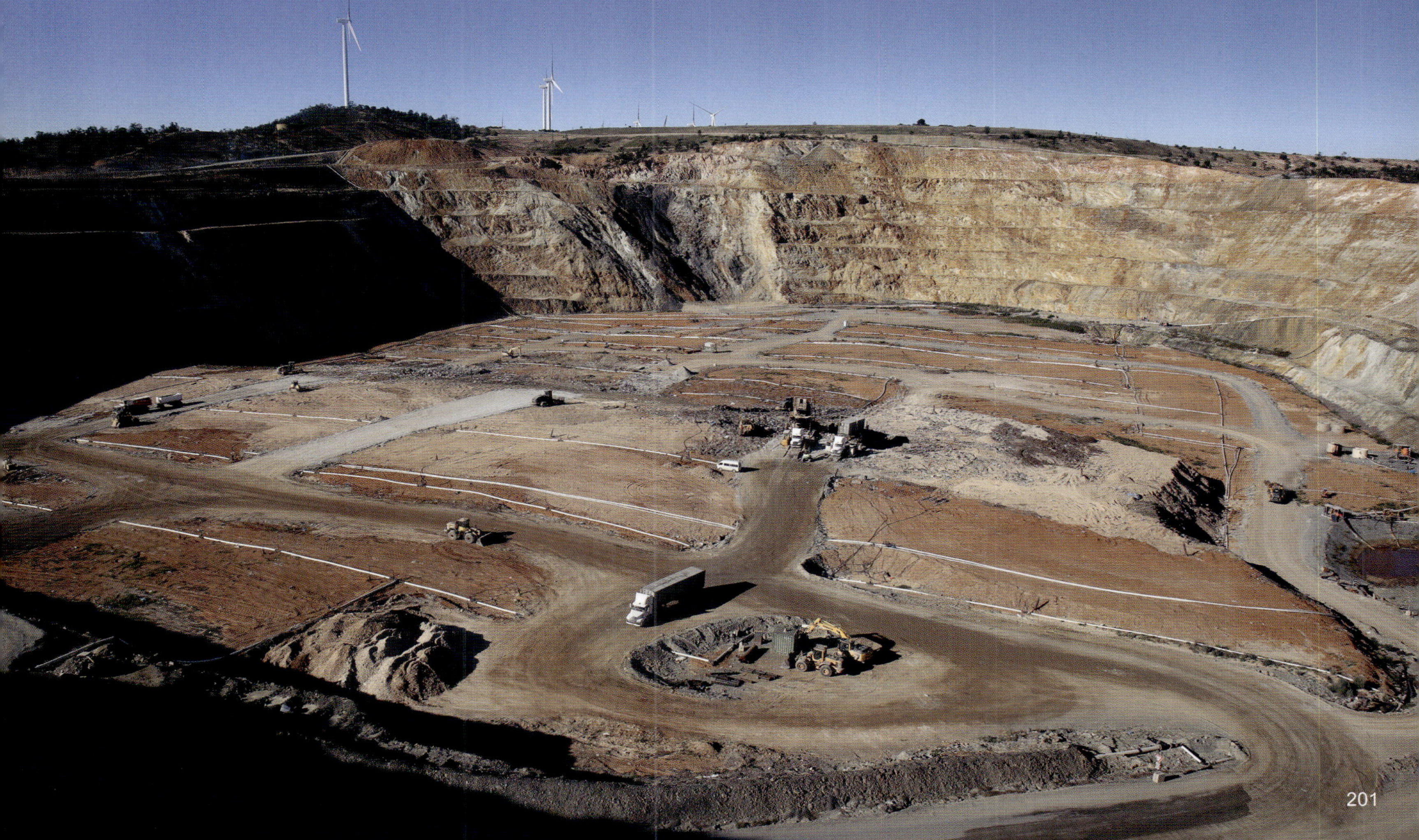

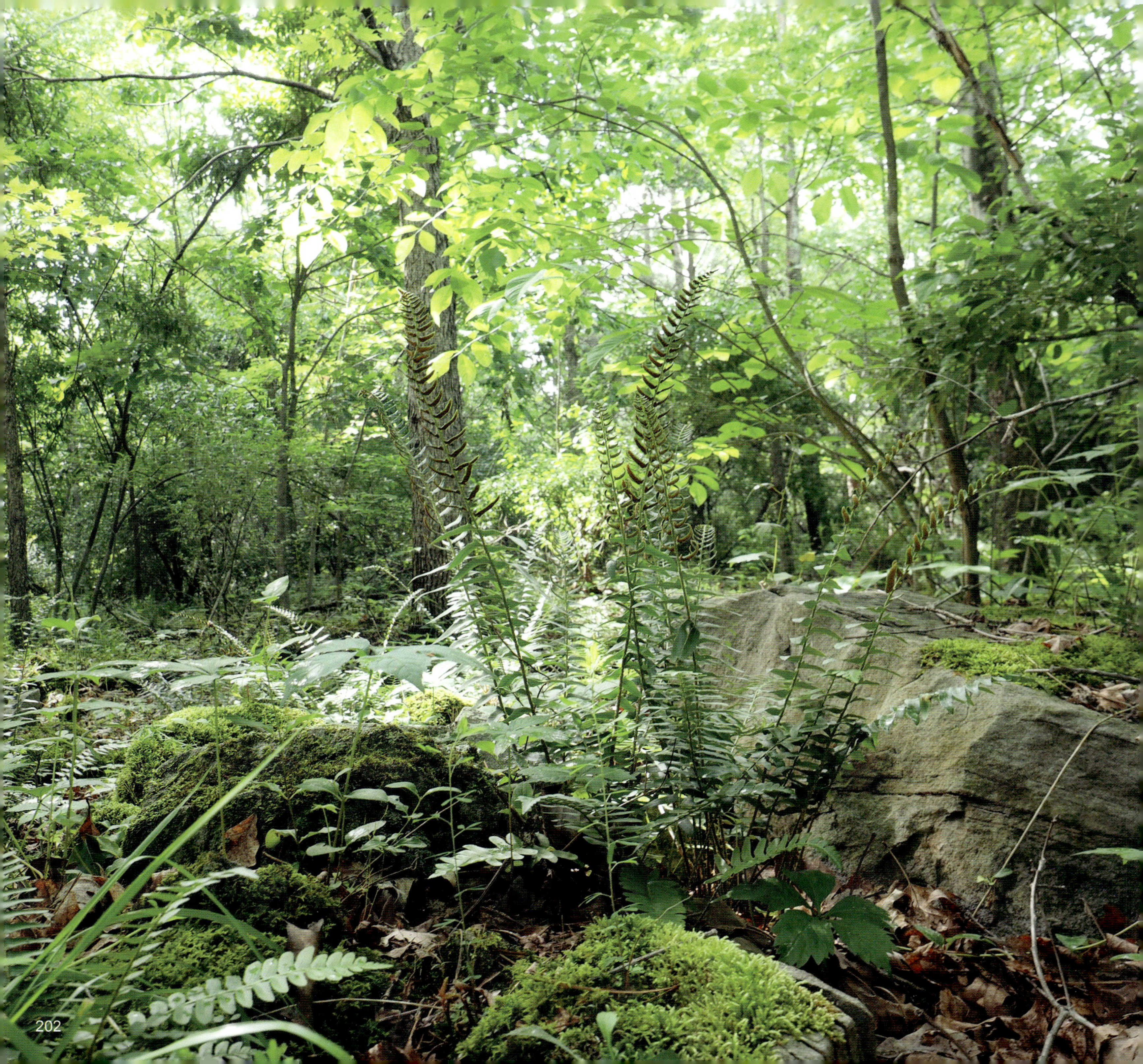

89

Restoring Appalachia's Forests

USA

The forest's cool, dappled shade offers a welcome retreat from the midsummer sun. Its mosses and ferns and tree seedlings growing in shallow but rich, earthy humus over angular, fragmented rock imply an ecosystem in repair. But just 20 years ago, this place near Three Forks, Kentucky, USA, was a vast expanse of compacted land and tough, dense grasses remaining after the mountaintop removal (MTR) coal-mining juggernaut had rolled through. MTR is a highly destructive mining method in which Appalachian forests are stripped and their mountain tops literally blown apart to extract the fossil forests beneath. It is highly controversial for many reasons, not least because of its impacts on surface water courses and the region's precious hardwood forests, which are unique, ancient and among the most biodiverse temperate forests to be found anywhere. They have supported human existence for millennia.

The 1977 federal Surface Mining Control and Reclamation Act (SMCRA) sought to address the worst depredations of MTR mining in which loose mountain rubble (aka overburden) was tipped into neighbouring steep-sided, narrow valleys and left. Although ideal for tree growth, these piles were unstable and hazardous, with obvious risks to downstream communities. SMCRA ended this practice by obligating mine operators to create reclamation plans and pay bonds to restore the land in a way that encourages future land use – usually upland plateaus (flat land is uncommon in these parts), with a quick-growing, herbaceous groundcover to protect the new surface and support a pasture end-use. However, the tough, non-native grasses (which were also a fire risk), combined with compaction, severely impeded natural forest regrowth. The once characteristic forested ridge and valley geography became utterly transformed into a seemingly boundless upland plateau blanketed with a 'biological desert'.

It is the job of the Office of Surface Mining Reclamation and Enforcement (OSM) federal inspectors to assess the reclamation progress till – all being well – the bond is released five years after mining ends. At the turn of the century, OSM inspectors Dr Patrick Angel and Scott Eggerud recognised the shortcomings in the SMCRA-mandated practices they were enforcing and vowed to find a more progressive approach to achieve its forest recovery requirements. Simultaneously, University of Kentucky scientists were raising concerns about poor forest recovery on reclaimed sites and persuaded the OSM that the twin challenges of surface stability and forest recovery could be solved. In response, they formed the Appalachian Regional Reforestation Initiative (ARRI). Its core team of federal and state regulators were housed in the OSM. It aimed to encourage mining companies in Appalachia to use improved techniques for establishing forests on active mining sites and abandoned mine lands. It became a diverse and effective collaboration between federal and state agencies, academia, landowners, civil society groups, and the coal industry across the Appalachian coal states.

ARRI developed a simple, bespoke methodology for overcoming the technical barriers to establishing native hardwood trees on MTR sites. Known as the Forestry Reclamation Approach, the methodology included alleviating compaction by the tipping of loose rock piles, or deep ripping with a robust steel shank mounted on the rear of a bulldozer; planting a groundcover compatible with tree establishment; planting both early successional tree species that enhance soil stability and support wildlife and commercially important trees; and to use proper tree-planting techniques. The team created demonstration sites showcasing the approach and produced advisories aimed at the stakeholders who conduct and influence coal mine reclamation and reforestation practices. Federal and state regulations allowed this approach under certain post-mining land uses and some states, like Kentucky, even widely adopted it. As a consequence, many millions of trees have been planted on mined land otherwise destined to become biological deserts.

ARRI worked initially with active mines, but a further 400,000 hectares of 'legacy mines' (pre-Forestry

Reclamation Approach, post-SMCRA bond-released sites) also needed similar repair. The ARRI team suggested that a non-profit entity would be best placed for raising the necessary funds for restoring native forest on legacy mines. The scale of tree planting necessary, coupled with the need to stimulate related socio-economic opportunities, required a different strategy and a new organisation. Set up as a non-profit tree-planting organisation, Green Forests Work, under the leadership of the University of Kentucky's Professor Chris Barton, grew out of ARRI to take forward the work on these legacy mines.

Green Forests Work activities began with mass volunteer tree-planting events, eventually attracting accolades from the United Nations and start-up funding from the Appalachian Regional Commission. It modified the approach to jump-start natural succession on legacy sites and non-mining habitat enhancements. Its two main aims are to stimulate employment through reforestation and to enhance the environment by eradicating non-native species and restoring ecosystem services. Since its formation in 2009, Green Forests Work has planted over seven million trees on nearly 5,000 hectares of legacy mined land, while the active mining industry – under ARRI – has planted well over 100 million trees on more than 60,000 hectares. Green Forests Work also tracks the number of jobs created, the financial contribution to local communities and measurable benefits to ecosystem services.

There are many organisations concerned with restoring the forests and waterways of Appalachia and collaboration is central to success. The fabled American chestnut once grew in its billions in Appalachia's forests, supporting thousands of generations of people, before being wiped out by a foreign fungus. A resistant hybrid produced by The American Chestnut Foundation is being planted by Green Forests Work by the thousands on formerly mined lands. Over the coming decades, it is hoped that this iconic tree will once again rule the forest.

Elsewhere, the destruction of non-native, mature Norway spruce and red pine once planted on mined land in the Monongahela National Forest's Mower Tract in West Virginia looks, at first sight, like an environmental disaster, but the fresh greens of native trees planted by Green Forests Work between their fallen, grey skeletons show how quickly the indigenous forest can return. The fallen dead provide shelter from herbivores and extreme weather while protecting the soil. Plantings include the red spruce – an important but declining element of this high-elevation mixed forest. Red spruce tree communities support hundreds of at-risk animal and plant species. This post-mining land was identified as a high-priority conservation corridor and is now at the heart of efforts to restore this key Appalachian forest component, which will enable sensitive species to move northwards along it as climate change renders the southern extent of their ranges inhospitable.

Green Forests Work's portfolio of environmental enhancement is increasing beyond purely mine site tree-planting. They are creating wetland areas and small pools to benefit the area's special biodiversity and slowly rebuild pre-mining peatlands, as well as more public access trails and campsites to entice more people into the forests.

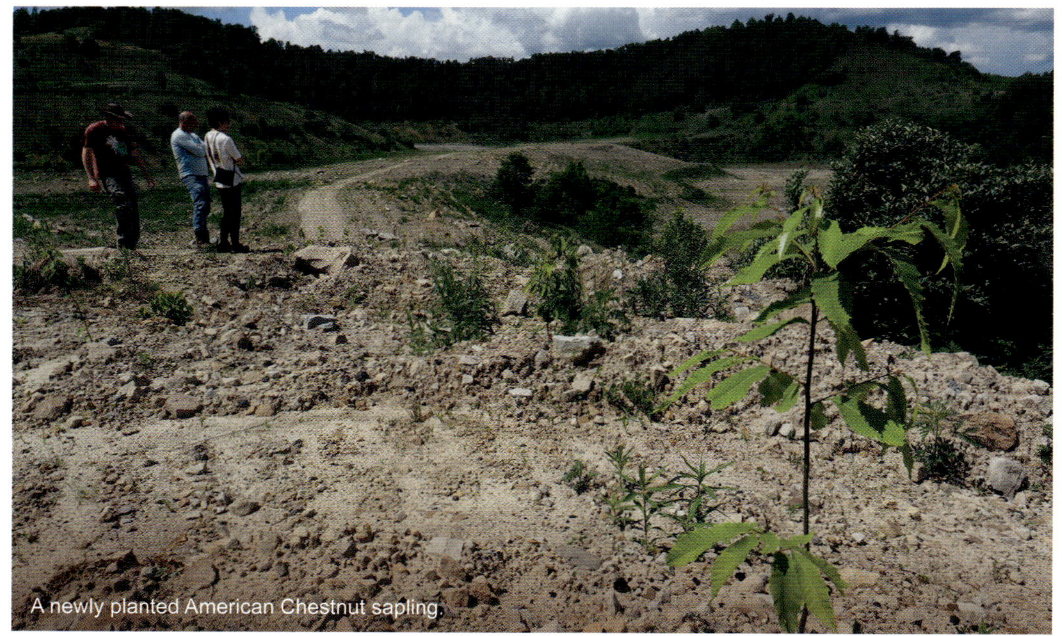
A newly planted American Chestnut sapling.

90
DeepStore
UK

Winsford Rock Salt Mine is the largest and oldest operational mine in the country. Owned by Compass Minerals, it supplies halite – rock salt – that is used for de-icing Britain's roads. As the mining continues, it increases the space for another business to operate: DeepStore, the UK's largest underground storage facility.

In 2003 DeepStore took enough paper records from the National Archives to fill 25km of shelves. The mine's natural ambient conditions of 15°C and 55% humidity make it ideal for storing paper documents like these, while its location means that records are easily retrievable. The National Archives is just one of many clients, and there's plenty of room to grow – it's estimated to be big enough to house all the of the UK's archives in its depths.

It's not the first time the mine has been used to hold government records. In 1940, almost 45,000 volumes were sent by the India Office to what was then known as Meadowbank Salt Mine before being returned to London in 1947.

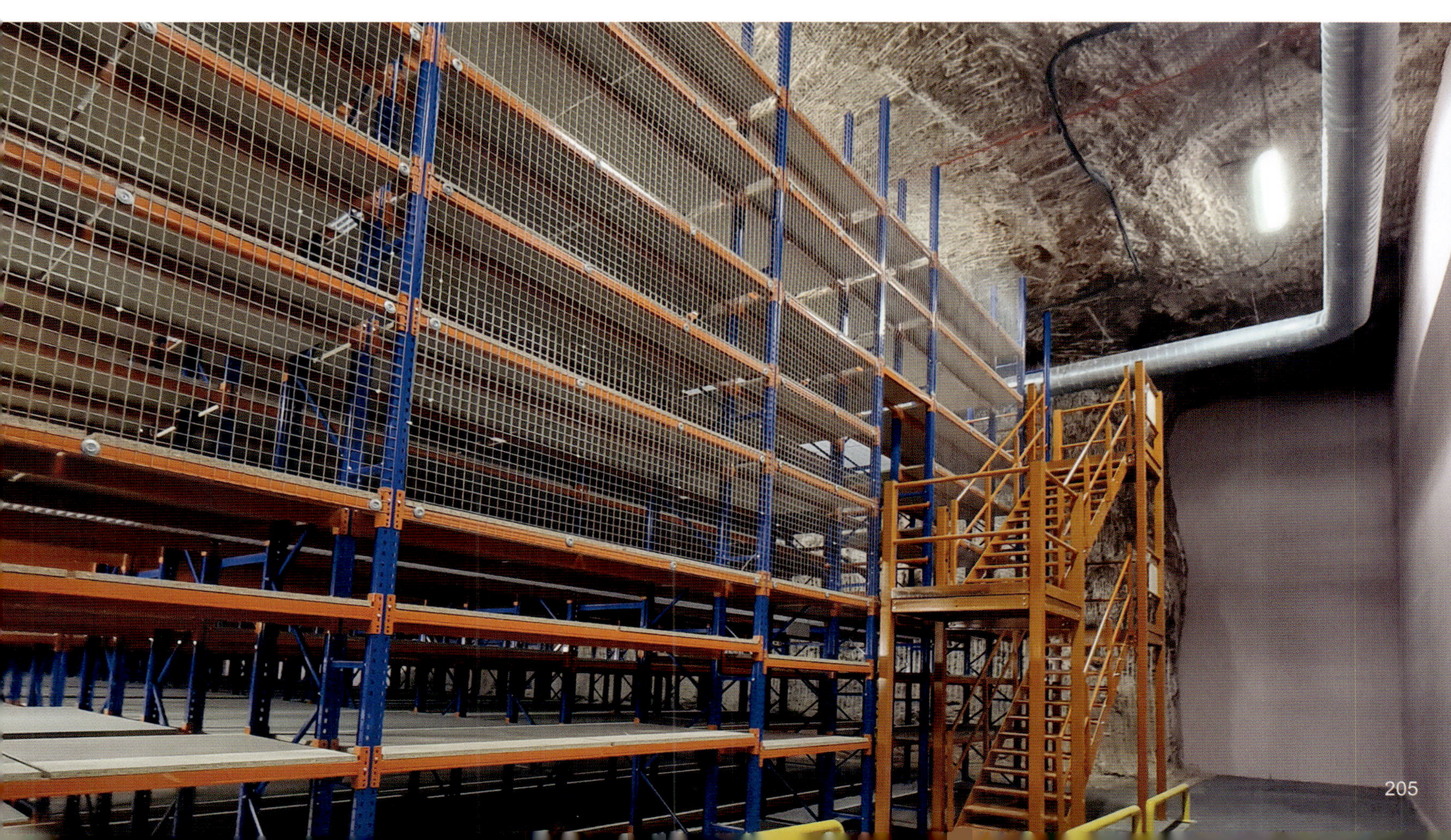

91
Coober Pedy
Australia

It's a long drive through nowhere to reach the Outback town of Coober Pedy, the Opal Capital of the World. The town is nevertheless one of the Outback's more accessible locations and, as well as being a mining centre and a focus for essential services, is also a major stop-off on the Stuart Highway between Adelaide and Alice Springs and a tourist attraction in its own right. Unlike many mining townships, the Opal Capital's subterrain is the very essence of the place. In summer it is searingly hot, so popular underground attractions include 'dugouts' (houses), the museum, hotels, eateries and churches.

Although very ethnically diverse, with over 45 nationalities represented in the population of around 2,500, the region is the traditional land of the Arabana (Ngarabana) people (other traditional groups are associated with some ceremonial sites). The name 'Coober Pedy' derives from Aboriginal languages, meaning roughly 'Whitefella – hole in the ground'. In 1975, the Aboriginal Community started using the name 'Umoona' (meaning long life) for the town, after the umoona or mulga tree, which is common in the area.

About half of Coober Pedy's people live in underground dwellings known colloquially as 'dugouts', mostly burrowed into the low hills at the edge of the plateau rather than repurposed mine workings. The practice is thought to have been imported by First World War soldiers returning from the trenches of France. The shallow geology is sufficiently competent to allow the excavation of large ceiling spans and, when joined to neighbouring underground properties, can form extensive dwellings.

Explosives were easier to come by in the past, so homeowners would expand their living space in the hope of finding opal – sometimes blasting through into the neighbour's dugout! Today, mining is banned in residential areas, but 'expanding the living space' can sometimes be a euphemism for mining…

Opal mining itself is kept small-scale by limiting claim sizes to slightly more than 15m^2. Individual mines might be small, but the cumulative effect is very apparent: there are over 250,000 mine shafts and entrances in the surrounding land, with horizon-filling panoramas of low orange and white molehill-like waste piles across the flat landscape. These are mined using repurposed machinery and vehicles reminiscent of *Mad Max 3*, which was filmed here, as was *Priscilla, Queen of the Desert*.

The town sports a range of underground accommodation for the intrepid visitor, from short-term dugout house rental to bed and breakfasts and hotels. Unlike many, the Comfort Inn Coober Pedy Experience Motel was actually once an underground opal mine which closed in the 1960s. Re-modelling into a motel began in 1990. During later

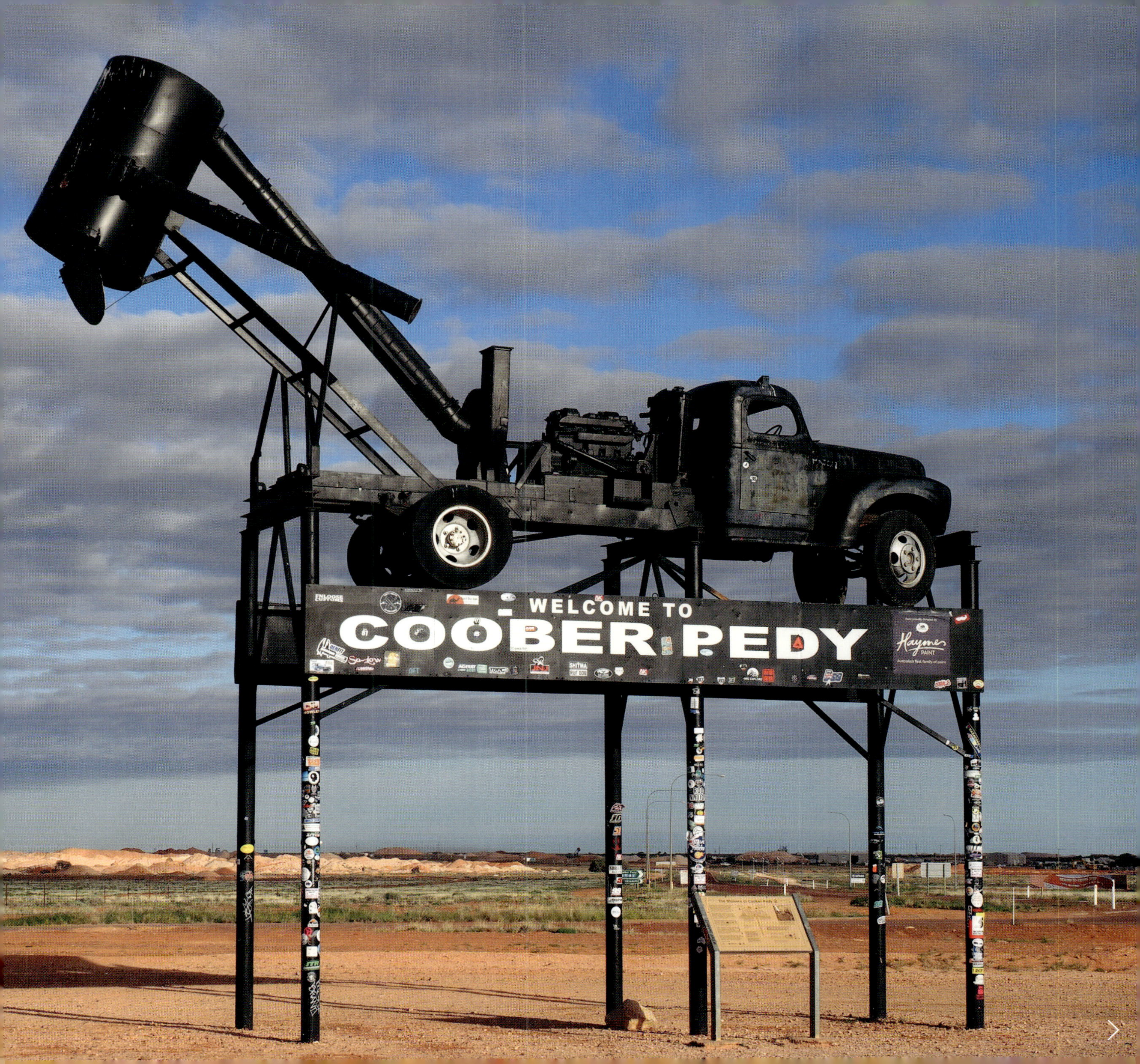

WELCOME TO
COOBER PEDY
Haymes
PAINT

excavations to extend the hotel, the owners unearthed over AUD 100,000 worth of opals and opaline shells missed by the miners, which they kept on display to tempt guests into a purchase.

Coober Pedy's underground churches serve several different denominations. One of the most visited is the Serbian Orthodox Church's Church of Saint Elijah the Prophet, which was hewn from the sandstone by volunteer Serbian/Australian opal miners in 1993. The complex comprises a community hall, parish house, religious school and a church excavated between three and 17 metres below ground. The church's scalloped ceiling bears the grooves from the teeth of the tunnelling machines that excavated it. The daylight streaming through the ground-level stained-glass window and the light radiating from the backlit painted glass iconostasis create a muted golden ambience of ethereal calm.

The disorientating tunnel complex of the Umoona Opal Mine and Museum tempts you in wrong directions as your underground guide, an 86-year-old former opal miner, leads you through the old workings. As well as narrating his first-hand mining experiences, he

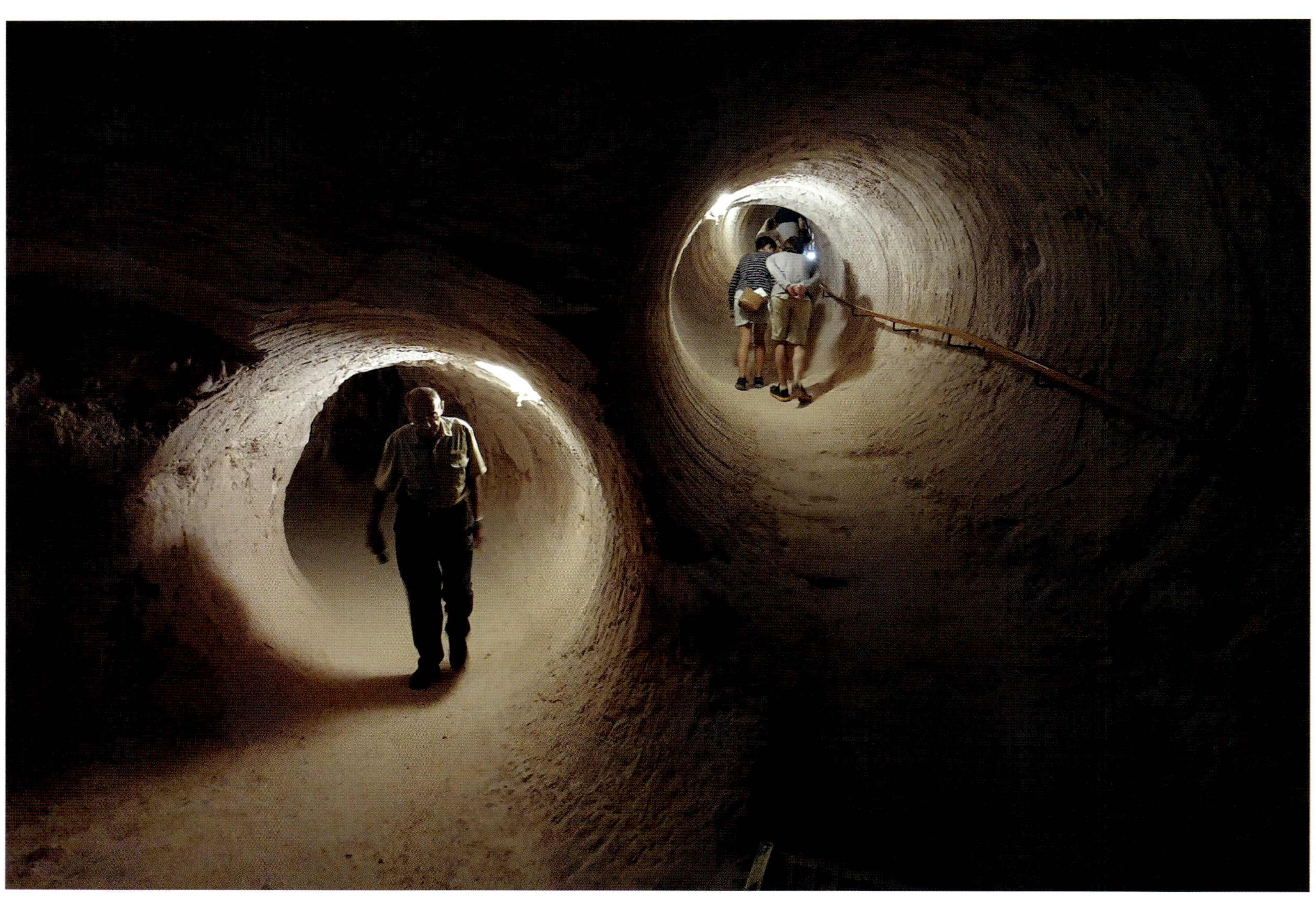

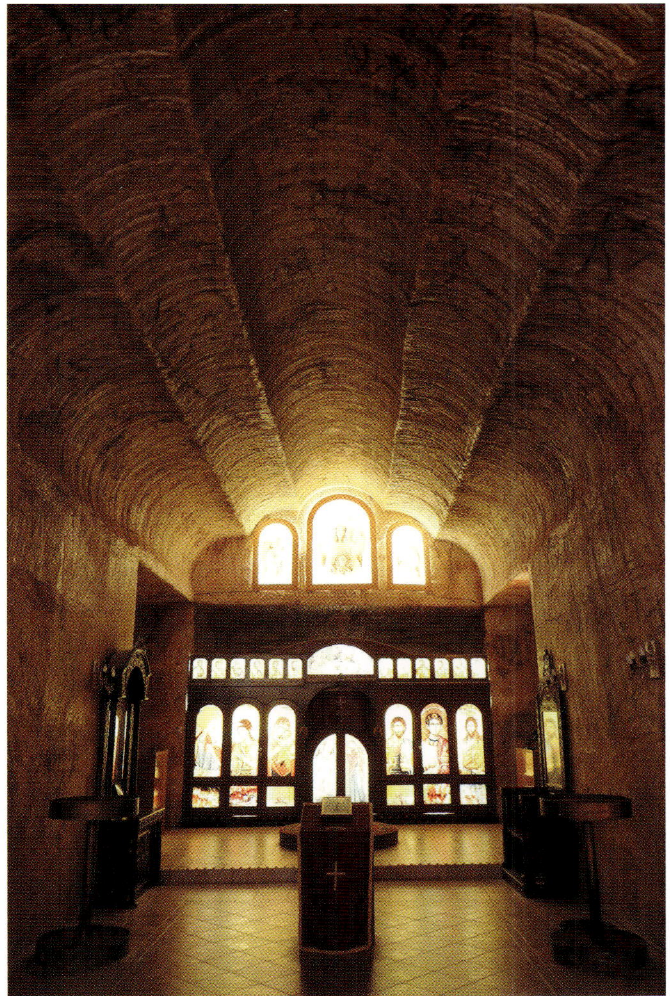

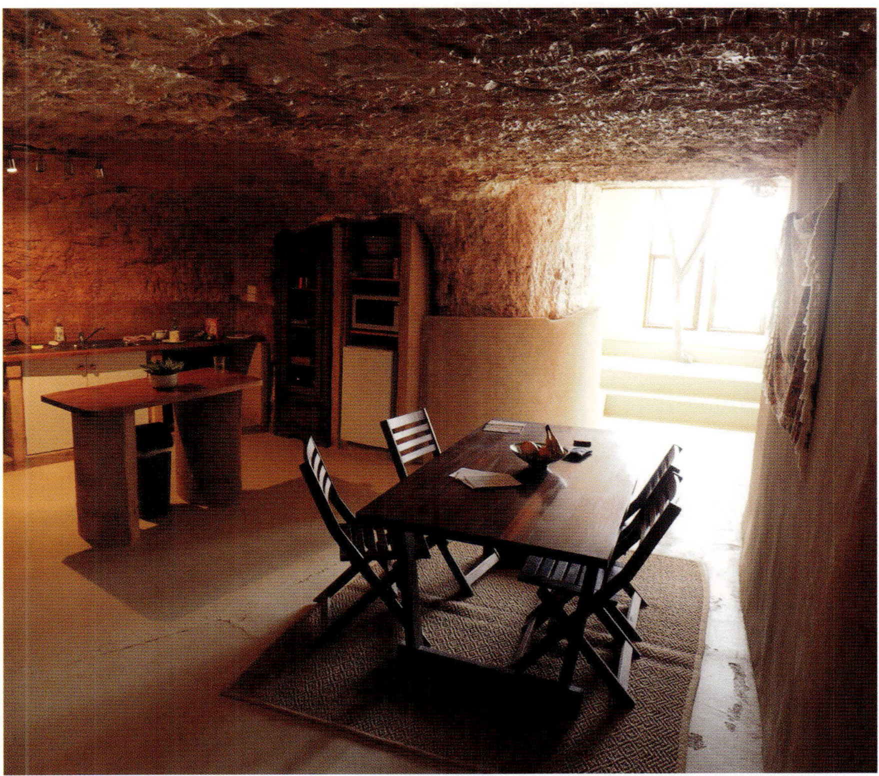

relishes regaling his captive audience with colourful stories of life in 'Wild West' Coober Pedy – the drinking, fighting, gambling and other adult pastimes – fascinating younger visitors while possibly embarrassing their parents.

Not only is this award-winning subterranean museum complex a repurposed opal mine, but it also presents the deep history of the place in numerous exhibits – its Aboriginal culture, semi-desert landscape and natural history and the history of Coober Pedy. A preserved dugout once inhabited by miners illustrates the town's underground past. The museum also houses a café, an opal shop and an underground cinema that shows an award-winning documentary about Coober Pedy, opal mining and the museum.

Living underground for a few days, even in a fully fitted-out dugout, is a strange experience. The night is velvet dark and creepily quiet. Critically, in the summer, when the surface heat soars above 40°C, or in the winter when a chill sets in, the equable 20–25°C temperature below ground removes the need for expensive air conditioning and heating. The comfort soon evaporates when you emerge from your burrow, blinking in the bright sunlight before your vision focuses on the stark, shattered landscape before you.

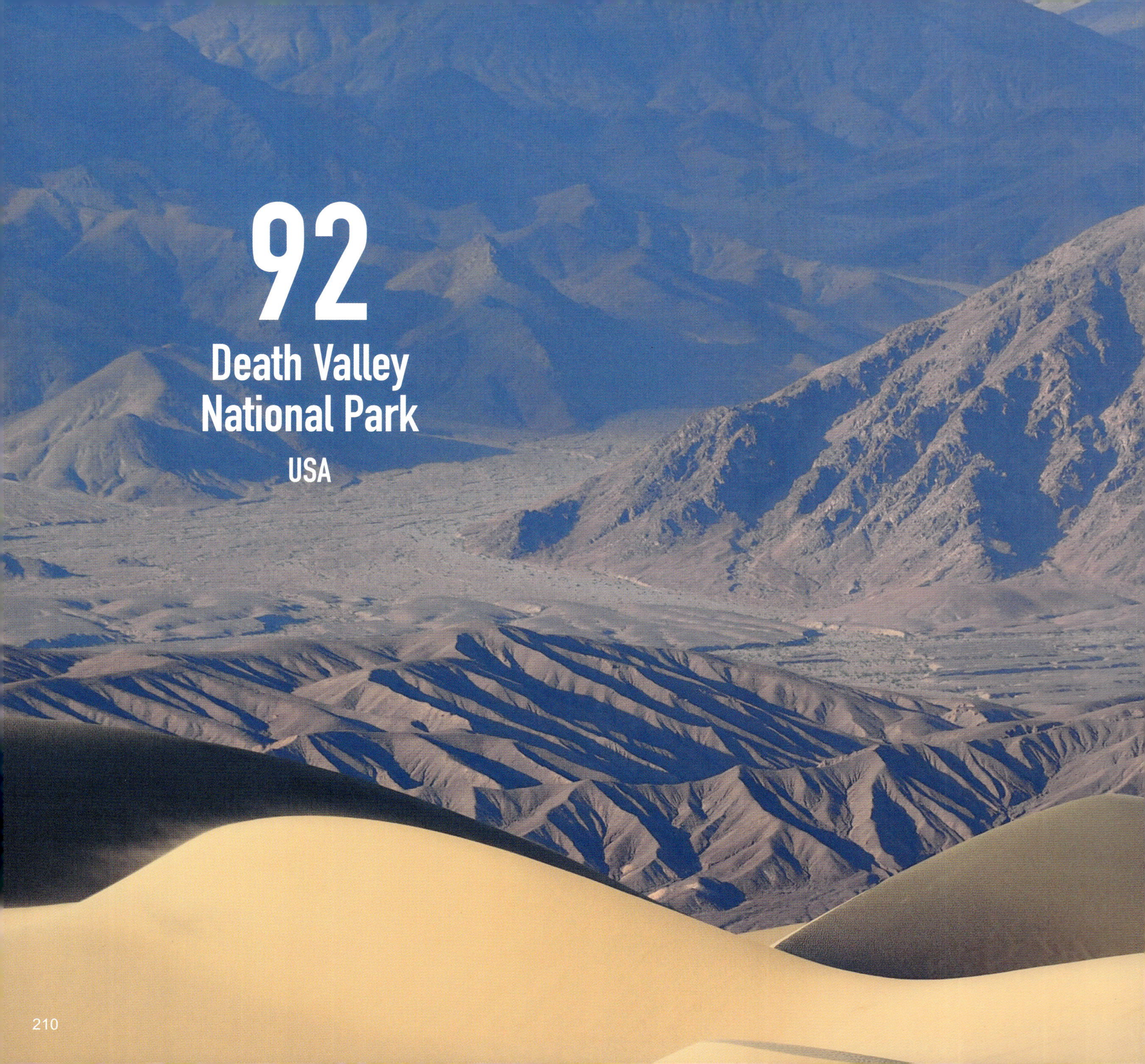

92
Death Valley
National Park
USA

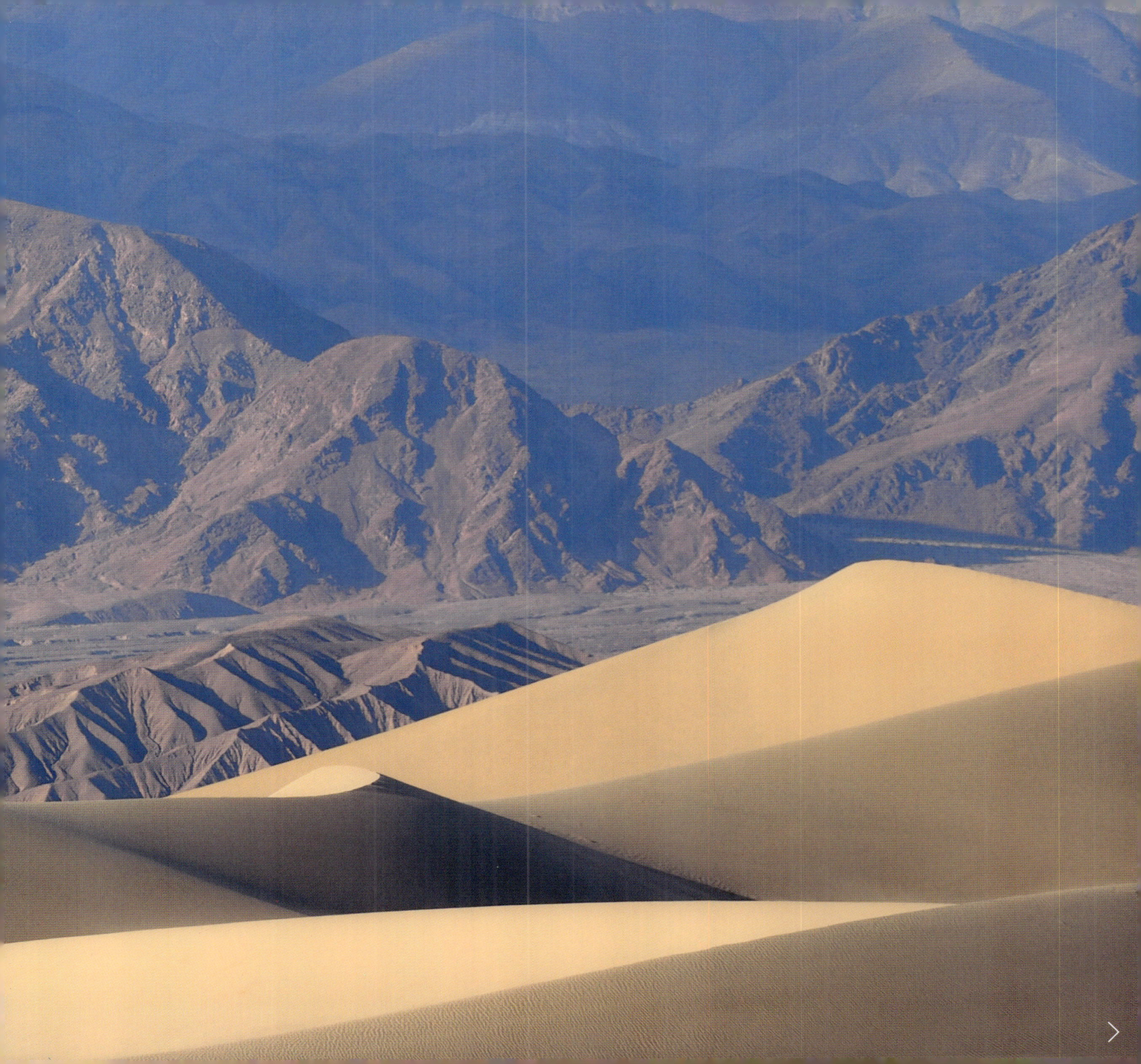

The row of giant masonry beehives is a strange sight after the lonely drive along a dirt road ascending from desert to forest. At over 2,000 metres elevation and surrounded by trees, Wildrose Canyon's charcoal kilns produced fuel for Death Valley's mining operations – a haunting, smoky aroma persists on venturing inside the structures. In this spectacular part of North America's western deserts, there were few other fuel sources close to the mineral riches being mined in Death Valley.

Named by Forty-Niners seeking an ill-advised shortcut en route to possible fortune in the 1849 Californian gold rush, Death Valley is entwined with mining and, surprisingly, the development of the nation's national parks. It's a starkly beautiful, elemental world, rising from salt pans below sea level to lofty peaks piercing an azure sky, with a unique ecology and cultural history, and boasting the nation's hottest, driest

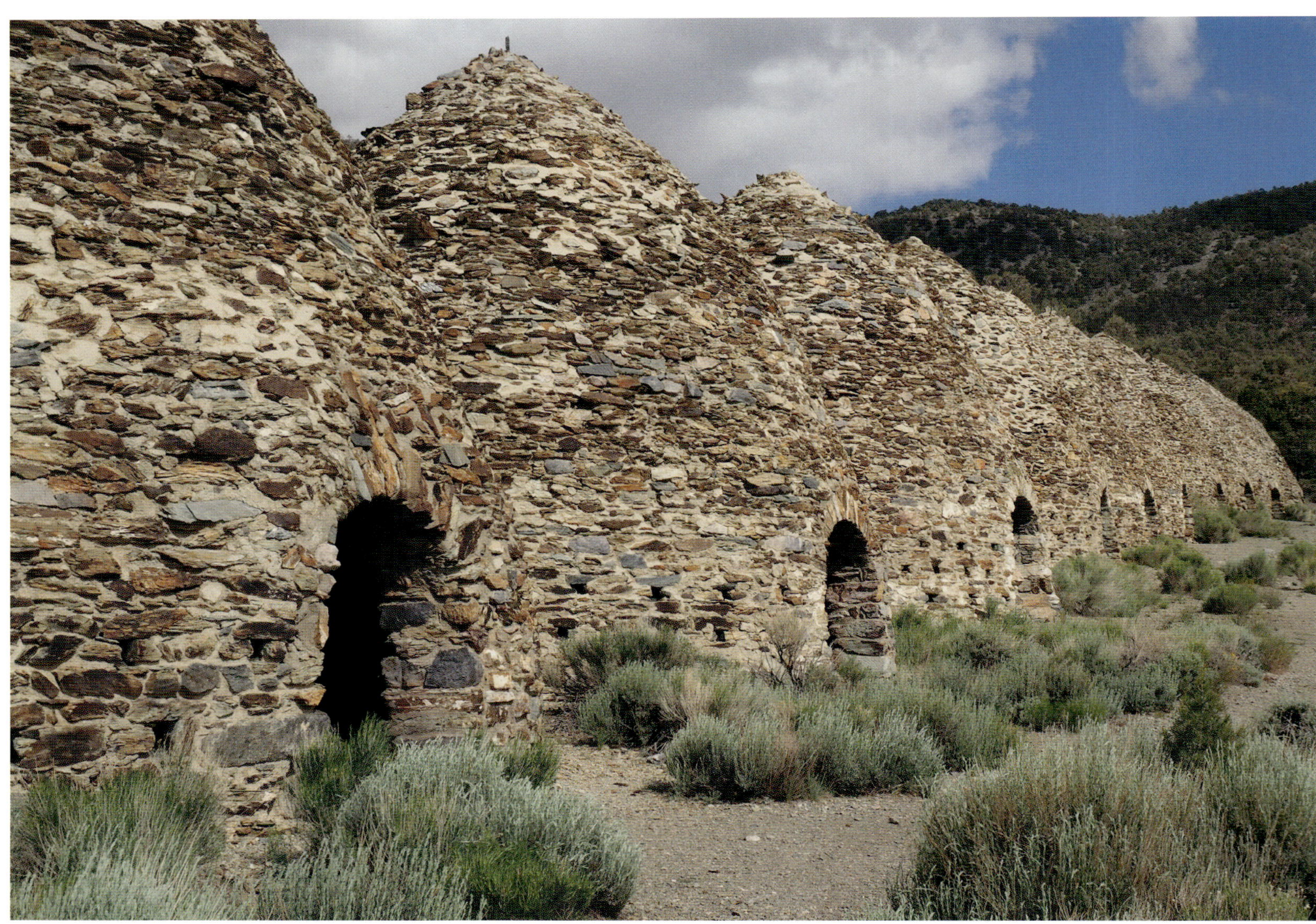

and lowest extremities. The blistering heat and aridity make survival difficult, although the Timbisha Shoshone have lived in their Death Valley homeland for generations, connecting the place to a deeper past.

A range of metals, including gold, was mined here, but the main mineral was borax extracted from Death Valley's dry lake beds. Borax was traditionally used in gold refining and for preserving meat, then later soaps and laundry detergents, amongst many other uses. Several operations were established, including, famously, Harmony Borax Works. Operating for only a few years in the 1880s, the Harmony Works left a lasting imprint on American culture with its 20 Mule Team operation hauling borax from the Valley. Two enormous wooden wagons, filled with nine tonnes of borax and a 4.5-tonne water wagon weighing a total of 33 tonnes, were pulled by 18 mules and two horses on a gruelling 10-day, 265-kilometre desert trek to the Mojave railhead. This venture tweaked romantic notions of the American West, featuring in radio and television series and became emblematic of the borax industry – '20 Mule Team Borax Soap' could be found in almost every household in the country. The 20 Mule Team is celebrated every year in Bishop, California's Mule Days event, with the team hauling three replica wagons through cheering crowds.

After its acquisition by the Pacific Coast Borax Company (PCB), the Harmony Works closed, and the focus of borax mining moved outside of Death Valley around the turn of the 20th century. Now on the National Register of Historic Places, Harmony's industrial remnants can be visited today, an experience that includes, uniquely, an original, weather-beaten, 20 Mule Team wagon set-up. Originally known as *Tümpisa* by the Timbisha Shoshone people who already lived there due to its spring water supply, the Greenland Ranch area had been developed by the Harmony Works for growing alfalfa for animal feed and supporting borax workers. It was crucial in opening up Death Valley and was subsequently converted to tourism and renamed Furnace Creek Ranch, which today hosts a petrol station, hotel and campground, shops, golf course, solar farm and a very welcome ice cream parlour! The national park's Furnace Creek Visitor Center is next door.

In 1914, PCB established its Ryan Camp operations, which ran until 1927. Then, along with the successful Stovepipe Wells Resort developed by others, in an early example of what we might now call 'sustainable development', PCB repurposed its miners' township, mineral railway and dirt roads, converted Death Valley Junction to a hotel and built the luxurious Inn at Death Valley to serve tourists eager to experience first-hand the Wild West Death Valley wilderness of 20 Mule Team fame. This denoted an irrepressible shift in the Valley's economy from mining to tourism.

Mining and Death Valley's intimacy extended into US politics too: Stephen Mather, a former PCB executive who appreciated the West's sublime landscapes and conceived the 20 Mule Team Borax brand, turned his attention to Congress. Together with a young Bishop lawyer, Horace Albright (friend of PCB executive Christian Zabriskie after whom the park's iconic Zabriskie Point is named), he advocated for and created the National Park Service in 1916. Mather became its first director. As PCB lobbied for park status, Mather felt it would be a conflict of interest given his past association with the company. In 1933, Horace Albright, as the NPS's second director, convinced President Hoover to declare it a national monument. Finally, in 1994, Death Valley was elevated to National Park status via the Desert Protection Act. Rio Tinto acquired the company in 1967 and the head of Rio Tinto Borax spoke at the park's dedication ceremony in 1998.

The Death Valley mining story took an ironic twist in the 1970s when the controversial impacts of Tenneco's Boraxo open pit mine became the poster child for changing the law on mining in national parks. The 1976 Mining in the National Parks Act led to all national parks (and national monuments) being closed to new mining developments (although it is possible to exploit pre-1976 mining claims). Death Valley's last operating mine, American Borate Company's Billie Mine, closed in 2005. Its rusting headframe stands as an impromptu monument to the park's industrial foundations.

US national parks exist to 'preserve unimpaired the natural and cultural resources and values of the national park system for the enjoyment, education, and inspiration of this and future generations'. Death Valley has become one of America's most popular national parks, welcoming 1.7 million visitors in 2019. Its mission to 'preserve' inevitably conflicts with other priorities for maintaining visitor infrastructure in a landscape regularly affected by flash flooding, heat and aridity – which are becoming more extreme. In the last seven years, three 1:1,000-year storms have caused USD 200 million of damage to critical service infrastructure. Already arguably the planet's hottest place, it's getting even hotter with temperatures now reaching the mid-50s Celsius with increasing regularity.

Within this brutal world of extremes, the national park has around 16,000 abandoned mining features – including eerie ghost towns – dating from 1870, narrating a profound human story of resilience and ingenuity. Many arid canyons contain emotive remnants of the area's mining history – a dilapidated shack, adits, waste piles, or defunct

engineering; adobe buildings dissolve imperceptibly back into the desert – not everything can be saved. As Wildrose Canyon's masonry beehives illustrate though, some things can, enabling Death Valley's mining story to continue being told.

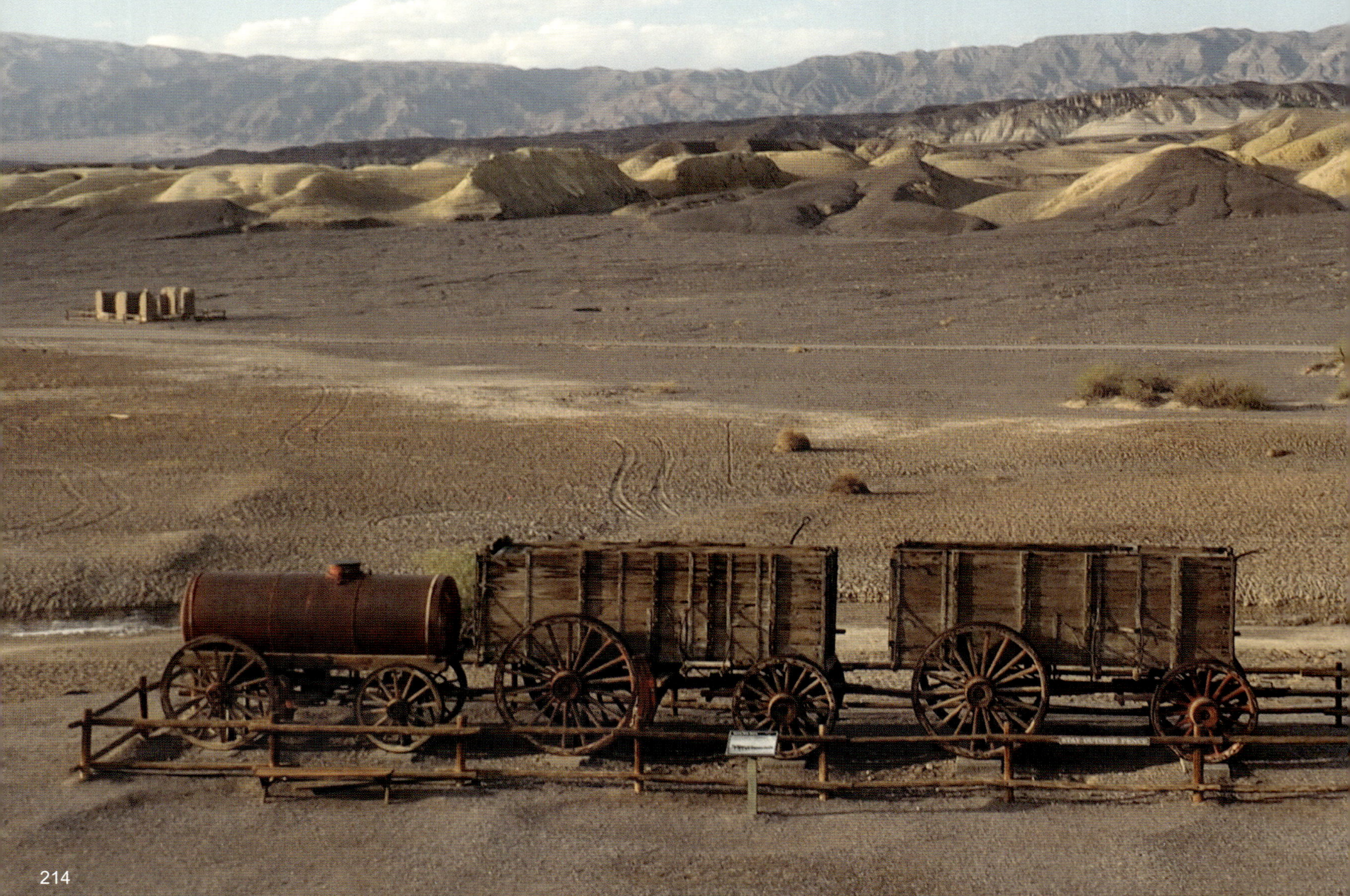

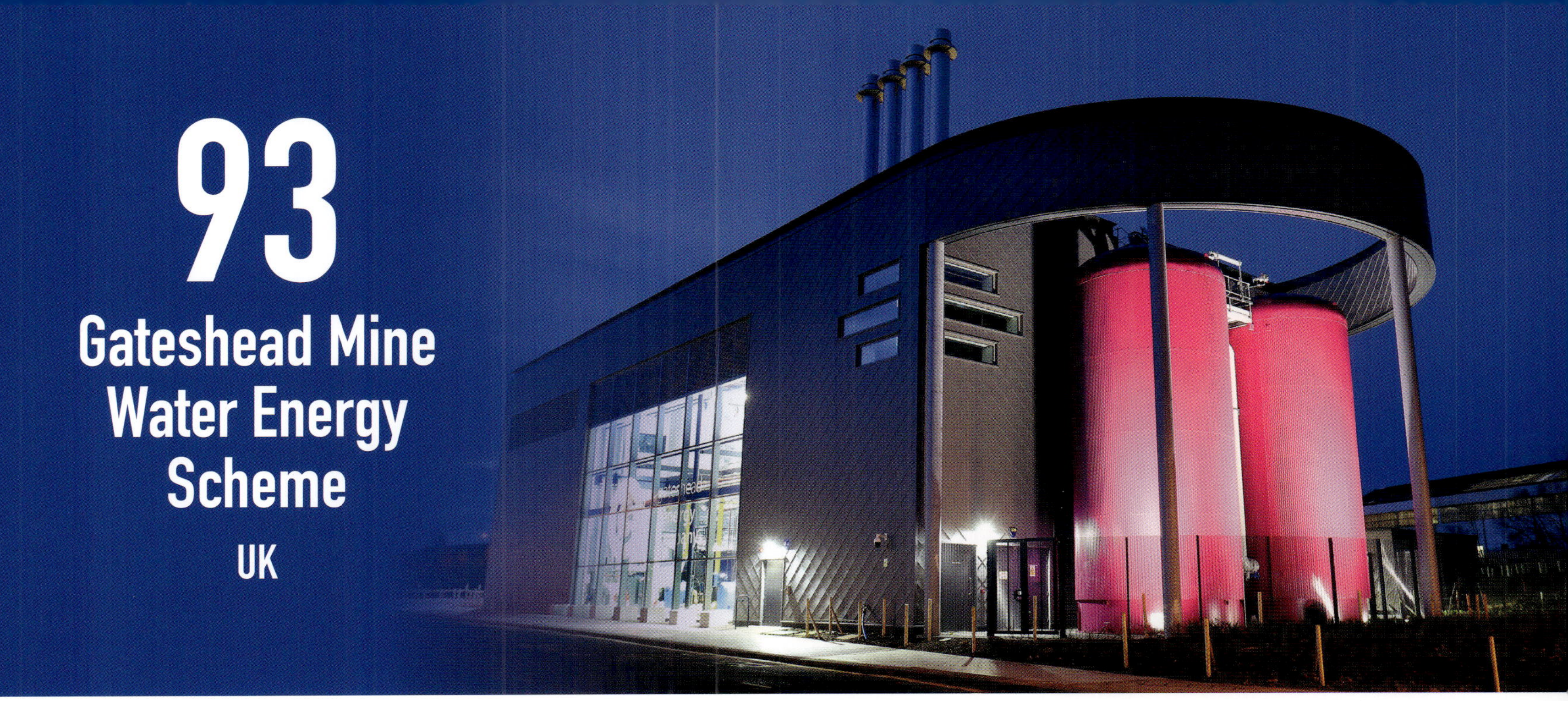

93
Gateshead Mine Water Energy Scheme
UK

The town of Gateshead in North East England was a coal powerhouse feeding Britain's industrial revolution; its last coal mines closed in the 1960s. When the pumps were turned off the mines flooded.

In 2023 the town turned on its 6MW Mine Water Energy Scheme, the UK's largest mine water geothermal project, to extract heat from water in its abandoned mines 150 metres below the town centre. Mine water is pumped at 15°C via bespoke boreholes to surface heat pumps and heat exchangers, which concentrate the heat to 80°C. The hot water is piped 350 metres from the drab Mine Water Heat Pump building to the District Energy Centre (DEC), which opened in 2017 with its eye-catching pink storage tanks.

The DEC's gas-fired, combined heat and power (CHP) engines supply district heating, now augmented by the mine water system, through five kilometres of pipework to town centre offices, businesses and public buildings, including the Baltic Centre for Contemporary Art and Gateshead College, and 350 council-owned homes. District heating systems are relatively unusual in the UK. Still, there are plans to expand the Gateshead network to include 270 new-build private homes, an existing low-rise council estate, a new hotel and a conference centre.

The Mine Water Energy Scheme is estimated to save about 1,800 tonnes of CO_2 emissions per year. Central to the low-carbon concept, nearby solar PV farms help power the heat pumps, although the network still requires electricity from the rapidly decarbonising UK grid. As well as CHP generation, the

DEC will increasingly focus on energy storage, grid services and low-carbon energy, leading to reduced costs for the town centre area.

This award-winning project received UK government grant assistance and is supported by the UK Coal Authority. The DEC itself is owned by Gateshead Council and run by the wholly owned Gateshead Energy Company.

The Authority estimates 300,000 offices and six million homes in Britain sit above old coal mines! High capital costs have limited the exploitation of this post-mining resource, although the running costs are lower than those of conventional systems. The UK – and much of the world – is just scratching the surface of this underground, low-carbon heat source, although there are many schemes in the pipeline.

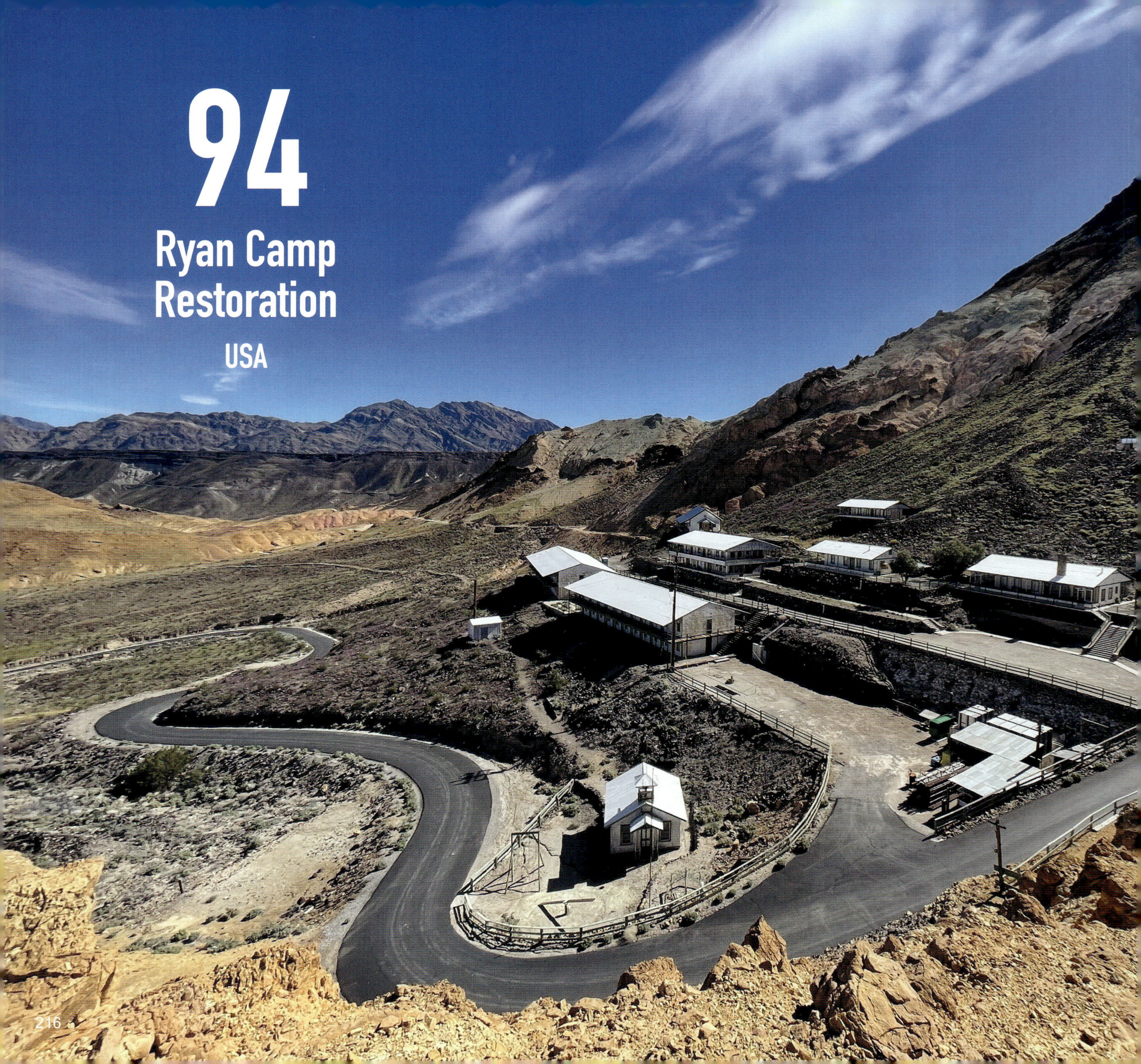
94
Ryan Camp
Restoration
USA

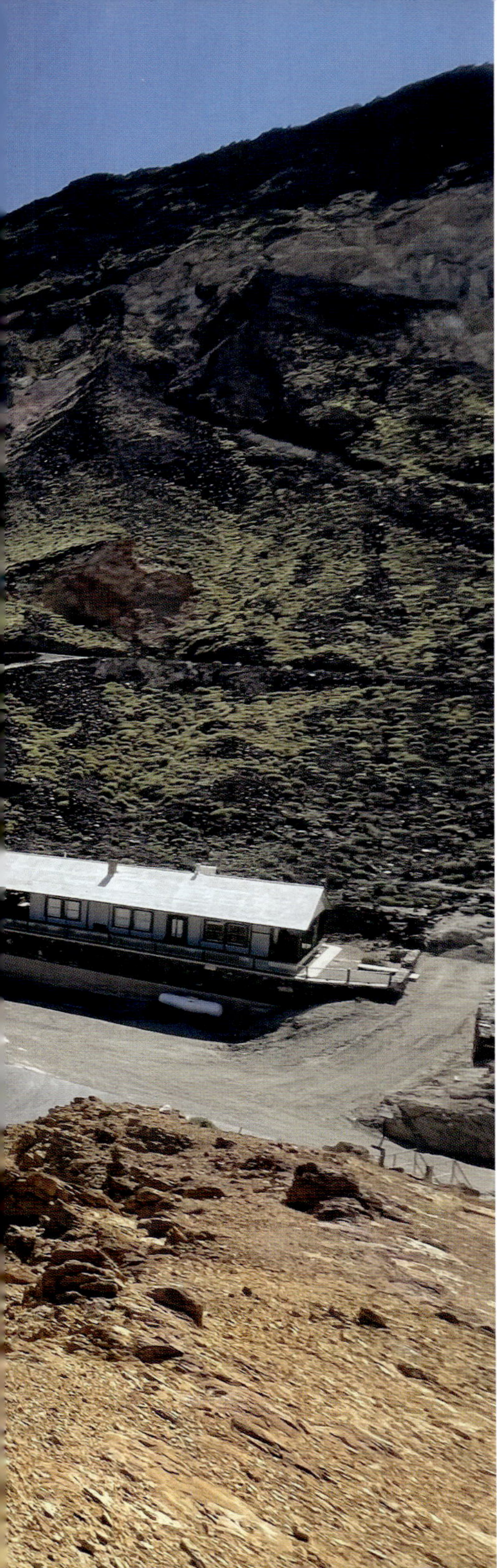

From halfway up a Death Valley mountainside, ominously surrounded by black basaltic scree, beckon the shiny metal roofs and green and white woodwork of the mining ghost town of New Ryan Camp. In 1914, the Pacific Coast Borax Company (PCB) established the township as the centre of its Death Valley borax mining operations by relocating buildings from nearby Death Valley Junction to their current location on a specially constructed, narrow-gauge railway. It expanded to house 300 miners and several families in relative luxury with all mod cons – power, heating, refrigeration, school, hospital, post office, general store and a recreation hall that also served as a cinema and a church.

After the company found more lucrative deposits elsewhere, PCB realised the railway offered the opportunity to remodel Ryan Camp for tourism, and the Death Valley View Hotel opened in 1927. In the 1928/9 winter over 10,000 visitors rode their railway to Ryan, then ventured beyond. The rapid rise of the car and the improving road network caused the railway and Ryan Camp to fall out of favour in the 1930s, leading to its demise and ghost town status, although with occasional use as overflow accommodation when PCB's two Death Valley hotels were full. PCB sold its Valley hotel properties in the 1960s but kept an onsite caretaker to maintain Ryan Camp township (periodic strong winds can peel off metal roofs like a sardine can!).

Ryan is a unique Californian mining ghost town in that it is undeveloped and has restricted access. Over time, through company mergers and acquisitions, it became part of Rio Tinto's legacy sites portfolio, which owns thousands of hectares of land, buildings and mineral rights in and around Death Valley National Park.

PCB executives were instrumental in forming not only the national park, but also the National Park Service itself, and today Rio Tinto keeps regular relations with Congress and the park. It has donated some of the mineral rights and associated surface rights to the park (including the land upon which the national park's visitor centre sits), and in 2013 it handed over land, heritage assets and some mineral rights in and around Ryan Camp to the not-for-profit Death Valley Conservancy (DVC), with the mandate for managing and restoring the site. It also financed DVC's endowment and works funds, which were sufficient to ensure long-term preservation and restoration activities, obviating the need for visitor revenues, and so no need to compromise the vision.

As well as collaborating with the national park in other parts of the Valley, DVC plays a custodial role at Ryan Camp to purvey an authentic, tangible story of a Death Valley mining community. Its mantra is to 'preserve what's left and restore where possible (and what is appropriate to restore)'. Since 2013, DVC has been returning Ryan to its 1927 hotel heyday. The 'arrested decline' model of heritage preservation that some organisations use has limitations for old wooden structures in harsh environments; with limited and competing funds, there is a persuasive argument for the adaptive reuse of old buildings to enable their upkeep. In this vein, DVC will not allow unrestricted public access to Ryan (strictly speaking it lies outside the national park) to protect the site's historic value, and its limited water supply will not support large visitor numbers. Instead, they facilitate public access in smaller groups on guided tours, specialist visits, and mine closure workshops. Their work initially focused on restoring the site's infrastructure (water, sewer, electric, communications). Since completion, the focus has been on renovating spaces for presentations, gatherings and the like.

The first building to receive attention was the former Death Valley View Hotel complex, including its lobby, dining hall, kitchen and workshop. The current priority is restoring the historic old church with its buckling walls. Originally built in Rhyolite, Nevada, it was moved across the border to Ryan and originally secured to the mountainside by cables. Doubling as a performance space, it was the social focus of camp life and still has great potential. The former miners' bunkhouse and other buildings will also be partially renovated as temporary accommodation for visiting students, researchers, craftspeople and others in 1927 'comfort'.

DVC is sensitive to Ryan's archaeological riches beyond the buildings and is, unusually, supported by a resident archaeologist (who happens to be married to the resident site superintendent). These 'riches' extend to erstwhile trash – cookers, fridges, bits of machinery, cans – unceremoniously discarded down the mountainside throughout the 20th century, forming a discordant stream of rusting debris. Other still-visible mountainside rubbish hails from the set of the opening sequence of the 1960s Stanley Kubrick film *Spartacus*, starring Kirk Douglas, in which Ryan Mine served as a slave trader camp and forced labour quarry. The 'riches' go beyond mining too: some of its tunnels were used as Cold War nuclear shelters to store essential canned foods and medical supplies. Although decommissioned in the 1970s, these supplies remain in-situ and offer a captivating, unexpected chapter to the story. The process to register Ryan as a National Historic District is underway.

Lessons can be learned from the DVC Ryan Camp programme. Critically, controlling the land and resident occupation discourages outsiders from entering. Also, ownership by an organisation with a purpose to protect and restore a place supported through independent funding allows fulfilment of a dedicated mission to preserve it for future generations.

Below: Restoration of the old church is underway.

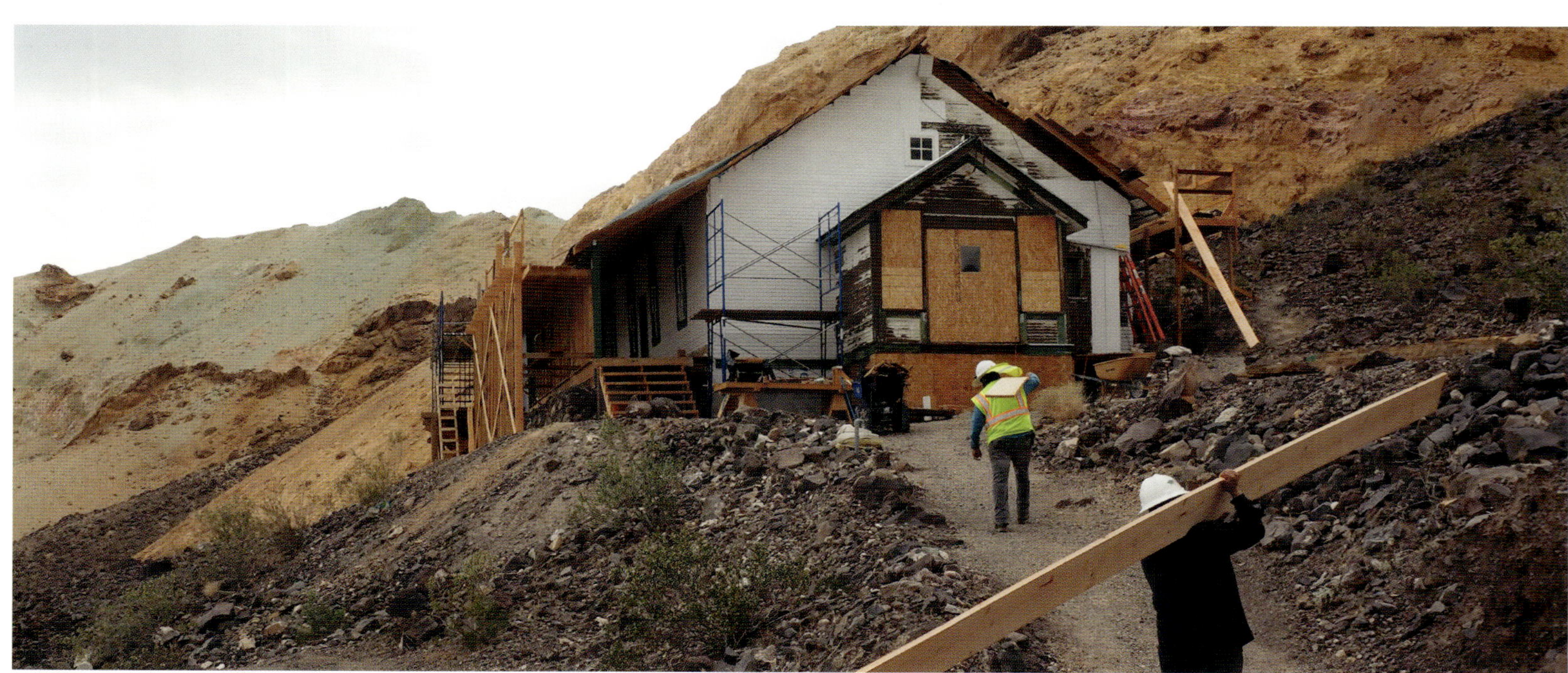

95

Salina Turda

Romania

The world's only subterranean Ferris wheel sits inside a Transylvanian salt mine. Mined by hand from the Middle Ages until the early 1930s, Salina Turda salt caverns now contain an amusement park. It's one of the few such attractions to appear in a Michelin Travel Guide; over 680,000 people visit the mine every year.

A lift offers panoramic views of the cone-shaped Terezia mine as you descend. At the bottom of the cavern, a round pier glows on the dark lake.

Ferris wheel aside, the amusements are more low-key; no dodgems or helter-skelters here. Instead, visitors can play billiards, table tennis or mini-golf, and go boating on the salt lake. There's also a 180-seat amphitheatre (with heated seats) that hosts classical concerts.

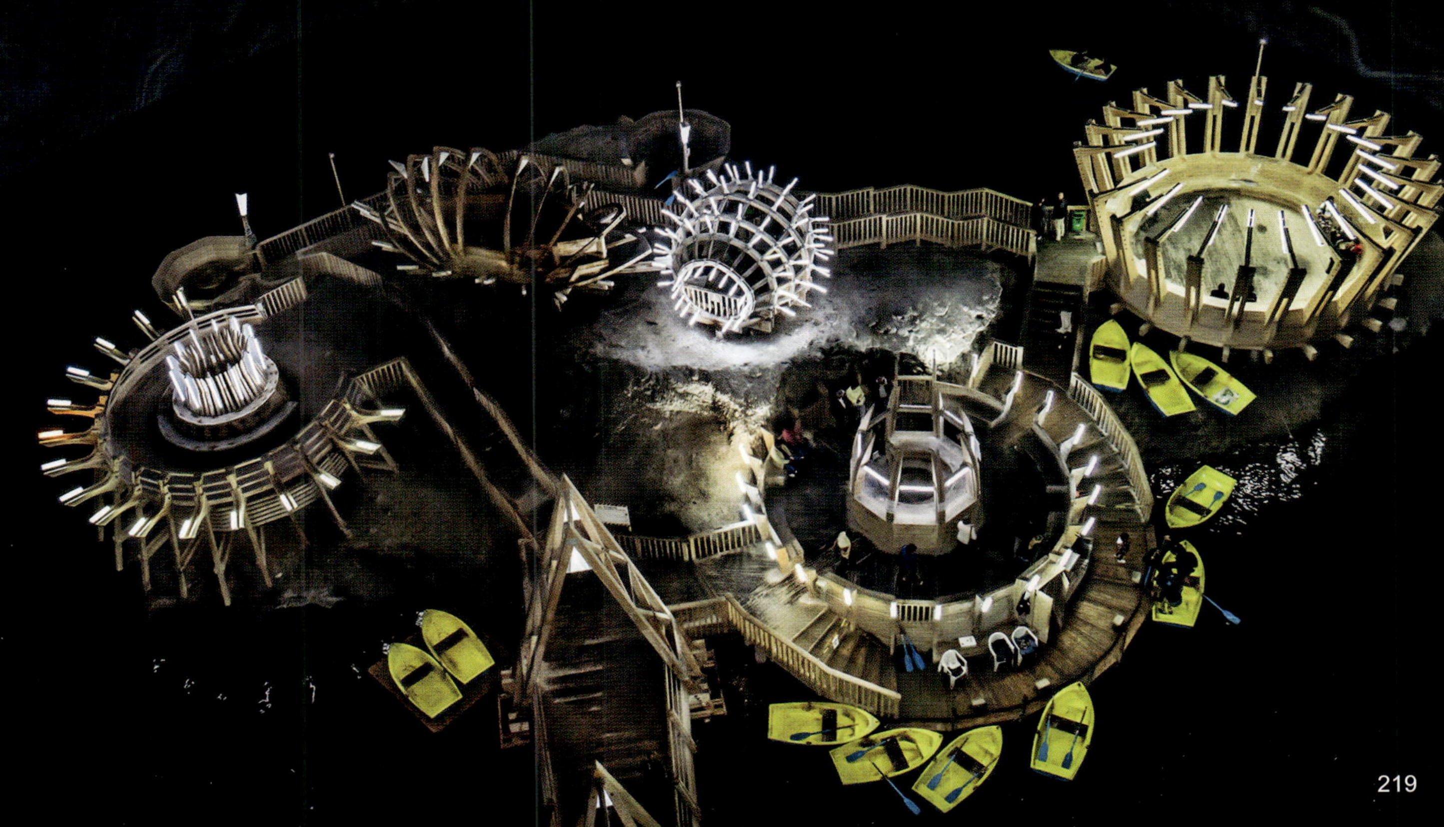

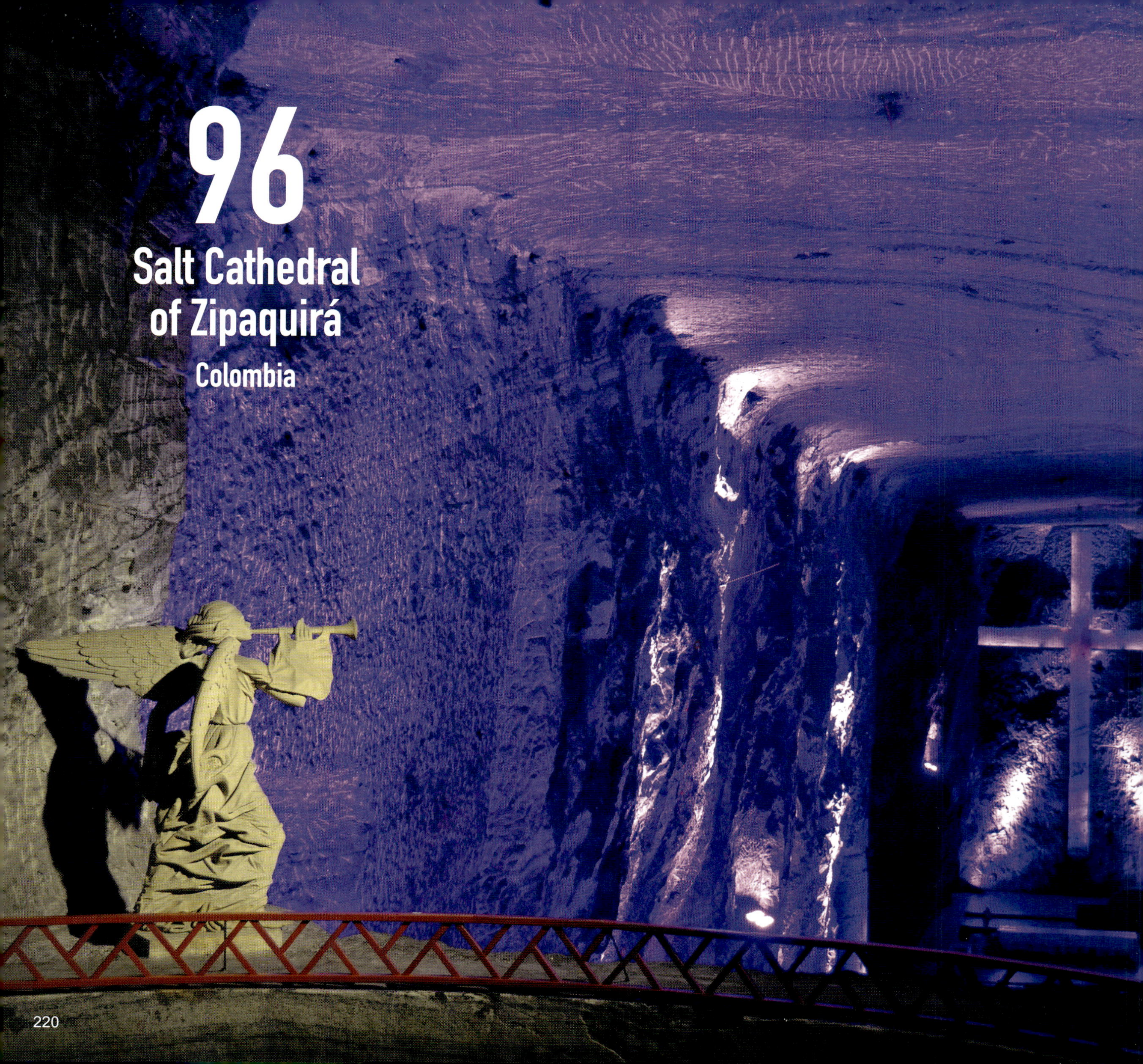

96

Salt Cathedral
of Zipaquirá

Colombia

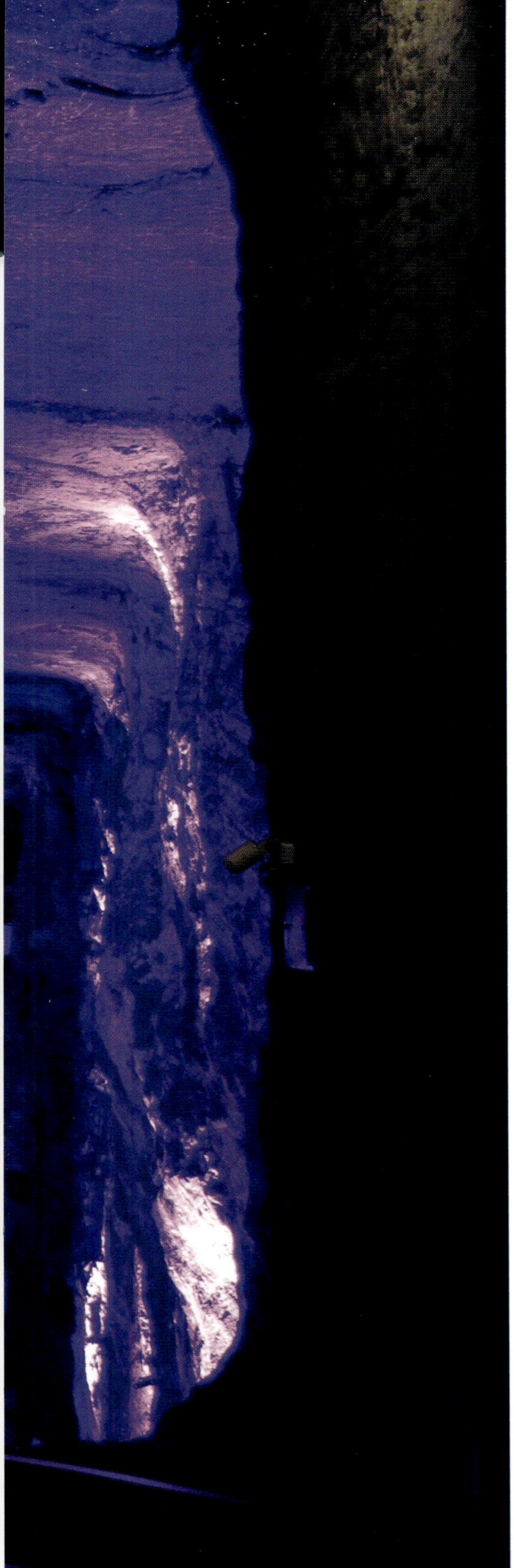

The Salt Cathedral of Zipaquirá is not merely a cathedral; it's an architectural masterpiece, a cultural treasure and a religious sanctuary that tugs at the heartstrings of Colombians. This place of worship plays host to thousands of pilgrims every year, including 6,000 visitors a day during Holy Week.

The roots of this extraordinary place stretch back to pre-Columbian times when the indigenous Muisca people considered salt more precious than gold. Muiscas worship the goddess Nygua, who they believe dwelled in the midst of the dark salt mine in Zipaquirá and from her shadowy palace provided them with protection and maternal shelter. The Muisca apparently dedicated the salt mines to this deity. During the rainy season, they distilled brine extracted from salt rocks, meticulously boiling it until it transformed into a solid crystalline form known as *pan de sal*, or salt bread. This valuable resource served a multifaceted purpose, acting not only as a preservative for food and a treatment for textiles but also was considered the currency used to obtain commodities such as gold, emeralds, textiles or food.

When mining began in the early 19th century, another religious figure came to dominate the mine. Catholic miners began to bring a statue of their patron saint the Virgin of the Rosary of Guasá (*guasá* means rock salt in the Muisica's language) into the depths of the mine.

In 1930 the miners carved out a small shrine dedicated to the Virgin in gratitude for her protection, and in response, the Colombian government agreed that part of the mine could be transformed into a more formal place of worship. In 1954 the Salt Cathedral of Zipaquirá was inaugurated, an act that would soon bring tourists from near and far.

However, the cathedral's structure suffered degradation from water damage, which progressively weakened its pillars and led to its closure in 1992. Three years later, a new cathedral was inaugurated 60 metres below the original. Like its predecessor, it is entirely carved from salt.

The pilgrimage into the heart of the Salt Cathedral is a mesmerizing two-kilometre journey. The Way of the Cross features fourteen stations adorned with small altars hewn from salt rock. As you venture deeper, reaching a depth of 180 meters, the central nave and its side aisles reveal themselves in all their grandeur, occupying a 1,200m^2 sinkhole. An imposing 16-metre-high cross stands as a beacon of faith within. The Chapel of the Virgin of the Rosary of Guasá holds a special place in the hearts of miners, a sanctuary where they entrust their labour to the Virgin. For adventurous souls, the Cathedral presents the 'Miner's Route' – a plunge into the very bowels of the earth that lets visitors explore the tunnels to dig and drill, delving into the intricacies of mining and geology. Over half a million visitors come here each year and the money they bring is the main source of income for the city's government, paying for its roads and parks amongst other things.

221

Cecil Rhodes, Ernest Oppenheimer and others brought organisation and industrialisation to Kimberley's mining and gave birth to the diamond behemoth, De Beers Consolidated Mines, in 1888.

vast geological powers and timescales embodied by diamonds.

Located on the north western perimeter is a collection of heritage buildings known as the Old Town.

houses the Butler's Hotel School, which teaches catering and culinary arts and management. Its courses are accredited by City & Guilds of London. Several refurbished period buildings also offer comfortable visitor accommodation in the heart of the Old Town. Unrestricted free entry makes it a valued social and cultural asset for the city and its streets host a monthly open-air market – weather permitting.

The complex directly employs 33 staff, which doubles when on-site business tenants are included. It is a national educational resource too and is plugged into the national curriculum and marketed to schools. It has won numerous awards and attracts around 90,000 visitors per year, including many international visitors who arrive on the luxury Cape Town to Johannesburg railway services.

Kimberley (population about 190,000) retains many of its historic buildings, but it's fair to say the city itself has seen better days. There are plans by the Big Hole to develop further as well as stimulate new community and tourism developments in the city centre in the hope it will entice more tourists to come and spend more time in the city.

Given its urban location, the Big Hole is constantly monitored. Its water level is creeping upwards, and its pit walls are slowly eroding outwards. Urban Kimberley itself still carries the scars of old mining activities as craters and hills of tailings (some of which are protected as

cultural heritage). Small scale operators are reprocessing old tailings tips to remove their residual diamonds and placing the reprocessed mineral wastes into old open pits where financially viable.

The Big Hole is on De Beers formative Diamond Route – a collection of the company's extensive conservation and heritage sites across South Africa and Botswana – encouraging intrepid travellers to visit. Given the profound history and international significance of the place, there are long term plans to apply to UNESCO for World Heritage Site status too.

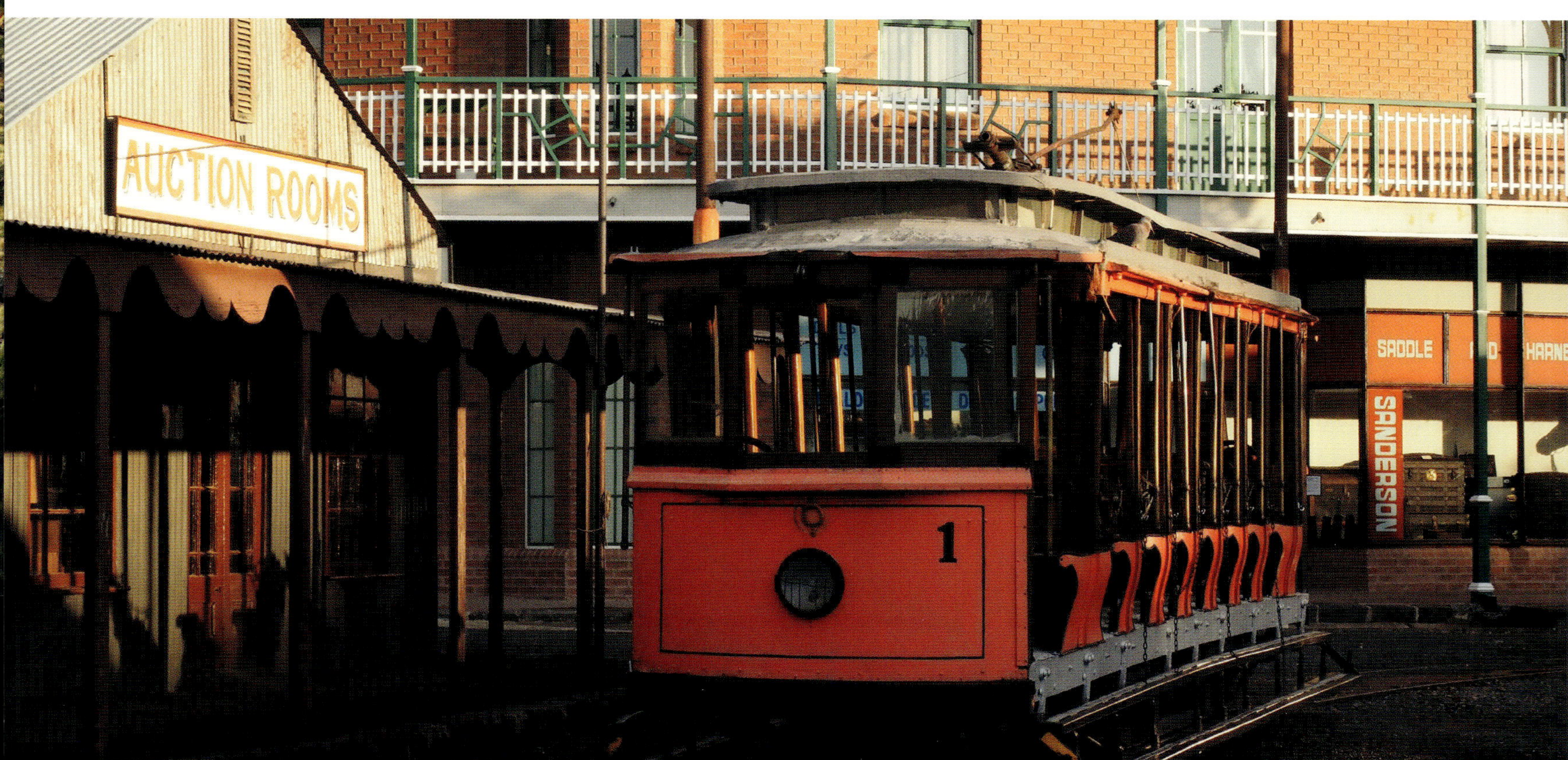

98
Kinlochleven Community Trust
UK

In 1907, when the North British Aluminium Company set up a smelter at the eastern tip of Loch Leven in the Scottish Highlands, it needed two things – power and people. The electricity came from Blackwater Reservoir, 10km away; the people came from the newly founded Kinlochleven, created from two hamlets on either side of the River Leven, the river that feeds the loch. British Alcan bought the North British Aluminium Company in 1982, and closed the smelter in 2000. Seven years later, Alcan was purchased by Rio Tinto.

At its peak, the smelter employed 800 people, but by the time of its closure this had dropped to just under a hundred. The loss of even 100 jobs in a small community can destroy it – especially when all of those jobs depend on a single employer. But Kinlochleven wasn't about to let that happen.

The village formed a community land development trust and organised the transfer of ownership of 31.5 hectares of land from British Alcan – with a further 1.5 hectares and some buildings leased for a peppercorn rent for 99 years. The trust, now called Kinlochleven Community Trust (KCT), manages these assets for the benefit of the community. Most of the original smelter was demolished, but its carbon store became the National Ice Climbing Centre. At the time, it was the world's largest indoor ice climbing wall. In addition to ice-climbing walls and bouldering facilities, it offered a bar, a restaurant and a sauna, catering for climbers and the many walkers visiting the area – Ben Nevis and Glencoe are nearby.

The centre closed in early 2023 while KCT looked for new tenants, and it's hoped it will reopen soon.

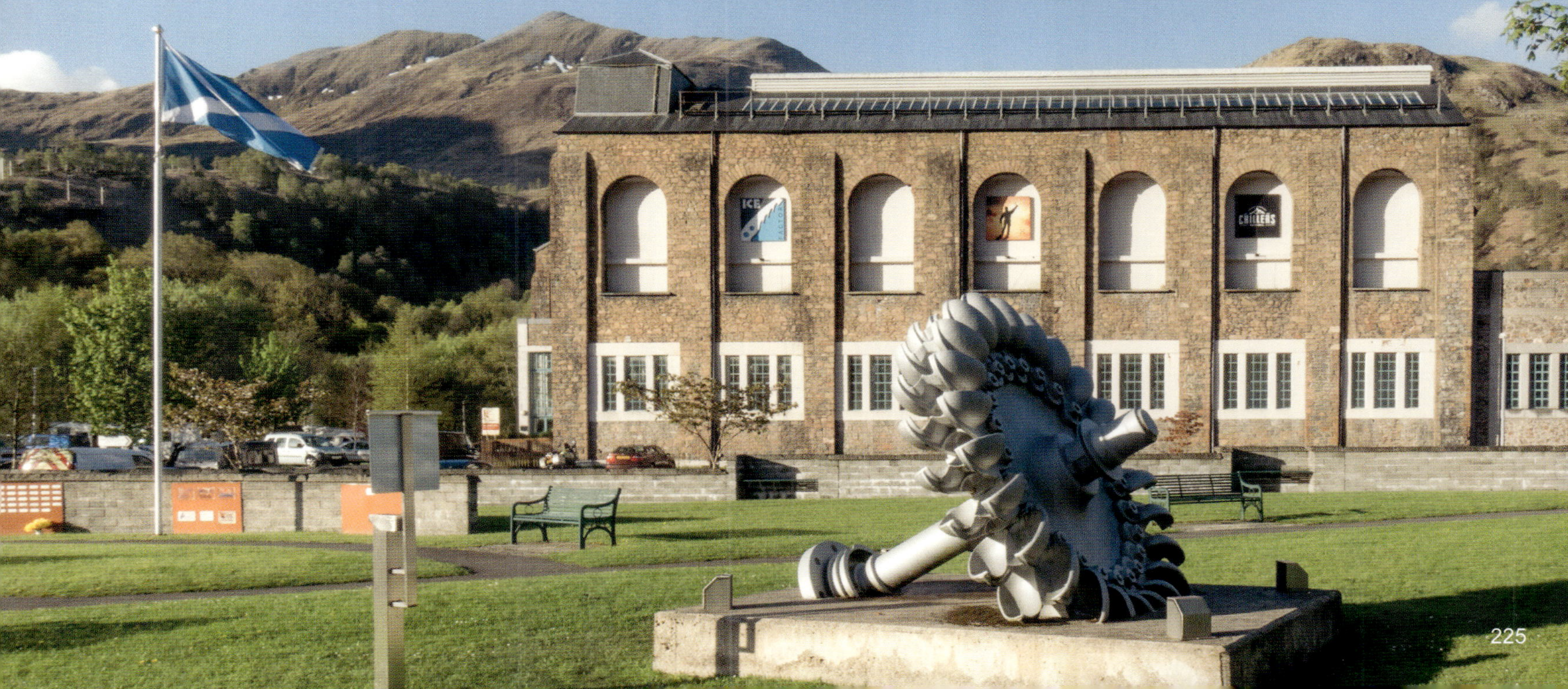

99

Peace Diamonds Restoration Initiative

Sierra Leone

Across Sierra Leone, many communities rely on artisanal and small-scale diamond mining (ASDM) for their livelihoods while conversely contending with its legacy of abandoned, steep-sided, flooded pits, contaminated water and damaged soils. The resulting deterioration undermines agricultural productivity and poses extensive risks to neighbouring communities' safety and health. This situation is also rooted in Sierra Leone's historical context, considering how the spectre of 'Blood Diamonds' was a major driver in the 1990-2002 civil war. Its lasting impact on the context and practice of ASDM is reflected in current challenges, including addressing land degradation.

The Sierra Leone team of RESOLVE's Peace Diamonds Restoration Initiative works in partnership with community leaders, government agencies and miners to restore mined-out areas.

It focuses on building knowledge and skills, provides tools and supports community resilience while restoring and repurposing damaged artisanal diamond mining land, and – importantly – helping to consolidate peace through collaboration and shared interests.

Launched in 2022, Peace Diamonds evolved from a 2019 Anglo American Foundation pilot project to regenerate old ASDM pits. It partnered with traditional leaders, government bodies and local communities to test the concept, which RESOLVE developed into Peace Diamonds, supported by The Tiffany & Co. Foundation, Brilliant Earth and the Gemological Institute of America.

Peace Diamonds works at a grassroots level to address the ASDM legacy by training miners and their communities to backfill flooded pits to help make the land more suitable for agriculture. This also enhances the land's potential for forestry and fisheries, benefiting public health and creating new food supplies and economic opportunities. Once an area is restored, female-led co-operatives use the land to grow subsistence crops, selling any surplus in the local marketplace. It is also closing the socio-economic development circle by supporting entrepreneurship and job creation and collaborating with public agencies to pass on the benefits of a restored environment to neighbouring communities. As the approach has developed, RESOLVE is working with community leaders and government

agencies to focus on community-led land planning. Using the country's land-planning policies, the group identifies areas for agriculture, conservation and other land uses.

Between 2019 and 2023, 110 old pits were reclaimed to create 83 hectares of land suitable for various new uses, but mostly farming. Peace Diamonds provided seedlings, tools, food support and agricultural extension services. Five hundred women grew, processed and marketed groundnuts, rice and various vegetables, including pepper, okra and more. Improved crop yields have led to an increased food supply, some of which is processed to add value for sale at higher market prices. Two thousand people directly benefited from the scheme, with a further 20,000 indirectly. The goal is to restore 1,500 hectares across three districts by 2030. There was also considerable progress in involving stakeholders in collaborative land-use planning.

Peace Diamonds has contributed to Sierra Leone's post-conflict peace-building efforts by fostering and sustaining positive outcomes in a region of increasing pressure for land. RESOLVE plans to share what it has learned to inform national programmes and help others restore mine sites and biodiversity, adapt to climate change, produce sustainable food, and create alternative livelihoods.

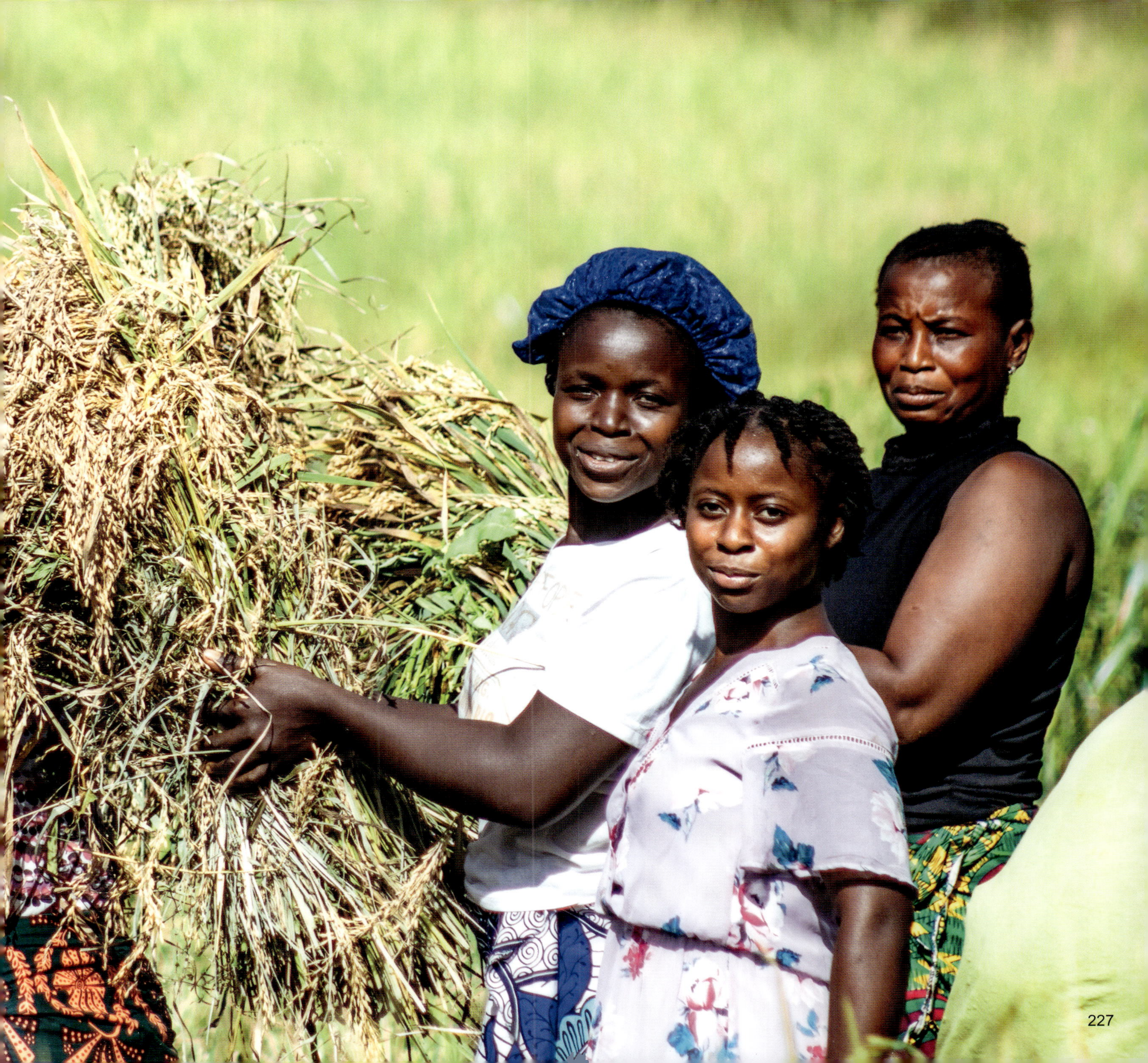

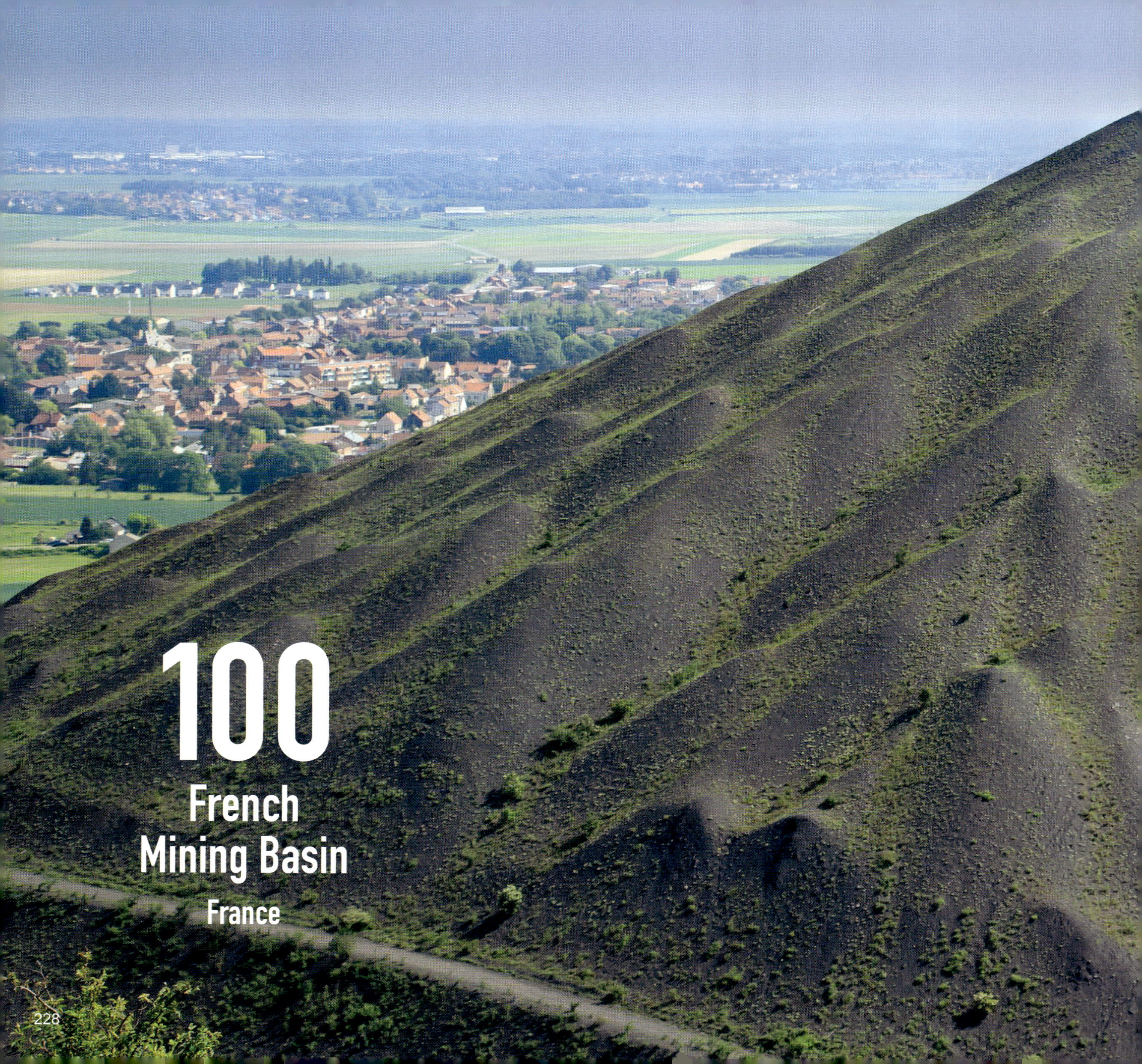

100
French
Mining Basin
France

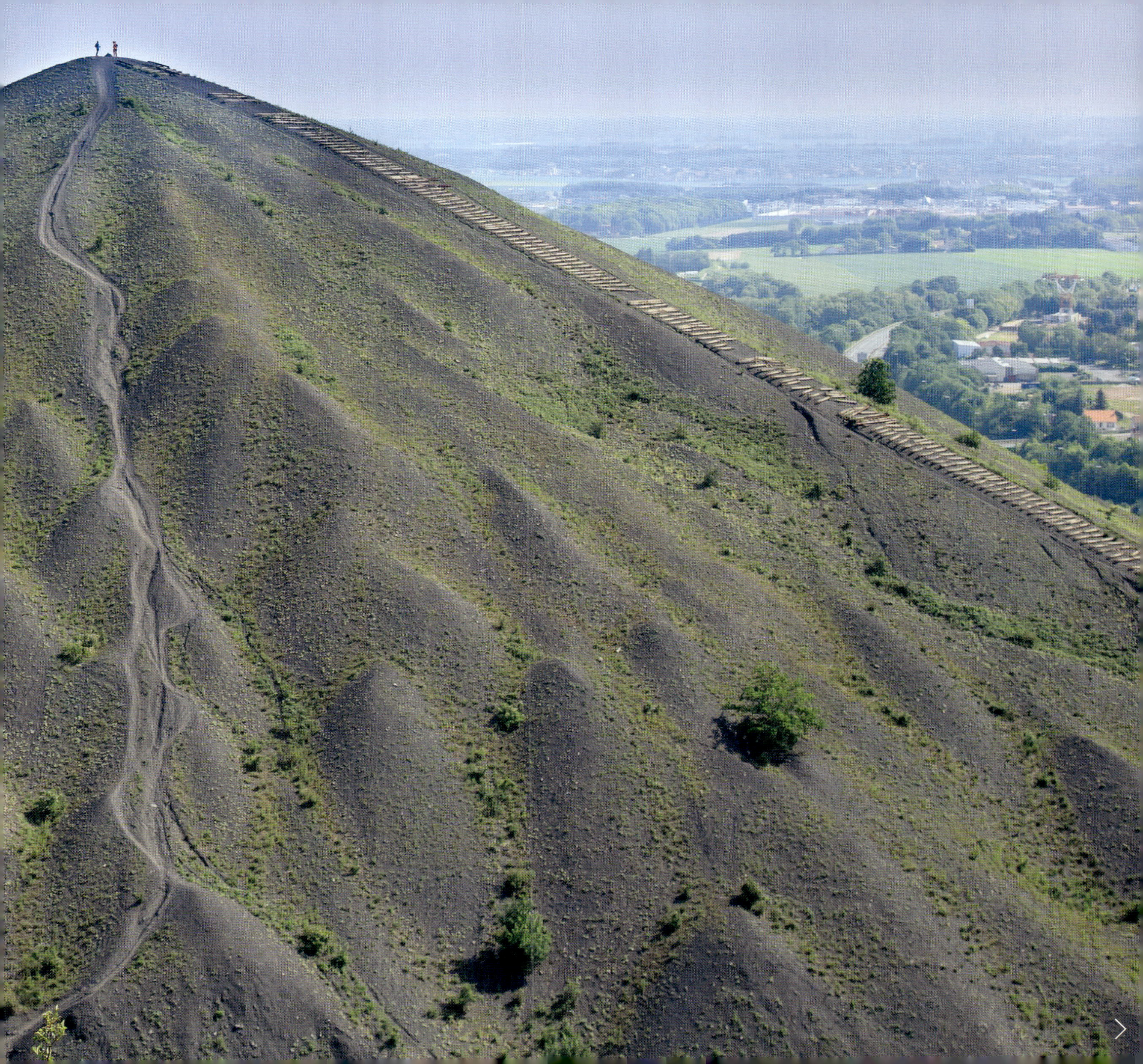

The Nord-Pas-de-Calais Mining Basin was once France's biggest coalfield, supplying half of the country's coal, producing 2.4 billion tonnes in 300 years. The industry went into decline in the '60s and '70s, mainly due to cheap imports, and the last mine closed in 1990, leaving behind a devastated economy.

Two hundred vast conical spoil heaps called *terrils* still tower over the region's otherwise flat landscape. At 198 metres, the tallest of these would dwarf the Great Pyramid of Giza. Regional rejuvenation has been led by Mission Bassin Minier (MBM), which campaigned for UNESCO World Heritage Site status, which was granted in 2012. Around a quarter of the *terrils* are protected along with mining infrastructure and workers' accommodation. MBM has continued to spearhead initiatives in the region and supports many others, but some are down to the creativity of the small communities that live among the *terrils*.

The Black Pyramid Trail (previous page)
Held every May, the *Trail des Pyramides Noires* race offers six routes ranging from 22km to 124km across the region's *terrils*.

Loisinord Ski Slope (below)
At 129m high, this artificial ski slope outside of Noeux-les-Mines in Nord Pas de Calais is officially the lowest mountain resort in France.

The idea of a former mayor, the 350m run comprises three slopes from beginner to expert and was created in 1996 for local people as much as tourists.

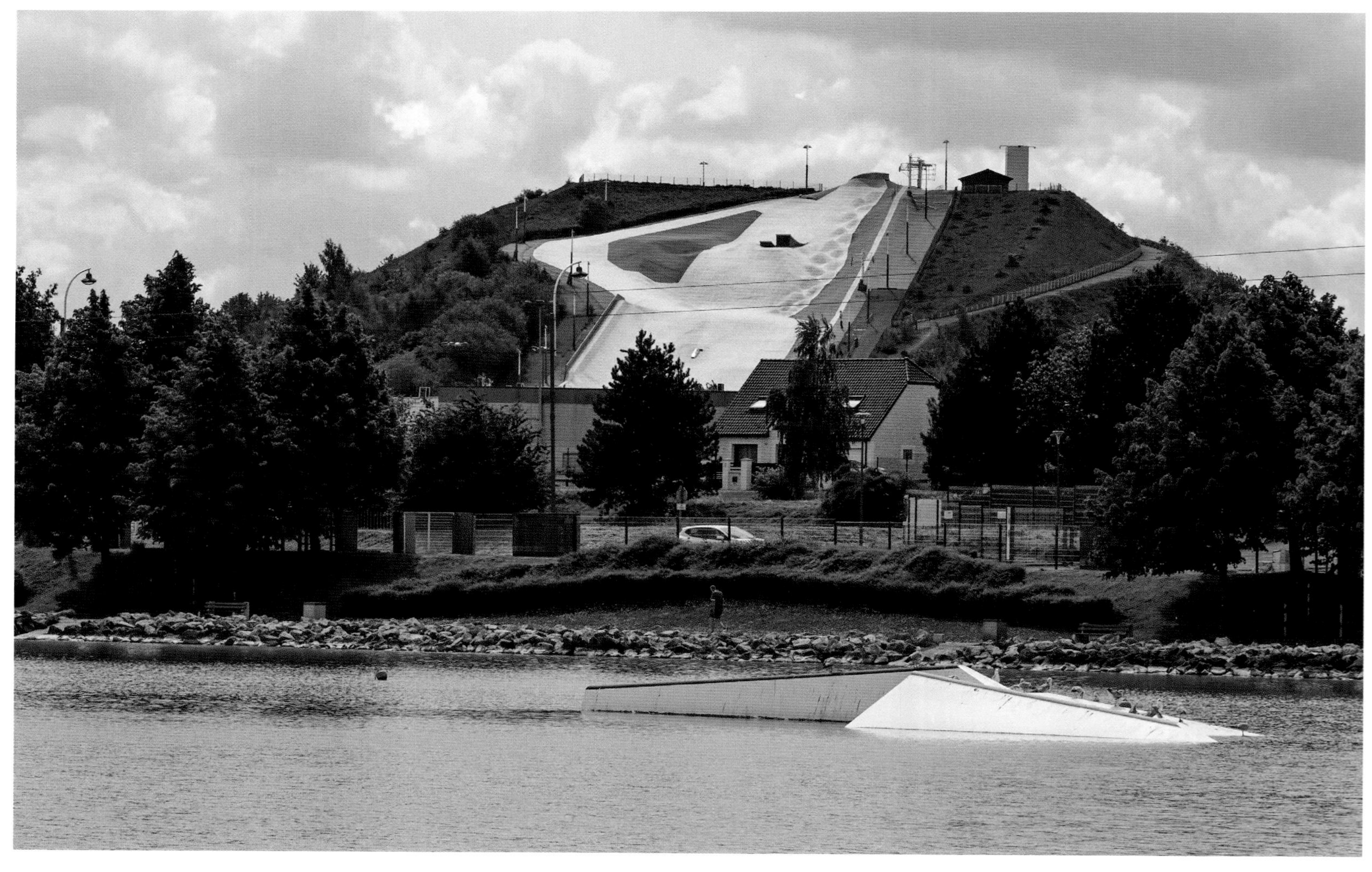

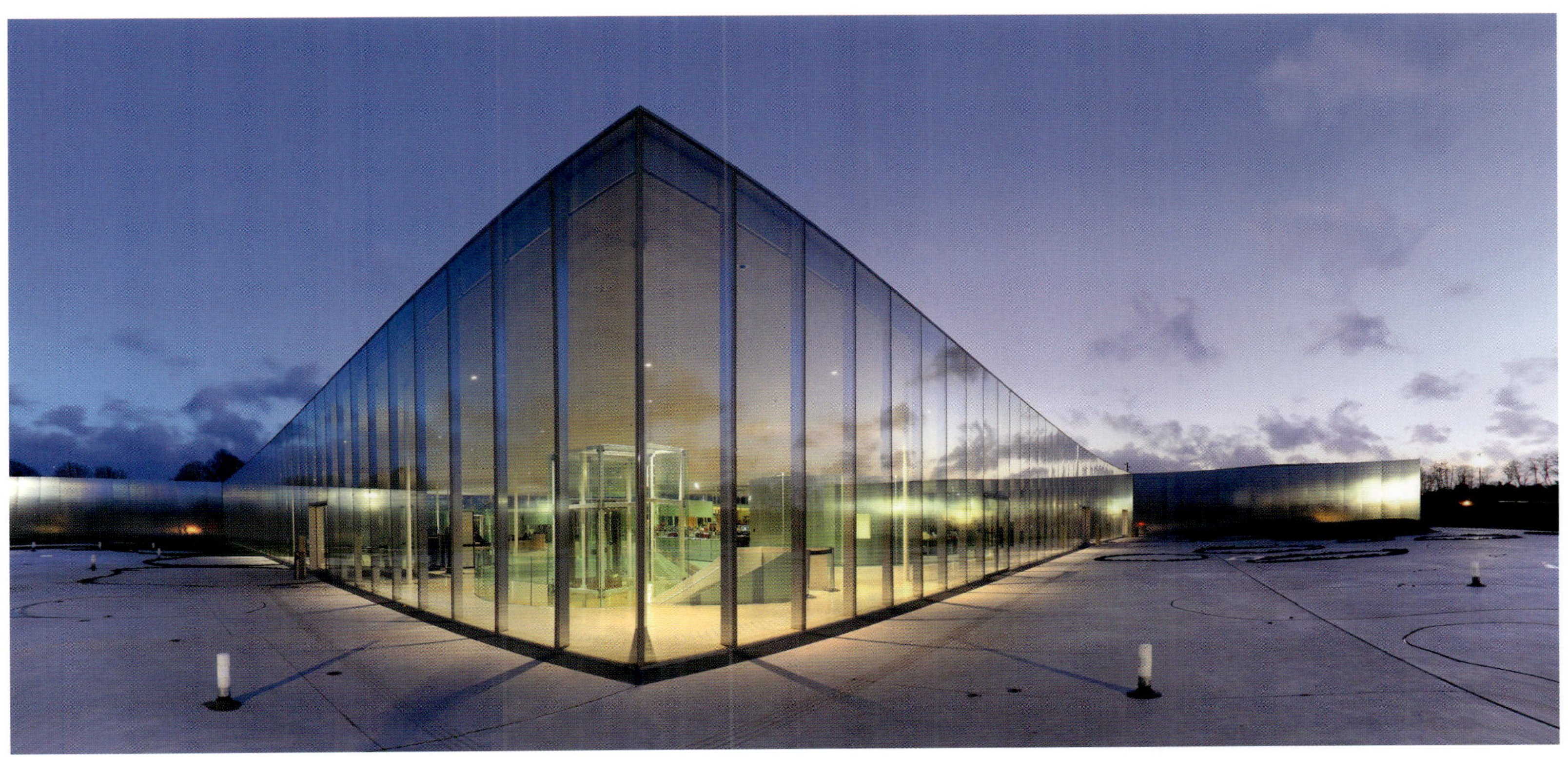

Louvre-Lens

When the Louvre in Paris declared it was going to open a regional outpost, only one regional département responded to the challenge: Nord-Pas-de-Calais (as it then was, now Haut-de-France). The département submitted six candidates, including Calais, but it was the small former mining town of Lens that won.

Winning the bid to host a regional Louvre represented a kind of recompense for the town's many hardships. Before the arrival of the Louvre, Lens was mainly known as the site of Europe's biggest mining accident – the Courrières mining disaster in 1906 – which cost the lives of over a thousand miners. The town had endured the Battle of Loos in 1915, occupation during both world wars and another mining tragedy in 1974. The last mine closed in 1986 with the remnants of its economy in ruins.

Louvre-Lens was built on a 20-hectare mining yard that had been abandoned since the 1960s. Two wings project at right angles to a central building in a deliberate nod to the Louvre in Paris. The museum offers 28,000m² of exhibition space, including the *Galerie du Temps* ('Gallery of Time'), a long open gallery that features over 200 works spanning 5,000 years loaned from the Louvre in Paris. The Louvre-Lens also hosts temporary exhibitions and has its own theatre that hosts an annual Muse & Piano.

The museum cost EUR 150 million to create, but the majority of the running costs are met by the regional government. It attracts 450,000 visitors a year on average, making it one of the most popular outside of Paris. Its demographic is different from the majority in France, however – its visitors are generally younger, and over half don't usually visit museums. Entry to the Gallery of Time is free – part of the collection is refreshed every year – and popular with local families. The park it sits within is also part of the draw.

The Louvre-Lens is the centrepiece of the Nord-Pas-de-Calais's regeneration, bringing EUR 20 million into the region every year. Perhaps more importantly, it has changed the way people look at what was once a heavily industrialised part of France.

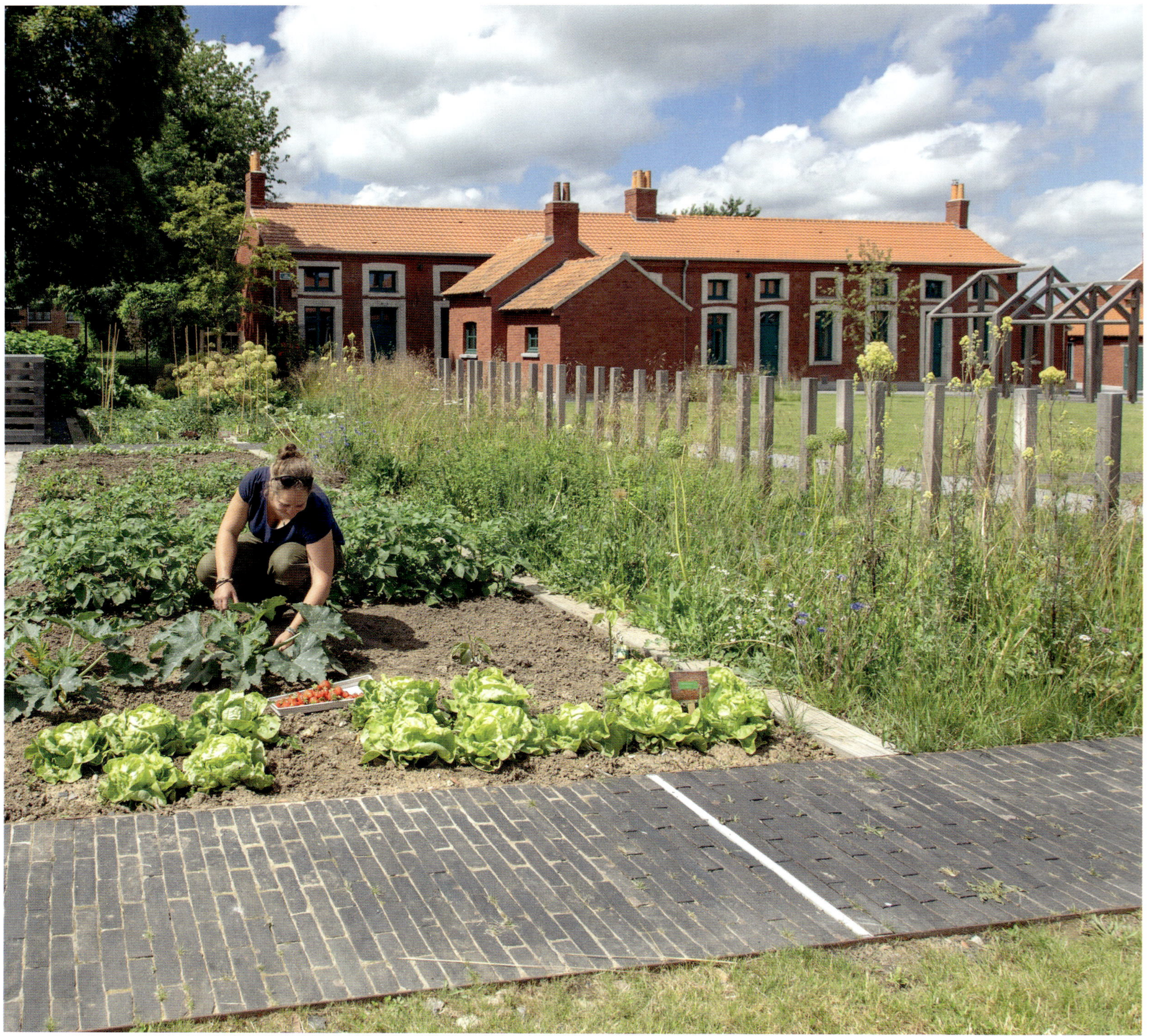

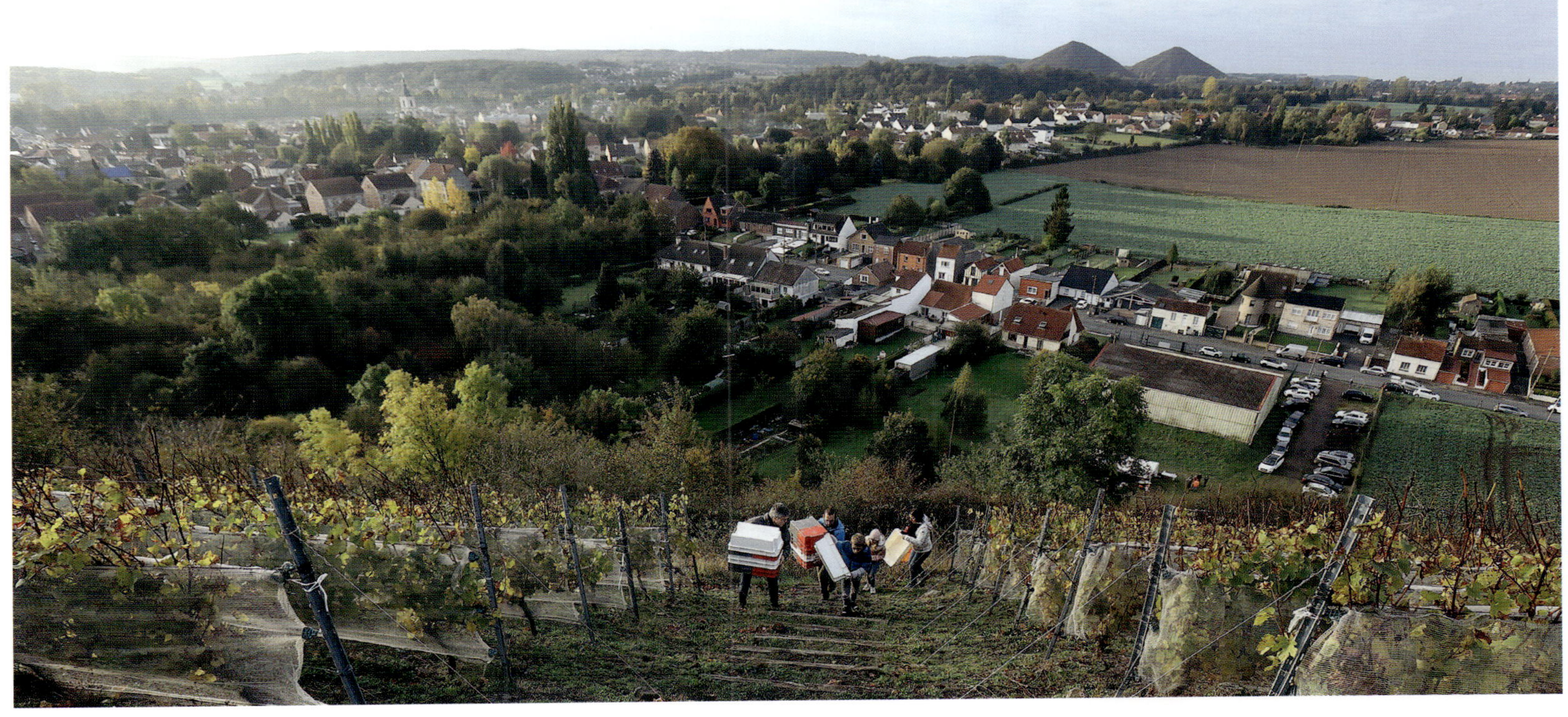

City of the Electricians

La Cité des Électriciens (left) was one of many purpose-built mining villages. 'City of the Electricians' was a nickname that stuck. Founded in the 1850s, the village predates the profession of electrician by at least 20 years, but the tiny village's streets were named after some of the field's pioneers, Ampère, Edison and Marconi. The local mine closed in 1979, and by 2008, *La Cité des Électriciens* was deserted. An artistic collective was hired to gather ideas from the public the same year. The village now has two exhibition spaces, one on the history of mining and the other for contemporary exhibitions. The former miners' cottages are also back in use, offering accommodation for visitors, artists in residence, and low-income households.

Haillicourt Vineyard

A still-combusting *terril* (above) probably isn't the first place you'd think of to create a vineyard, but for Olivier Pucek, it was a long-held dream. Pucek was born into a mining family in Bruay-La-Buissière in Nord Pas de Calais but had left the region over 30 years before for Charente, where he owned a small vineyard. In 2011, Pucek and the winemaker Henri Jammet set about planting 3,000 vines on the southern aspect of a 136m-high slag heap just outside of the village of Haillicourt. The steepness of the *terril* means that it can only be harvested by hand, but the 'soil' is fast draining, and the vineyard's altitude brings other benefits: the wind helps prevent spores from settling on the vines, which means it hasn't needed treating with pesticides.

Most of the cost of creating the vineyard came from Pucek and a few of his friends, with the rest coming from the mayor's office. The vineyard now produces well over a thousand bottles of 'charbonnay' every year – named for the Chardonnay grape grown on a coal – *charbon* – slag heap.

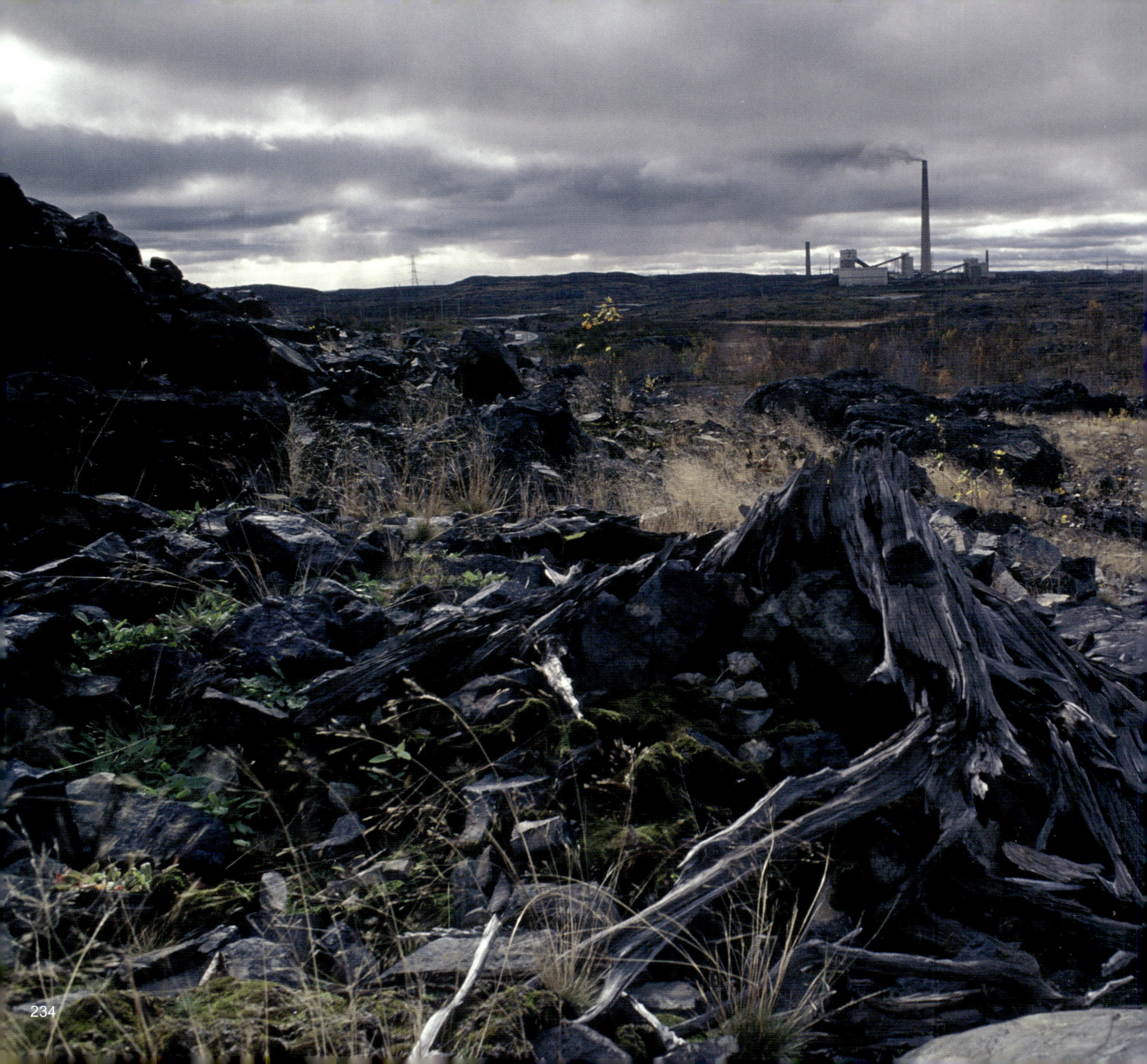

101

Sudbury's Regreening

Canada

For much of the 20th century amongst northern Ontario's seemingly endless forests lay a darkened, toxic landscape of black rock and lifeless lakes, bleached tree scraps scattered like driftwood, or sporadic, stunted survivors. Sudbury, in Canada's nickel mining heartland, was called a 'moonscape' by the media when NASA's Apollo astronauts arrived in the early 1970s to study the geology of its immense meteorite impact crater in preparation for their visits to the lunar surface. The description damaged the city's reputation for decades. Today the same place is a worldwide poster child for environmental renewal. The town's regreening has been praised by environmentalist Jane Goodall, who featured it in her film *Reasons for Hope*.

A combination of historical logging and early ore roasting and smelting activities had dispersed dense, sulphurous fumes and metal particulates over a wide area. The Canadian Shield's warm-coloured rock outcrops darkened, its forest failed, and its lakes turned to acid. The denuded, blackened landscape, known as the 'barrens', spread over more than 19,000 hectares close to the old smelters, surrounded by 64,000 hectares of 'semi-barrens' with severely diminished forests mainly of stunted white birch.

For the people who lived in this single-industry town, something had to change. Despite early environmental research, there was little drive for cleaner air until the late 1960s, when public health issues and acid rain moved up the political agenda, and the provincial government enacted the 1967 Air Pollution Act.

In the early 1970s the main nickel miner Inco, now Vale, was identified as the largest source of acidic emissions in North America. They constructed the iconic 381-metre 'Superstack', which incorporated scrubbers to reduce emissions and dilute and disperse the rest over a much wider area, improving local air quality but polluting further afield. Sudbury's old Coniston smelter also shut down, further enhancing air quality.

Distrust between dominant industry, community and environmental scientists stymied environmental progress. The provincial government took the lead, formally requesting scientific studies and mandating incremental reductions in smelter emissions. In 1973 the municipal government instigated the Vegetation Enhancement Technical Advisory Committee (VETAC), aka the Regreening Advisory Panel, made up of volunteer technical experts from government, academia, industry and community. Its brief was to revegetate key visible barrens around communities and major transport routes, employing recently laid-off miners and providing trees for planting on private property.

Laurentian University scientists identified acid generation from decades of sulphurous deposits as the obstacle to natural forest regeneration. Soil acidity was releasing toxic metals such as copper, nickel, arsenic and selenium from particulates (also from the emissions) and leaching aluminium from pre-existing soil minerals, poisoning both plants and soil. The acidity also degraded soil organic matter and caused nutrients to leach away. The scant, bare soil was vulnerable to temperature extremes, further hindering seedling establishment and erosion. Simply adding limestone reversed acidity and toxicity, triggering the conditions for revegetation. The barrens were particularly challenging; here, trees would not survive without an initial grass cover, so limestone, fertiliser and a grass-legume seed mixture were applied; the less-damaged semi-barrens were simply planted with trees and shrubs. This became known as the Sudbury Recipe. The first revegetation trials began in 1975 on half a hectare of barren hillside behind a school, upon which the children spread limestone, fertiliser and grass seed.

1978 proved an auspicious year; regional authorities gave high priority to enhancing Sudbury's landscape and image as 3,500 mineworkers were laid off and seasonal employment for students in the industry ended. VETAC's research became a major operation, transforming into the regreening programme, which employed summer students and out-of-work miners to implement the

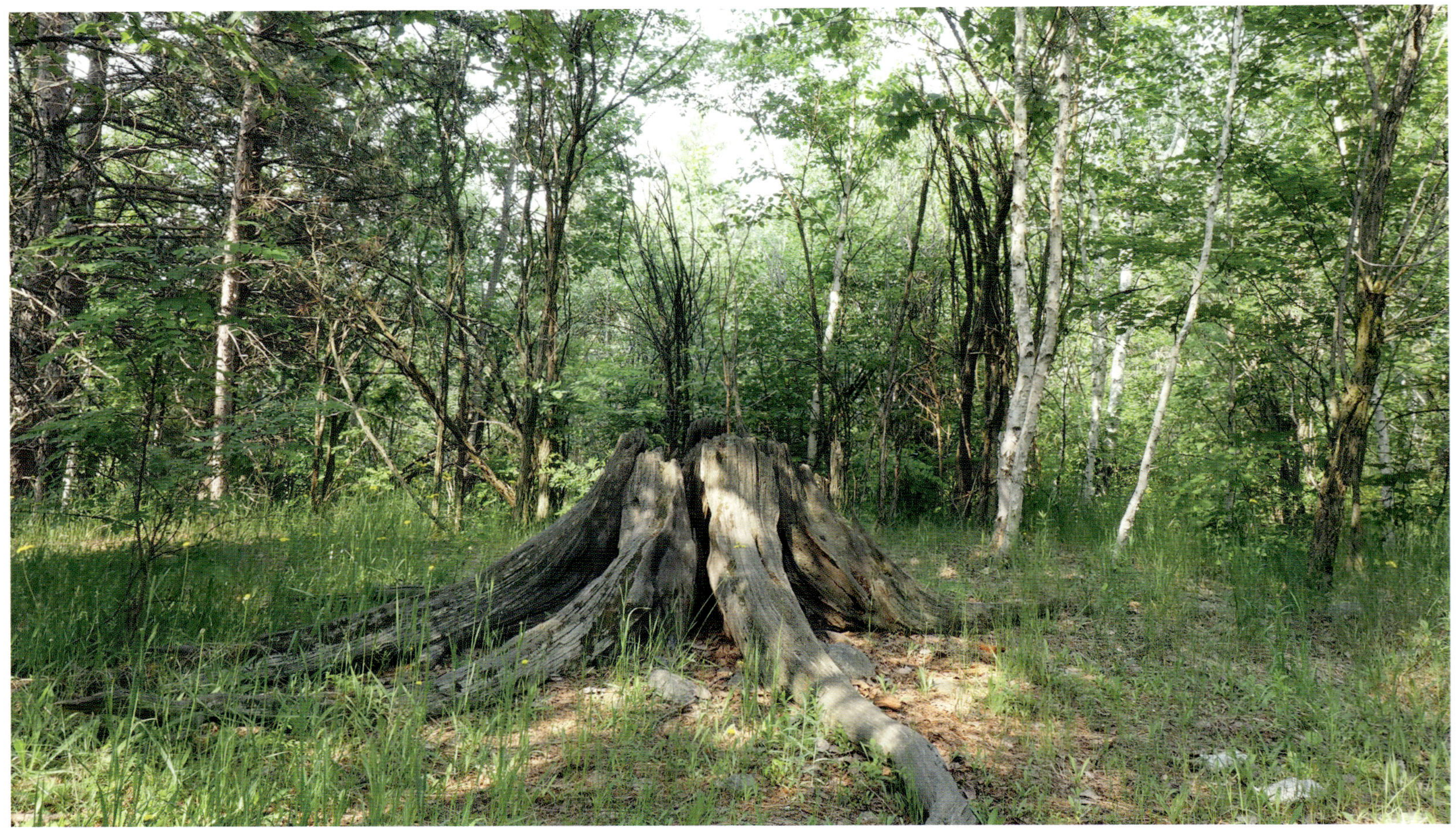

Sudbury Recipe. The first literal green shoots of recovery on a blackened barren were spotted in August 1978. So began possibly the largest-ever and longest-established community-based reclamation programme on industrially damaged lands. Ever since, VETAC has coordinated planting activities by small armies of community groups, volunteers and paid workers, funded by adaptively sourcing funds and in-kind support from industry and government.

The first trees were planted in 1979. By VETAC's 50th anniversary in mid-2023, 10,787,588 trees and shrubs had been planted and over 25,000 hectares had been regreened, including all of the easy-to-reach sites. In the early years, the only grass seed mixtures and trees available were commercial types. Refinements over the years mean the planting mixes now promote local, native species biodiversity.

Air quality improvements and terrestrial limestone applications have improved lake water quality in hundreds of acidified lakes since the 1970s, although aquatic ecological recovery has lagged behind chemical recovery. (Only a few lakes were directly limed, and they soon re-acidified if the surrounding land was not treated.) Lake recovery has been dramatic but is not yet complete. It is closely linked to how well the surrounding terrestrial ecosystem recovers.

Regreening has been so successful that the green arboreal carpet now confuses the memory of what went before. For this reason, some areas of barrens have been kept as reference points. The provincial government was persuaded to designate a mainly untreated, 620-hectare barren as the Daisy Lake Uplands Provincial Park. Ecologists argued that its metal-tolerant plants and mosses, and the self-seeded, small white birch amongst the bare rock, constituted a unique habitat that should be protected as a naturally recovering ecosystem for research purposes. They estimate that even with the regreening efforts, its environment will take about 200 years to completely recover; without intervention, it would take 2,000 years!

Planting trees alone will not create a diverse, resilient ecosystem. Tree recruitment occurs naturally in the older areas, but the understorey and forest floor remain impoverished and accumulating conifer needles present a fire risk. VETAC has been planting imported turf mats of forest floor vegetation, and their herbaceous layer is slowly spreading. Trials are underway to regenerate ecologically important lichens by broadcasting lichen fragments from aeroplanes; now, small patches of lichens are very gradually returning to places they last grew more than a century ago. It is estimated that the new forests have fixed two-thirds of a million tonnes of carbon, a major contribution towards Sudbury's 2050 carbon-neutral target.

Sudbury's regreening has received numerous national and international conservation awards over the years. The original partnership has expanded beyond all recognition and supports residents to improve their own doorstep environments. Within a single generation, the community's attitude has changed from one of apathy and resignation to a can-do attitude that takes pride in its achievement. The regreening of Sudbury has cost about CAD 40 million so far, but the reputational and socio-economic gains are huge, according to the city council. Investment in cultural attractions like Science North and Dynamic Earth, built in a former mine, have raised the city's profile. The admirable environmental effort has influenced economic diversification; anecdotal evidence suggests that visitors and workers are more readily attracted to Sudbury, resulting in a range of new commercial, tourism and cultural opportunities, including new hotels; an expanding regional medical centre and school, with a state-of-the-art cancer treatment centre serving north-east Ontario; major retail outlets; green tech; call centres building on the population's bilingual skills; and, of course, mining services.

This 50-year regreening story is one of applied science, community action, government regulation and leadership, and industrial collaboration. The restored forests and lakes are not perfect renditions of their progenitors, and they don't pretend to be – these are works in progress. Most significantly, the character and reputation of the place have moved on from a moonscape to an inspiring global symbol of people-led renewal.

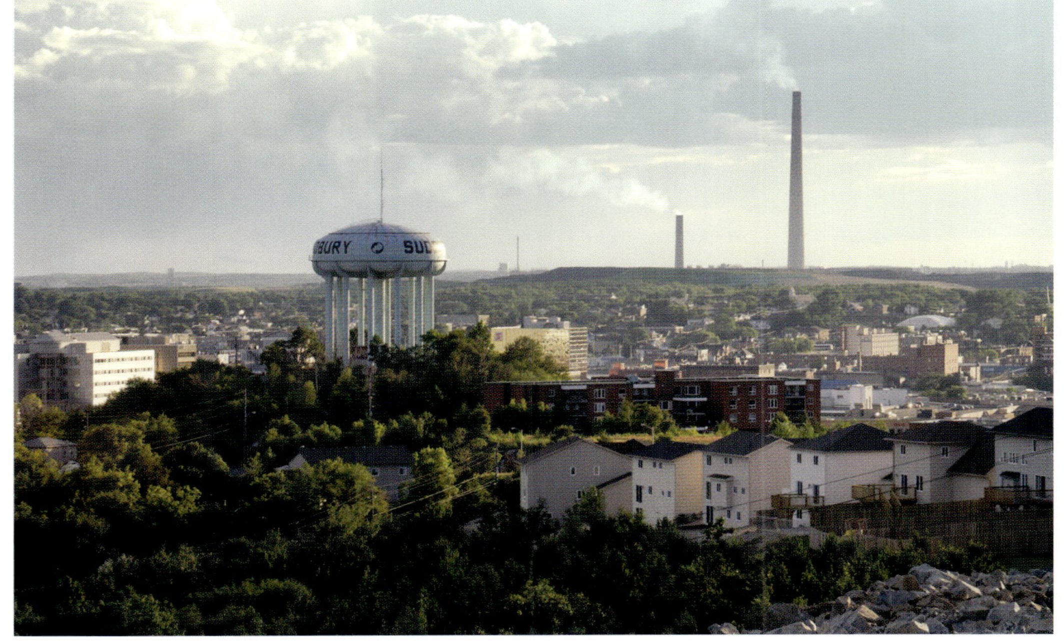

Previous page: Sudbury's 'barrens' in the shadow of its iconic Superstack. **Opposite:** Forest regeneration on the once poisoned footprint of the old. **Left:** Sudbury's verdant city centre with the now inactive smelter stacks behind.

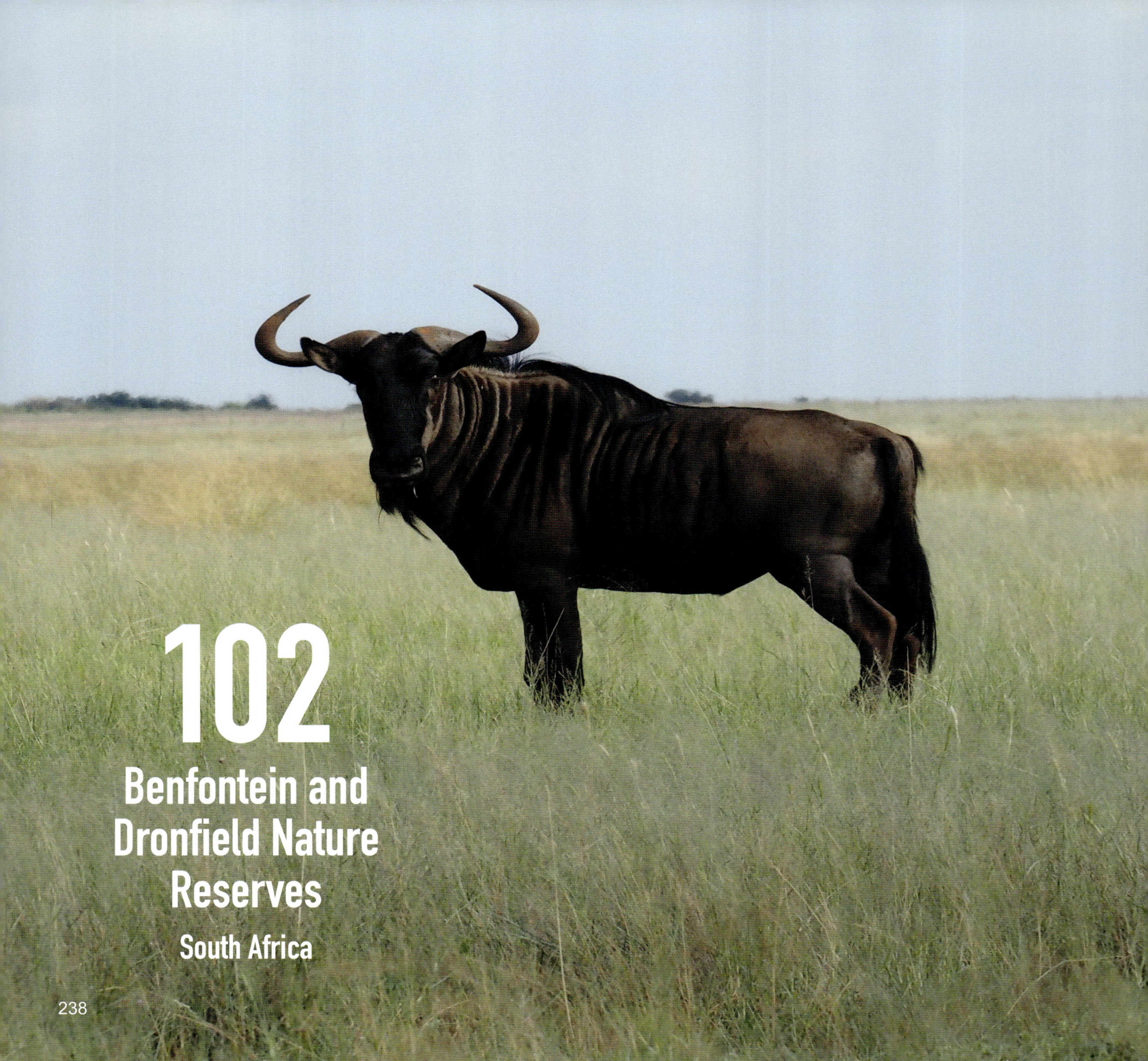
102
Benfontein and
Dronfield Nature
Reserves
South Africa

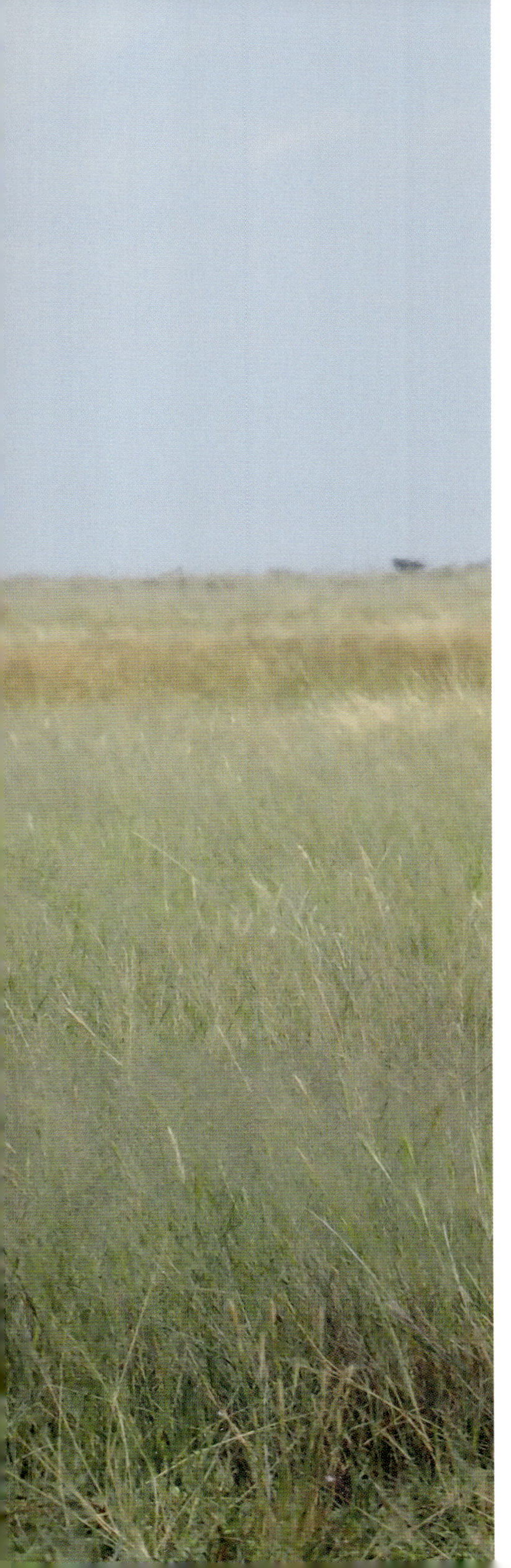

Not all land owned by mining companies is actually mined. Over time, this land may acquire new environmental, socio-economic and cultural values by virtue of its protection from disturbance or development. To enable access to land for mining, De Beers bought farms in key locations across southern Africa many decades ago. Many locations are now important wildlife and game reserves, such as Benfontein and Dronfield near the diamond heartland of Kimberley.

Originally bought by De Beers in 1891 for its diamond reserves, Benfontein's 11,000 arid hectares comprise a variety of habitats – karoo, thornveld, savanna and salt pan – resulting in the rich biodiversity and rarities for which they are famed. It is a designated Important Bird and Biodiversity Area (IBA) with 25 mammal and 260 bird species, including many endemics.

The focus is on ecological research – its long-term studies have influenced conservation practice in southern Africa for years. It is home to South Africa's purest breed of black wildebeest – the result of a captive breeding programme begun in the 1940s when the species' global population dropped below 100. Half of the remnant population was moved to Benfontein, where it was rescued from the precipice of extinction. Globally important research is also underway on Benfontein's aardwolves, its secretive black-footed cats (one of Africa's smallest and rarest cats) and other nocturnal mammals, which famously bring the place alive at night. Group visits allow bird watching, mountain biking and horse-riding.

To the north of Kimberley, Dronfield was purchased in 1888 by De Beers for game farming. It has a different character and purpose to Benfontein, with the flat-crowned camel thorn trees of the Kimberly thornveld partly open savanna being managed for ranching eland, gemsbok, blue wildebeest, and more. From 1994, it was farmed for cattle before becoming a nature reserve specialising in breeding white rhino, sable antelope and other high-value species. It is also an IBA, boasting 140 bird species, including the critically endangered white-backed vulture, which breeds here and has been the subject of conservation research since the 1970s. Unlike Benfontein, Dronfield hosts high-end wildlife tourism. Its six well-appointed thatched chalets in the shade of the camel thorn trees are a tranquil base for exploring the 12,000-hectare reserve or, further afield, Kimberley and its surroundings.

Alongside game breeding, conservation research and eco-tourism, Dronfield also hosts educational visits for students of all ages arranged by environmental education organisations. The Northern Cape Nature Academy uses the site to deliver training in wildlife tourism and management and game capture.

Benfontein and Dronfield, along with neighbouring Kimberley, form three of a group of eight culturally and ecologically important De Beers sites across South Africa and Botswana branded collectively as The Diamond Route. This network totals close to 160,000 hectares, but each site has a different character and use, and contributes to economic diversification and shifts in perspective.

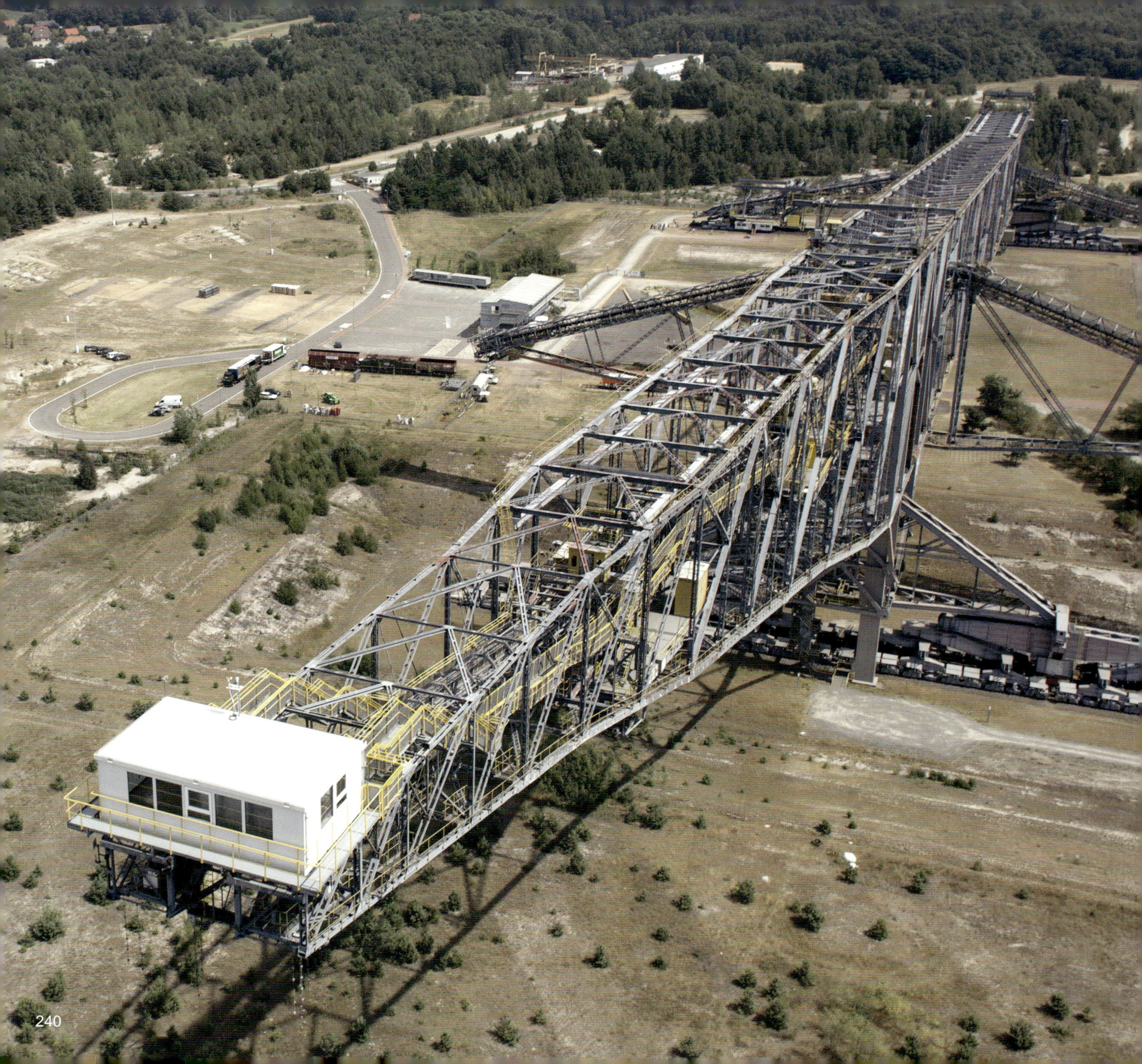

103

F60 Visitor Mine

Germany

Everything about F60 is enormous and impressive: like the skeleton of a massive prehistoric beast, it rises high over houses and trees, dominating its terrain. In its natural opencast mine habitat, it ate the earth to reveal the lignite beneath – a daily diet of 50,000 tonnes of overburden, enough to bury a football pitch eight metres deep! Relatives of F60 and its family of industrial behemoths still work today; as they slowly grind through their vast, devasted landscape, they are so massive that humans become almost invisible specks.

F60 is named after its 60-metre cutting height. Officially known as an overburden conveyor bridge, it is also known as the Reclining Eiffel Tower, although at over 500 metres long it would dwarf the Paris landmark if it were stood on end. The 200-metre-wide, 80-metre-high, 11,000-tonne structure is the largest and longest vehicle ever made, bettered in weight only by its in-pit relative, the bucket wheel excavator.

After the rapid collapse of the former East German lignite mining industry in the 1990s, these icons of the industry were no longer needed. They met their demise by being unceremoniously dismembered with explosives, leaving behind a poignant pile of twisted metal. F60 was originally slated to be similarly destroyed when the Lusatian Klettwitz-Nord opencast mine closed.

It was saved by a group of forward-thinking enthusiasts led by Lichterfeld community representatives and Elke Löwe, a landscape architect from Senftenberg. They understood the F60's heritage symbolism, but also that it signified the restructuring being experienced by the region and its role in pointing to a new future. There were initially more opponents than supporters, and there were the inevitable questions over who would pay to convert this machine into a visitor attraction and move it to a new site. Who would run the new venture? How would support from the local authorities be gained? An expert investigation by the German Institute for Tourism Research in Berlin addressed these concerns and convinced local political leaders that saving the machine was worthwhile.

F60 was saved from the scrapyard in 1998, with ownership passing to Lichterfeld-Schacksdorf municipality. It became a flagship project of the ten-year International Building Exhibition (IBA) Fürst-Pückler-Land, which helped develop the concept for an F60 Visitor Mine heritage attraction. In 2000, it was moved to its current resting place, an engineering feat witnessed by a 4,000-strong audience. Top-level political support was given when German Chancellor Gerhard Schröder visited. The F60 Visitor Mine management structure was formed out of the F60 Friends Association and the F60 Concept company that manages and markets the destination – initially with the support of the IBA.

The open pit adjacent to F60 began to be flooded in 2001 to create the 330-hectare Lake Bergheide, and the site now features beaches, a campsite and holiday accommodation, watersports and a solar farm, as well as space for nature. The following year, the site opened to visitors, attracting 70,000 in its first year, and in 2009, it celebrated its 500,000th visitor. It now attracts around 60,000 people annually. After dark, the F60's light and sound installation, created by light artist Hans Peter Kuhn, offers a different kind of visitor experience.

F60 has become an international icon of the Lusatian region. It is well-served by road and cycle routes and is an important stop on the European Route of Industrial Heritage. It hosts sporting events, such as the F60 Triathlon, and cultural events, including a popular outdoor cinema, as well as activities such as abseiling and 4x4 tours.

As one clambers through its metallic skeleton with a guide, the open structure offers little shelter against the elements and the ground, far below, is worryingly visible between one's feet. From the highest point at 74 metres, even through the rainy gloom, the views across the low-lying, post-industrial – and surprisingly wooded – landscape stimulate contemplation about change and what it can achieve.

104

Reclaimed Earth Colours

USA

Beauty is in the eye of the beholder – and so is opportunity. To see both in acid mine drainage (AMD), one of the most intractable and expensive mining-environmental challenges globally, requires a creative and activist insight. John Sabraw and Guy Riefler (respectively, Professor of Art and Professor of Civil and Environmental Engineering at the University of Ohio, Athens) saw the potential and have created a beautiful solution to dealing with AMD in America's Rustbelt: using it to make pigments and paint.

Obscured from view amongst the rural tranquillity of Appalachian Ohio's small fields and woods, the babbling Sunday Creek, near Athens, sounds idyllic, but on approach, the pungent sulphurous-metallic odour does not bode well. Up close, the bright orange streambed contrasts with the fresh green leaves of the overhanging branches; the lifeless water is crystal clear. The origin of this decay is evident just upstream as a bubbling seep of water known as the Truetown Discharge. On first appearance the acid-laden, iron-rich water is clear, and only by reacting with atmospheric oxygen at the surface do the smothering iron hydroxides, or ochre, precipitate out. Over 2,000 kilometres of streams and rivers are similarly impacted by coal AMD in Ohio alone.

The discharge drains the abandoned Truetown underground coal mine, which closed over 100 years ago. After closure its 96 km^2 of workings filled with groundwater, and the AMD broke out into Sunday Creek in the 1980s. The discharge issues 3,800 litres of pH 3.5 to 4, iron-laden water per day into the creek, part of the Ohio River catchment. This equates to nearly 1,000 tonnes of dissolved iron annually – approximately two cars-worth per day. The Creek is lifeless for over 11km downstream; the main impacts are due to acidity and ochreous smothering, although thankfully, toxic metal concentrations are very low. This discharge is one of the Ohio Department of Natural Resources' highest priorities.

While visiting similar Appalachian AMD sites with an environmental group around 2007, painter and environmental activist Sabraw was struck by the scene of decay and the disturbing bright hues of the AMD sludge coating the riverbeds. Aware that ochre has been used as a pigment for hundreds of thousands of years, he was inspired to take a creative response to this environmental challenge.

Trudging through waist-deep acidic water and orange sludge, Sabraw and Riefler collected ochre by hand – a filthy job! In the workshop they extracted the iron oxides and mixed them with oils and acrylics to make unique artists' paints imbued with a sense of environmental responsibility. A range of colours was created by heating the minerals to different temperatures to make yellows, oranges, browns, reds and violets. Credibility grew when Gamblin, the artists' paint manufacturers, agreed to a trial run of 500 tubes of 'Reclaimed Earth' paints. Sabraw uses the paints in his own innovative practice, for which he has received much media interest and many accolades, one highlight being invited to present his art at the UN Headquarters in New York.

Today a pilot-scale AMD Remediation Testing Site stands in a field adjacent to Sunday Creek, from which the ochre is now collected. Around 200,000 metric tonnes of iron oxide are used in the US each year in paints and to tint a range of products from cosmetics to concrete, most of which is imported, mainly from China. The pair, in collaboration with Michelle Shively of the citizen action group Rural Action, who brings environmental knowledge and marketing expertise, are building a commercial-scale treatment plant to treat all of the AMD emanating from the Truetown Discharge and make iron oxide pigments on-site as a feedstock for American industry. Millions of dollars have been provided by the State of Ohio and the federal Abandoned Mine Land Economic Revitalization Program for the capital outlay. Operational costs will be met by selling the pigment. It will be run by the bespoke social enterprise True Pigments, the aim being to reinvest any profits into cleaning up AMD-contaminated streams elsewhere.

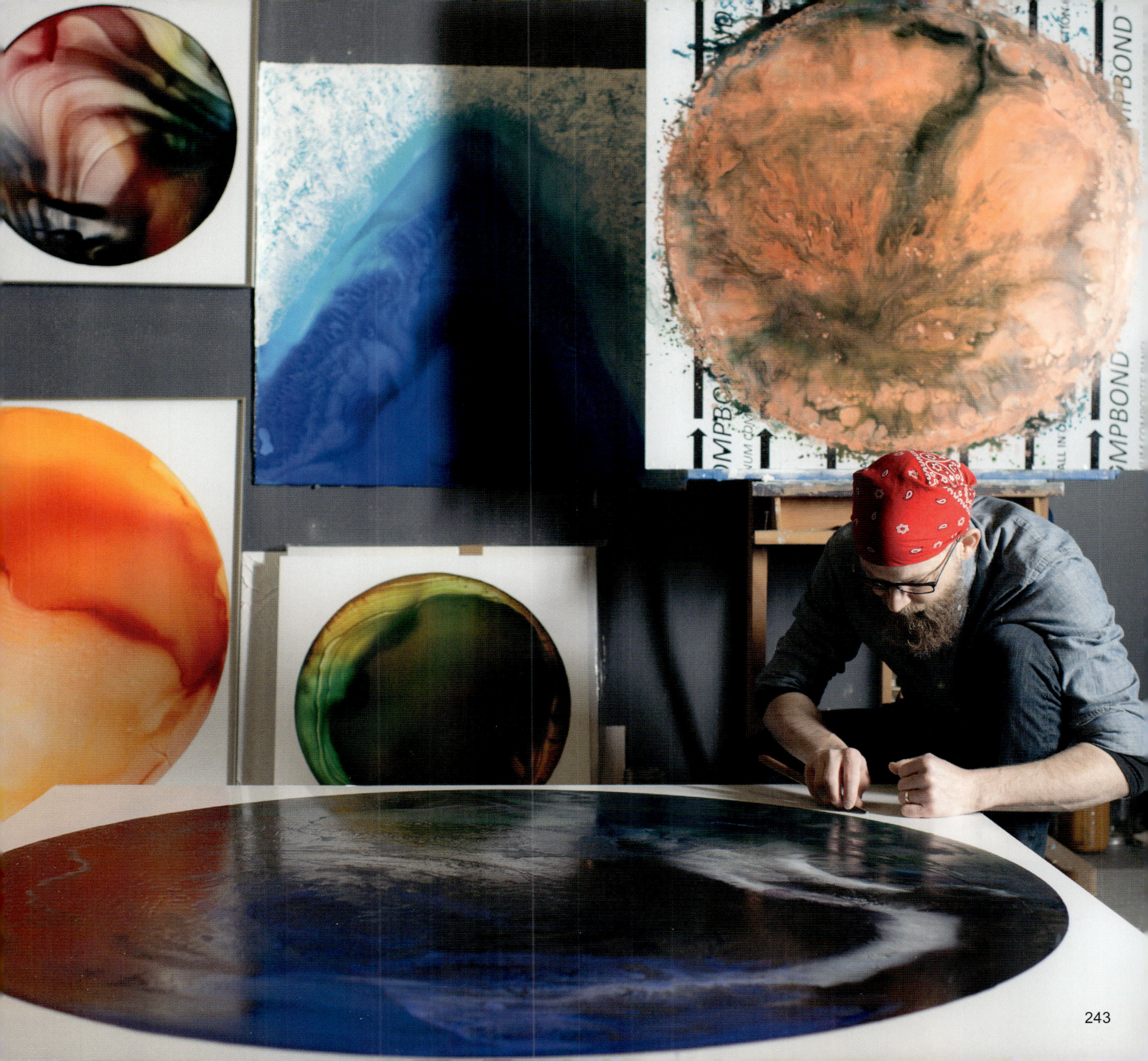

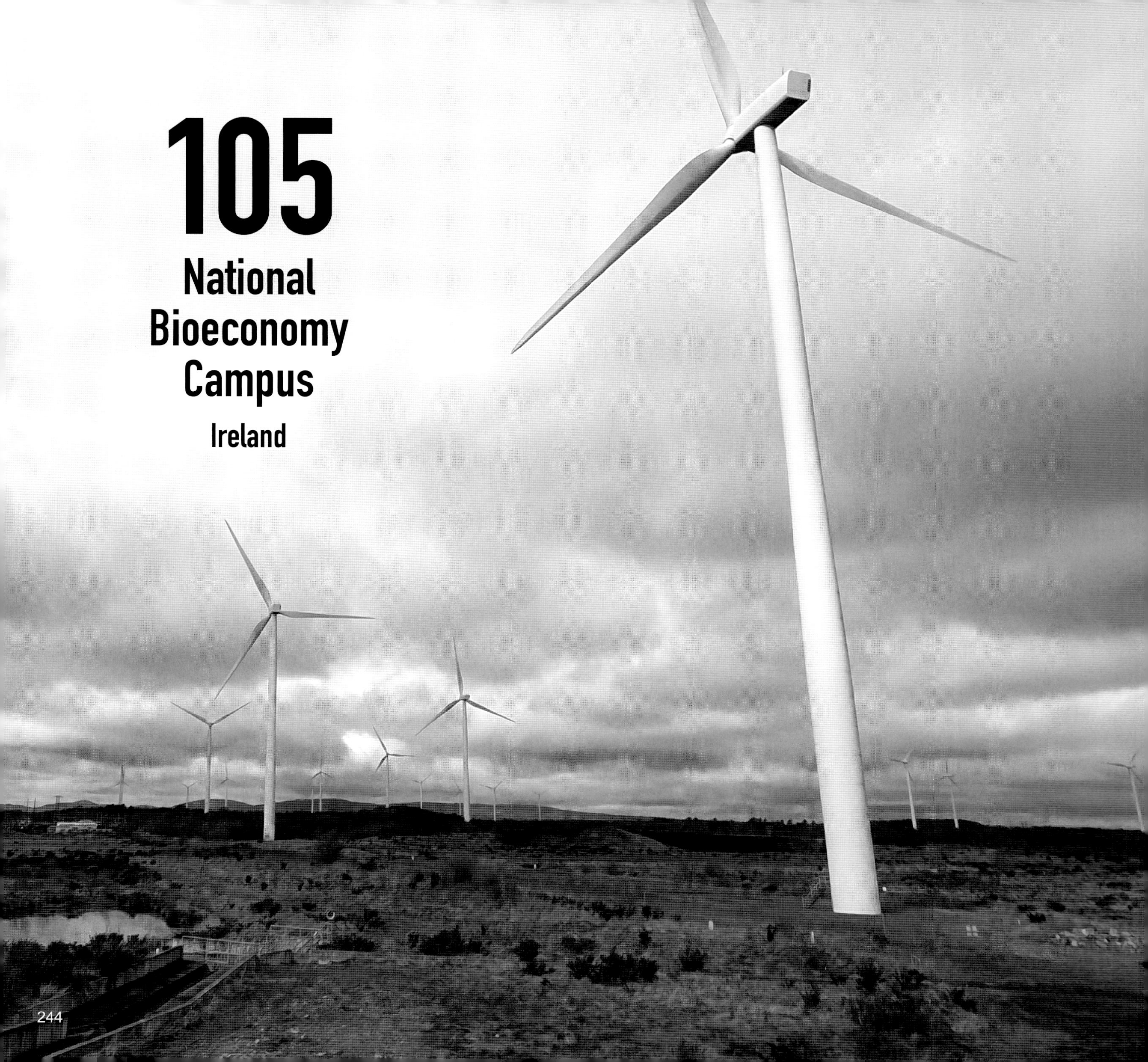

105
National
Bioeconomy
Campus
Ireland

It's a bleak place in winter with little shelter from the biting wind atop the steppe-like expanse of the closed tailings facility. Gazing over this part of Ireland's County Tipperary from the closed Lisheen mine site, the movement and vertical lines of its forest of wind turbines steal the view and hint at some innovative mine site repurposing activities over the last decade.

Zinc and lead ores were extracted from Lisheen's underground workings between 1999 and 2015. Initially operated by Anglo American, it was sold to Vedanta Resources in 2011. Anglo began planning the mine's closure in 2005 and implementation became Vedanta's responsibility after 2011. A closure task force was set up in 2013, with the full support of Tipperary County Council, to confer with stakeholders about alternative post-closure uses for the site. Most of the site's infrastructure was removed during the closure phase, except for a few buildings that might have a use for future economic activity. In 2021, after receiving a coveted mine closure completion certificate from the Irish Environmental Protection Agency, the site transitioned into aftercare.

Although the closure engineering challenges were complex, due in part to early planning and preparation they were relatively straightforward to handle. The social transition aspects were more demanding as there were many factors that could only be influenced – rather than controlled – by Lisheen. Again, early recognition of these challenges and prompt action were crucial; for example, the employee outplacement programme started two years before closure and was well received, with over 90% of employees finding new work within a year of closure.

A closure vision for a green energy hub was developed by Anglo in the mid-2000s. Lisheen successfully applied for planning permission to construct 18 wind turbines on-site during operations to help reduce the pumping costs for what was a very wet mine. Wind energy was a new industry in the country and one that divided opinion nationally. The Lisheen team worked with the local community to overcome anti-turbine sentiments, proving the value of wind power to the area. Once the community accepted the project, subsequent windfarm developments have had a smoother passage. On closure, the electrical infrastructure installed by Lisheen to support the windfarm was kept and subsequently used by others to develop further wind farms in the vicinity. The Lisheen substation is now connected to 52 wind turbines, installed in four phases, with a capacity of 131MW. Planning permission has also been obtained for a 122MW solar farm on the closed tailings facility. Discussions are underway regarding power storage projects, including batteries and – potentially – hydrogen production and storage.

As the green energy hub concept was being implemented, the vision morphed into one of a national bioeconomy hub to capitalise on the site's buildings, transport links and readily available renewable energy and water supply. The bioeconomy concept is based on low-carbon growth and resource efficiency that uses renewable biological resources, including wastes, to produce food, energy and industrial goods. A national bioeconomy hub at Lisheen was subsequently named in the 2018 government's National Policy Statement on the Bioeconomy. The Irish Bioeconomy Foundation (IBF), which arose as an outcome of the task force, is headquartered at Lisheen. Its mission is to promote the development of a sustainable bioeconomy. It is a not-for-profit organisation set up by the county council and various commercial and academic members.

The IBF provides practical technical support and advice for companies to solve issues related to the designing and scaling-up of bioprocessing engineering. Plans are underway to develop a state-of-the-art bioprocess pilot plant on-site.

The Irish Government is committed to providing EUR 4.6 million towards Lisheen's bioeconomy innovation and piloting facility. The IBF and its members have attracted additional investment and funding to the site, and work is ongoing to grow the campus and make the Lisheen site the fulcrum for Ireland's bioeconomy as the National Bioeconomy Campus. The campus is now a focal point in county and regional development plans; the area has been nominated as a Biomethane Demonstration Area in the county's recently published Draft Climate Action Plan. It is also at the core of the nominated Decarbonisation Zone.

Other related environmentally focused businesses with links to the bioeconomy are moving onto the site with interests in composting and the anaerobic digestion of organic wastes that will produce biogas. Some of the biogas will be used as a fuel in the combined heat and power plant to drive the facility, and the digestate will be used to fertilise agricultural land.

106 Agribusiness Development

South Africa

Kumba Iron Ore, an Anglo American subsidiary, has embarked on an ambitious initiative to create a major agribusiness hub at its massive Sishen mine in South Africa's Northern Cape. The project is part of the company's broader effort to drive sustainable economic diversification and community development across the mining area and stimulate prosperity that will outlast the lifespan of the mine.

The Sishen open pit iron ore mine, one of the world's largest at nearly 14 kilometres long, has roads, rail, extensive land holdings, power and water infrastructure that give Kumba a strong basis for agribusiness ventures. The vision is for the company's land assets, capital and convening authority to catalyse a commercial farm enabling agri-processing services and offtakes to independent small and mid-size farmers, with added agriculture expert services to promote job and small business creation. This will provide meaningful post-mining livelihoods and transferable skills to community members.

In 2022, Kumba commenced practical trials and studies to find high-value crops suited to the Northern Cape's climate and soils that also have large-scale farming potential. Crops being trialled include pistachios, pecans, pomegranates, raisin grapes and saffron. They have established two one-hectare test farm plots, one on flat agricultural land at the company's Lylyveld property and another atop a 100-metre-high waste rock dump called G80. The plots test growth rates, yields, quality, water usage and their potential for post-mining land use after rehabilitation. Plants are watered with drip irrigation using the mine's wastewater and are protected by windbreaks.

On G80's challenging calcrete rock surface, they pit-planted over 100 pecan trees to evaluate the agricultural reuse of mine waste areas. If they thrive, there is potential for up to 500 hectares of agricultural production on G80. Year 3's crop, the first, is being analysed for safe food use. At Lylyveld, alley cropping is being piloted; legumes grow between rows of tree crops like pistachio and pecan to enhance soils and enable earlier harvests before the trees mature.

Other promising trials include prickly pear as a protein source and livestock feed, and beekeeping for pollination.

Building local skills and small businesses is critical, so in partnership with commercial mega-farms, Anglo American/Kumba is investing in training programmes with an emphasis on work experience and apprenticeships for youth and women. The strategy aims to nurture independent farming businesses through coaching in agriculture, finance, and marketing to avoid dependence on mining subsidies.

With integrated planning, it's projected that profitable and sustainable farming can be established on over 1,400 hectares of the company's land. The target is for agribusiness ventures to generate a good return on investment and directly employ over 500 people within ten years, plus many more jobs in processing, transport and support roles.

In such ways, the company aims to drive sustainable prosperity for Northern Cape communities beyond the boom-and-bust cycles of commodity mining. As with progressive ecological restoration during mining, Anglo American is starting this economic diversification early to lower transition risks and costs as mining operations eventually wind down. The Kumba initiative faces substantial challenges but could provide a replicable model for mines globally.

Right: Pecan trees planted on top of Sishen Mine's G80 waste rock dump.

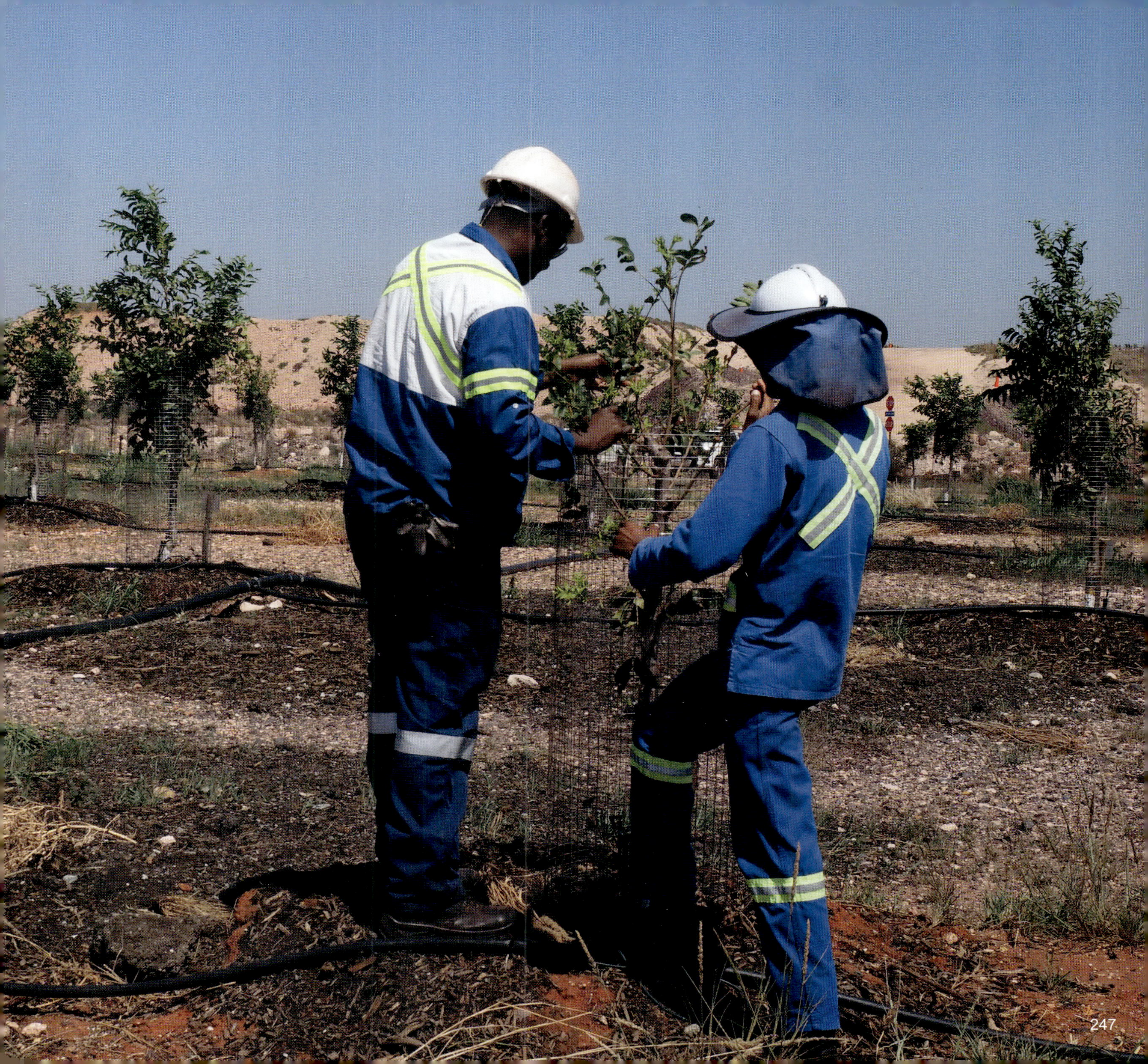

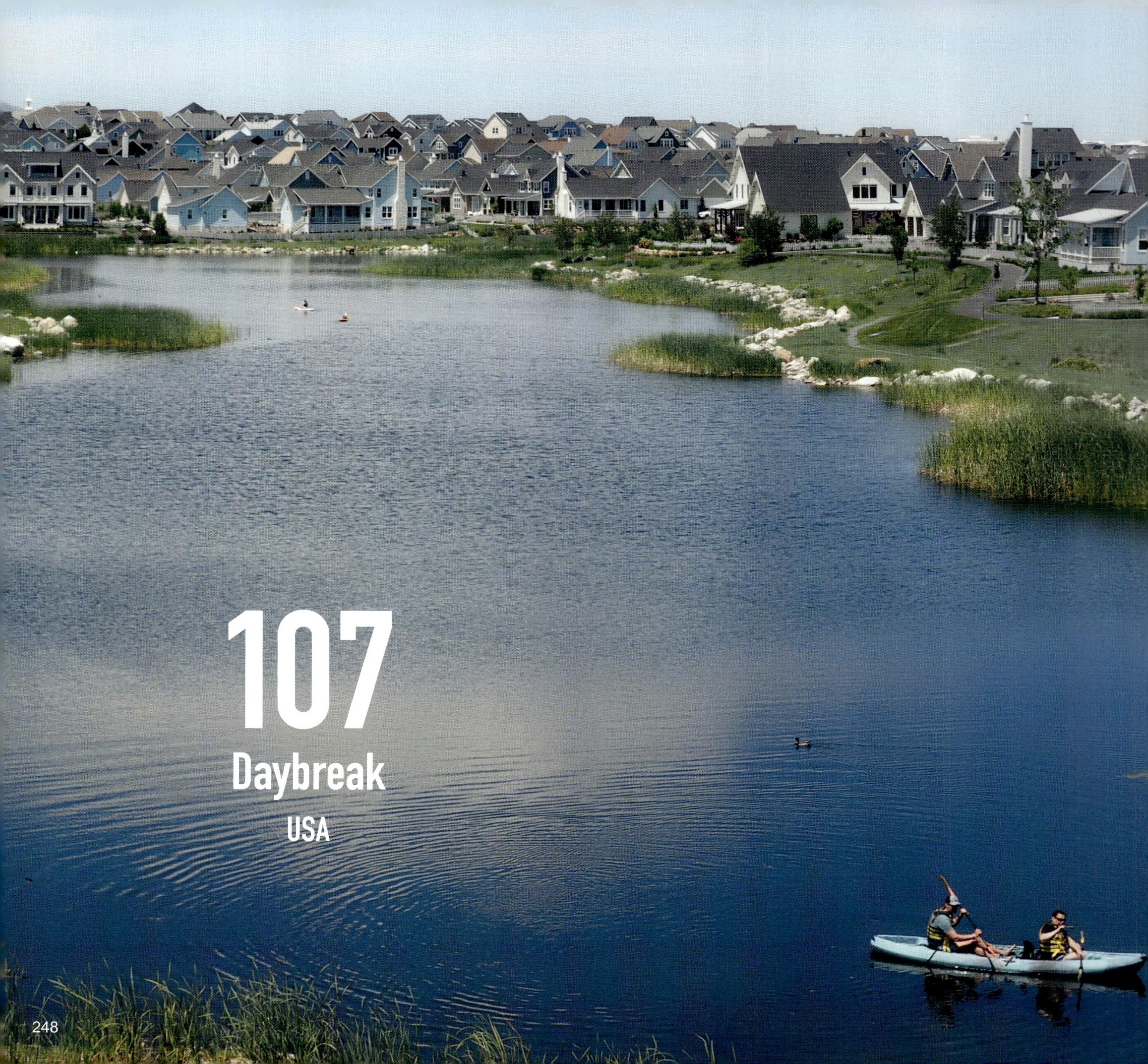

107

Daybreak

USA

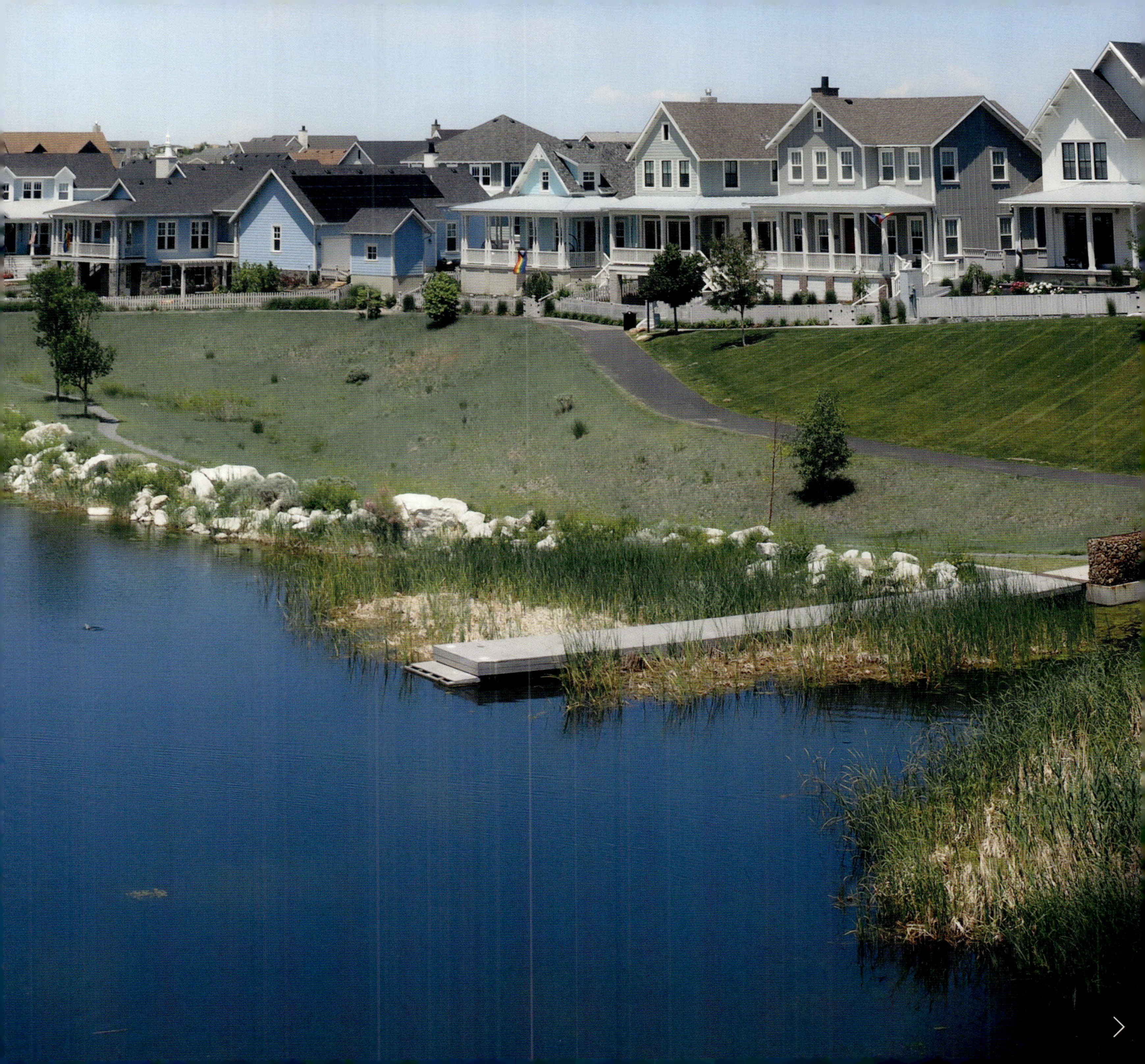

Utah's Oquirrh Mountains host one of humanity's greatest holes – Rio Tinto Kennecott's (RTK's) Bingham Canyon copper mine. It's over a century old, and its waste rock mountains loom above the western Salt Lake Valley, where 150 years of mining legacies have been the subject of expensive clean-up for decades. Meanwhile, the geography and the rapidly growing cities of South Jordan and Salt Lake conspire to increase the pressure for land for housing, services and jobs in an area already affected by water scarcity and impaired air quality.

When RTK acquired the landholdings of old mines in the Salt Lake Valley, it assumed liability for cleaning up their extensive areas of heavy metal contamination and an 18,600-hectare plume of sulphate-tainted groundwater emanating from old workings in the Oquirrh Mountains. Listing as a Superfund site (a contaminated site that the US Environmental Protection Agency [EPA] has ordered to be cleaned up with costs recouped from those responsible) was avoided by cooperating with the EPA and the State of Utah and spending USD 370 million on reclamation.

The most problematic contamination was on land occupied by evaporation ponds. First dug in the 1930s for storing and evaporating Bingham Canyon's mine water, they expanded over time until their use ended in 1987, leaving a legacy of contaminated soil. The EPA mandated that this be remediated – an enormous challenge that took years and required the removal and safe storage of the contaminated soils to federal and state standards.

The hundreds of hectares of newly cleaned land were in a prime location for residential development to accommodate Salt Lake County's urban expansion and to generate funds to offset the reclamation costs. Recognising this opportunity, in 2001, RTK set up Kennecott Land as a separate business unit, becoming one of the few mining companies to own a real estate arm. They focused on creating an exemplary sustainable community to be called Daybreak. Construction began in 2004, and a year later, Kennecott Land achieved ISO14001 status – a US first for a community developer.

The Daybreak masterplan covers about 16km^2 of the valley, incorporating 20% open space, 20,000 homes and 850,000m^2 of commercial development. From the outset, sustainable design and a progressive approach formed the central philosophy.

By 2020, about 8,000 homes had been built, housing 28,000 people. Daybreak offers diverse mixed neighbourhoods as well as distinct areas for families and over-55s. Construction is to a high standard with deliberate architectural variety and features like street-facing porches and balconies to stimulate interaction and neighbourliness, as well as garages located behind homes or accessed by alleys to separate cars from pedestrians. All homes are certified by the US Green Building Council for their sustainability and energy efficiency and by Energy Star for their indoor air quality, as are many commercial buildings. Photovoltaic and solar thermal panels were promoted years before they became popular, and most homes have fibre optic broadband.

The groundwater plume extends deep beneath Daybreak. Since 2004, RTK has pumped and treated the water in reverse-osmosis plants to supply drinking water in compliance with regulatory standards. It is predicted that the plume will have substantially decreased in size after 40 years of such treatment.

The Salt Lake Valley is severely water-stressed, so Daybreak is designed with water efficiency in mind. Its stormwater management system can retain rain even from a 100-year event to protect the groundwater. Secondary water runs into Oquirrh Lake from where it is used to irrigate large community parks, gardens and churchyards. Water consumption is carefully monitored and managed to maximise efficiency with limits on the extent of private lawns.

The artificial Oquirrh Lake at the heart of Daybreak's Lower Village allows non-motorised recreational use and fishing. The lake, plus about 13% of Daybreak itself, sits on the footprint of the former evaporation ponds. The recently constructed Watercourse is a 1.6-kilometre-long, publicly accessible system of waterways and a green corridor through the Upper Village.

An ambitious plan to plant 100,000 native trees will help mitigate the urban heat island effect, as will over 400 hectares of green space. Native, drought-tolerant grasses are used on public and private lawns to minimise irrigation, and wildflowers attract pollinators and feed the on-site beehives.

Poor air quality is another environmental concern in the valley. Sustainable alternatives to car use encourage clean air and healthier active travel. A designed-in 5-minute 'walkability' rule means easy, car-free access to green spaces, shops, offices, schools, public healthcare and public transport on nearly 50 kilometres of trails and paths. The urban design has moved workspaces closer to homes to reduce commuting; some major businesses, including Rio Tinto and eBay, have set up offices. These already employ over 2,000 people, many of whom live in Daybreak – a good omen. The development catalysed the construction of the TRAX light rail line to the University of Utah on the other side of the valley, with connections to downtown Salt Lake City and the international airport.

Alongside the recreational opportunities afforded by the trails and community parks, Daybreak's sports facilities accommodate most requirements, as do its abundance of shops, eateries, bars, cafés and farmers' markets selling local produce. Community cultural events, like summer concerts, holiday activities and educational and healthy living activities, are organised through the independent non-profit organisation LiveDaybreak. In a unique model for residential development, its community programmes are financed through the sale and resale of homes.

Kennecott Land was sold to Värde Partners in 2016, which agreed to continue development as a new company, Daybreak Communities. An undeveloped 500+ hectares was sold again in 2021, with the new developer focusing on a flagship downtown area, the centrepiece of which will be a new stadium for the Salt Lake Bees minor-league baseball team.

With its backdrop of the snow-capped Rocky Mountains to the east, abundant water features, softening green spaces and diverse architecture, and the lifestyle options these engender, the multi-award-winning Daybreak has been an influential post-industrial success story that offers valuable lessons to others.

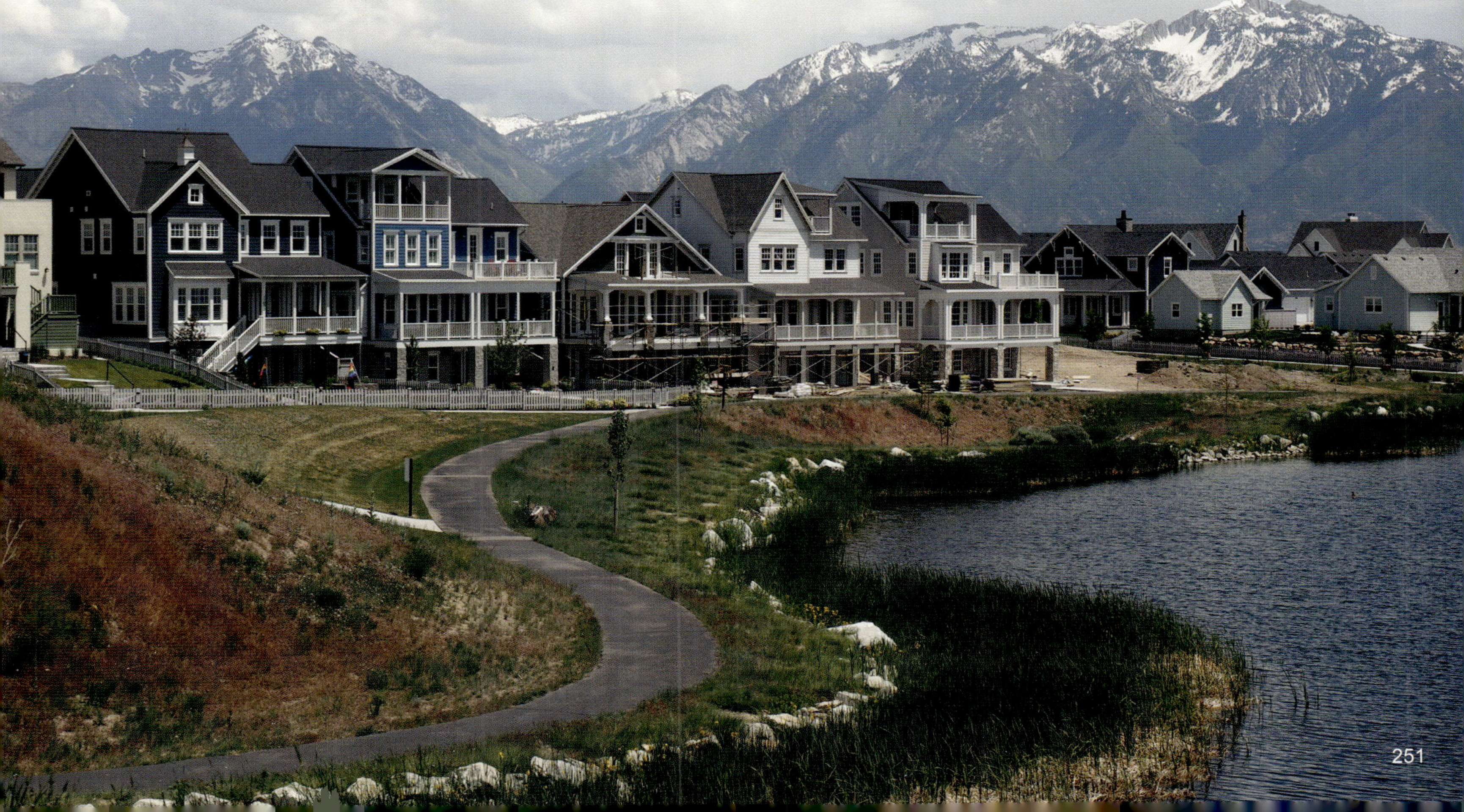

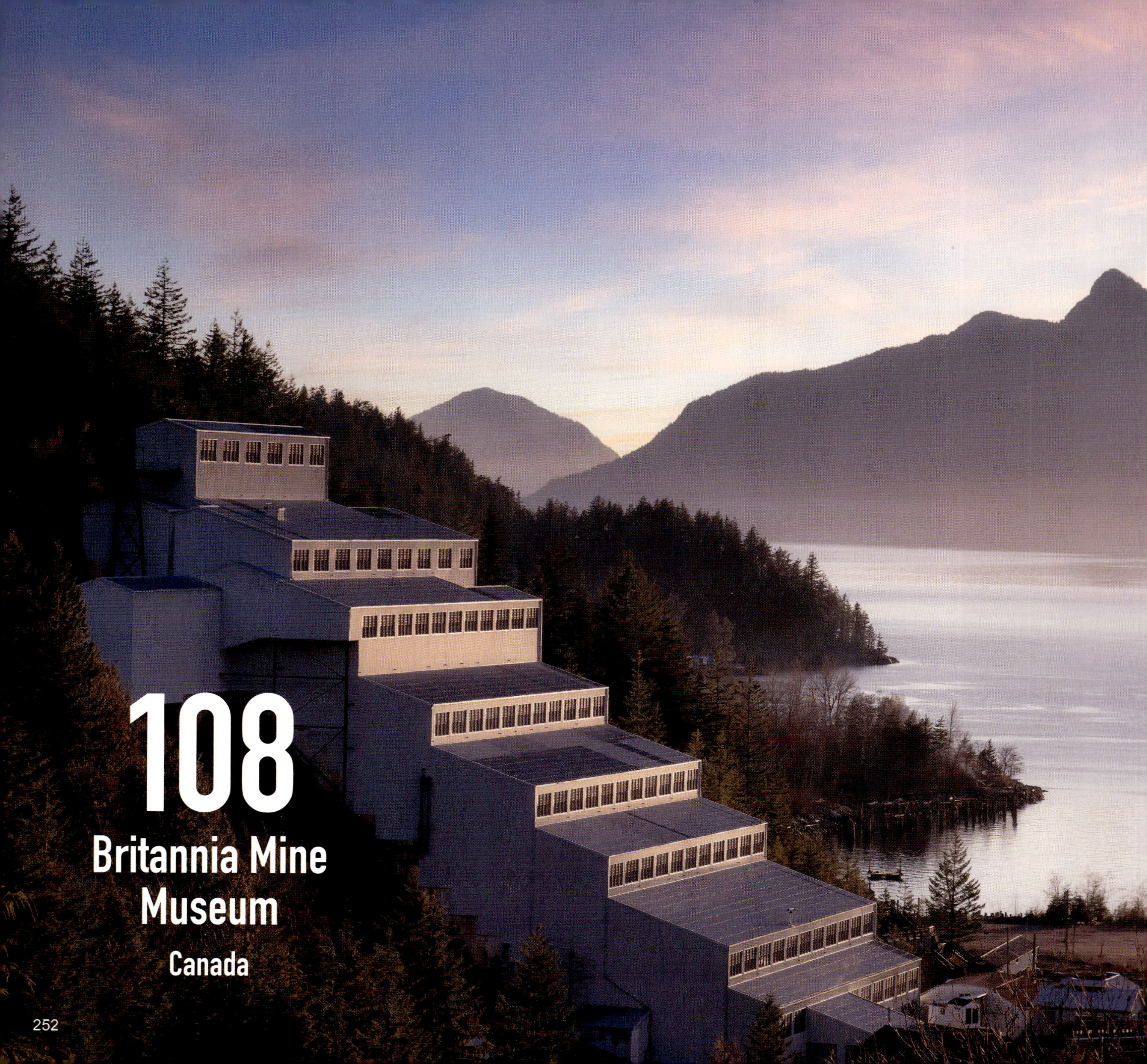

108
Britannia Mine
Museum
Canada

Successful visitor attractions often grab the attention with a 'big shiny thing'. Nestled between the narrow Howe Sound and the rugged mountains of the Sea-to-Sky Highway connecting Vancouver and Whistler, that 'thing' is the 20-storey Mill No.3 building of the old Britannia Mine, fixed to the mountainside like a frozen cascade. It embodies a century-long sense of place, as encapsulated by Joan, a former secretary for the Anaconda mining company:

'Can I tell you the first time it struck me that this place was gone? They decided in '74 to shut it down because it was costing too much to get the ore out and it wasn't profitable. One day I went to Squamish grocery shopping and when I came back, the Mill, the Concentrator was dark and I sat up on the bluff and I couldn't come home because I was crying so hard. It was the saddest thing. I'd never seen the Mill building without lights in it. And that's when I knew it was over.'

Located in the traditional territory of the Squamish Nation, Britannia Mine was once the largest copper mine in the British Empire, and it oozed cultural diversity; in its 70-year life, 60,000 people from more than 50 countries lived and worked there. Its 210 kilometres of tunnels produced over 50 million tonnes of ore and turned a mountain into a honeycomb. Completed in 1923, it was one of the last gravity-fed concentrator mills in North America, using the slope for shifting ore by gravity through each mineral processing step from the top of the building to its bottom.

The idea of using the buildings as a heritage asset was floated years before the mine closed in 1974, and the estimated costs of demolition and clean-up were greater than keeping and reusing them. The mine owner gave the buildings and land to a non-profit group, the Britannia Mine Museum Society (BMMS), which still owns the site. The closed mine site reopened as a museum in 1975 – a lot has changed since then.

The BMMS board is largely dominated by the mining industry, which has been effective in raising funds for essential capital works to preserve the National Historic Site. From 2005 to 2007, Mill No.3's roofs and sidings were renovated, its foundations stabilised, and its windows repaired – all 14,416 glass panes were painstakingly hand-puttied into their restored frames. In 2010, almost CAD 15 million was received from public and private (mainly industry) sources to redevelop the site into what it is today, using local contractors wherever possible. The well-designed facilities, including a visitor centre, interactive exhibition space, scenic boardwalk, revamped gold panning area and an outdoor events plaza, make learning by osmosis amazingly easy. Open year-round, the museum attracts around 70,000 visitors annually, including many from overseas, and about 120 kids per day on school visits – a figure limited by the capacity of the underground miners' train. It employs 46 people at the height of the season.

Mill No.3 looks enormous from the outside yet, somehow, it's even more overpowering from within; a cavernous, decaying, industrial cathedral. A pervasive damp, metallic odour; colossal concrete structures weeping stalactite tears; a muted kaleidoscope of rustic, tawny wood, cinnamon rust, iron stains and sugary efflorescences on a grey concrete canvas; then, flashes of turquoise minerals in the exposed bedrock while emerald green moss and algae flourish in patches of damp and light. The experience evokes an emotional response; it's easy to imagine the cacophony and smells and vibration of the heavy engineering in motion that once occupied this space.

It is a National Historic Site, so the BMMS are obligated to preserve it, which limits how much it can be modified. The museum team has responded to these restrictions with creativity. BOOM! is an award-winning and captivating audio-visual experience that narrates the mine's story while cleverly circumventing its limitations. Created by local companies VISTA Collaborative Arts and Dynamic Structures, the CAD 4.5 million cost was met by the federal and provincial governments and the museum's board. The visitors are seated while around them the dormant mill awakens with storytelling, theatrical lighting, deep, resonating industrial noises and a death-defying engineering experience as a finale. The show takes full advantage of the building's enormity and character to create a thrilling visceral experience. It includes Joan's quote above.

The photogenic architecture has been used as the backdrop to many TV programmes and films. Building on its totemic relationship with local people, the museum is also a valued community asset hosting a diverse range of events and travelling exhibitions, on issues as diverse as avalanche safety, the environmental recovery of Howe Sound and a high-profile Pride event, Old Town, New Queens, held in Mill No.3.

The museum also relates the extraordinary and tortuous 30-plus-year story of environmental renewal. Britannia Mine was infamous as Canada's worst point source of metal pollution: during its life it dumped 40 million tonnes of sulphidic tailings and released millions of litres of acid drainage carrying a daily average of 600kg of toxic dissolved metals into Howe Sound, decimating sea life. Legal action by the province eventually resulted in a pay-out of CAD 30 million towards the clean-up and perpetual water treatment costs.

The University of British Columbia's innovative Millennium Plug project installed two massive plugs, built to last 1,000 years, in key drainage tunnels to re-engineer groundwater flows within the honeycombed mountain and allow storage of up to 430,000 cubic metres of acid water in the old workings for controlled release into a water treatment plant. Golder Associates (now WSP) became the project manager for site remediation and closure planning for the provincial government; then utility company EPCOR built – and still operates – the plant to neutralise the acidic water and remove metals. Who knows for how long it will be necessary?

The tailings in Howe Sound are generally considered low risk because they have settled at depths with low oxygen levels and are being slowly covered by sediments from the Squamish River. Environmental monitoring shows improving shoreline water quality and ecology, with whales once again sighted in the area and salmon returning to the lower parts of the Britannia Creek.

The decades-long Britannia Mine story is one of cultural respect and environmental renewal. A wonderful asset has been created that offers many lessons to mining visitor attractions and that uses the past as an honest broker to catalyse new possibilities for people, former mining communities and their environment.

Previous page: Mill No.3 on Howe Sound.
Below left: The atmospheric interior of Mill No. 3.
Below right: Old Town, New Queens Pride event at Britannia Mine Museum.

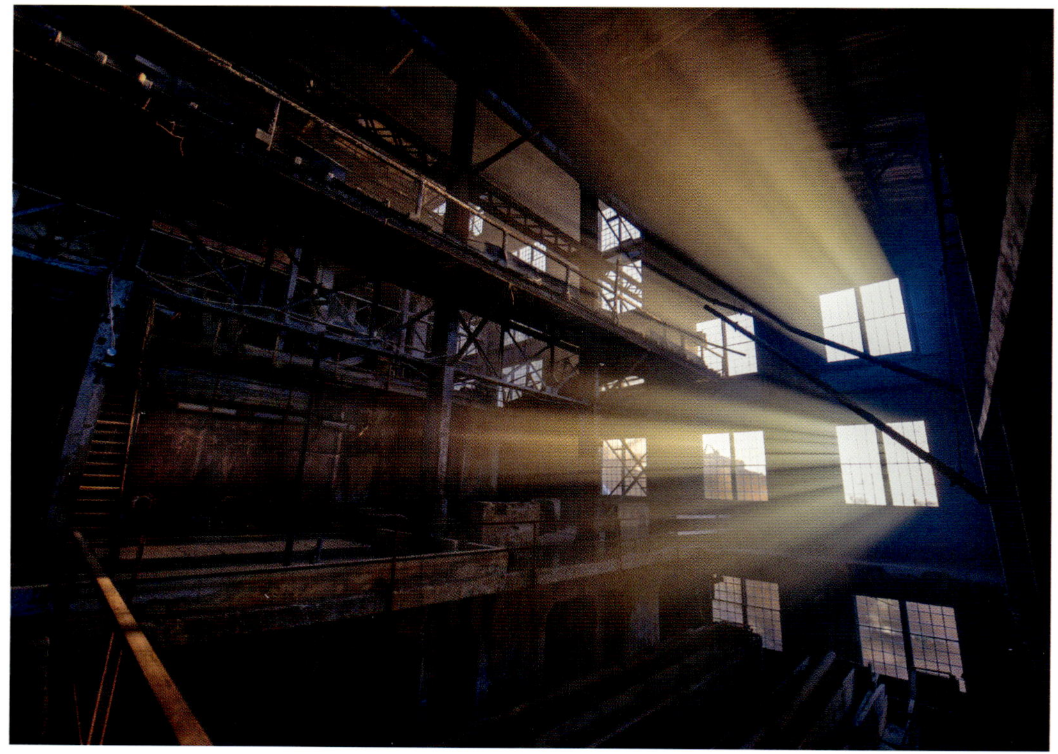

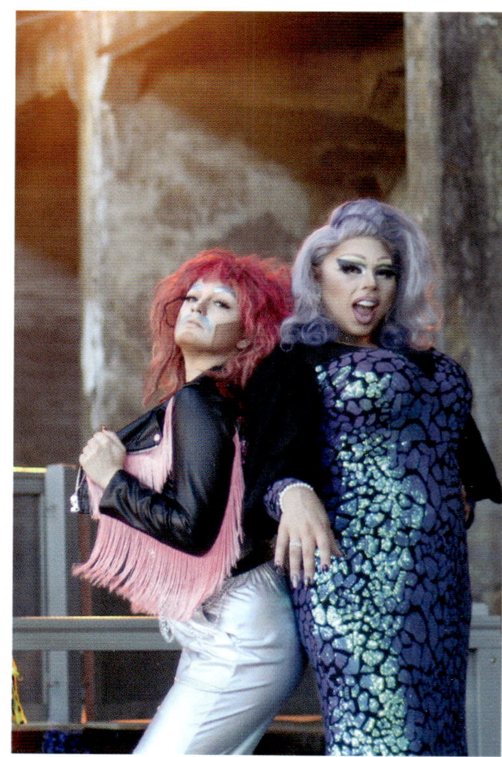

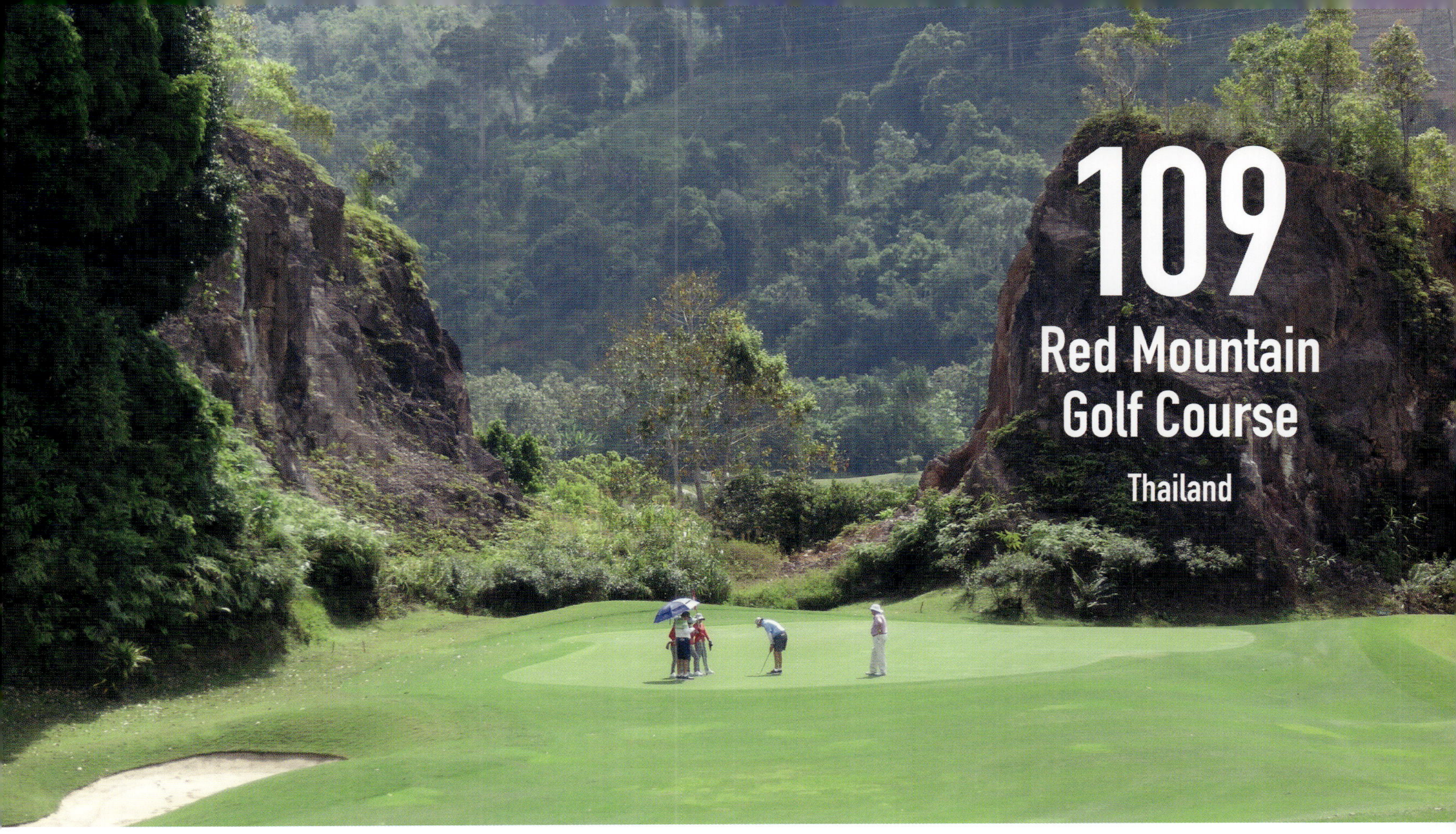

Many golf courses have been fashioned from former mine sites. Arguably one of the most spectacular is Phuket's Red Mountain Golf Club. It opened in 2007 and its name derives from the geology-imparted colour that attracted miners in the first place.

The 18-hole, par 72, 6,314-metre course was imaginatively painted on the harsh blank canvas of an old tin mine, creating a picture of emerald fairway greens, rusty geological reds, deep jungle greens, white sand bunkers and a brilliant blue sky. The experience is enhanced by orchids, ferns and other tropical plants that attract colourful birds and butterflies.

Red Mountain was cleverly designed and engineered to work with the undulating, underlying terrain to create its twisting fairways and dramatic landforms, with cliffs and lakes fashioned from the old mining pits. Strong elements of the original mined landscape feature: the first hole's immaculately manicured and contoured fairway leads to a green which lies in front of a remnant overburden dump; and the third hole capitalises on mining's disturbance with its treacherous approach to the green sculpted from deeply eroded mineral waste ridges and gullies, carpeted in grass – sometimes there is value in not levelling everything when rehabilitating a mine site to make a golf course. Alongside stunning views across the steaming jungle-clad hills of central Phuket, the seventeenth presents a different technical challenge with its steep 40-metre drop from the tee through lofty trees to what looks like a postage stamp-sized green far below.

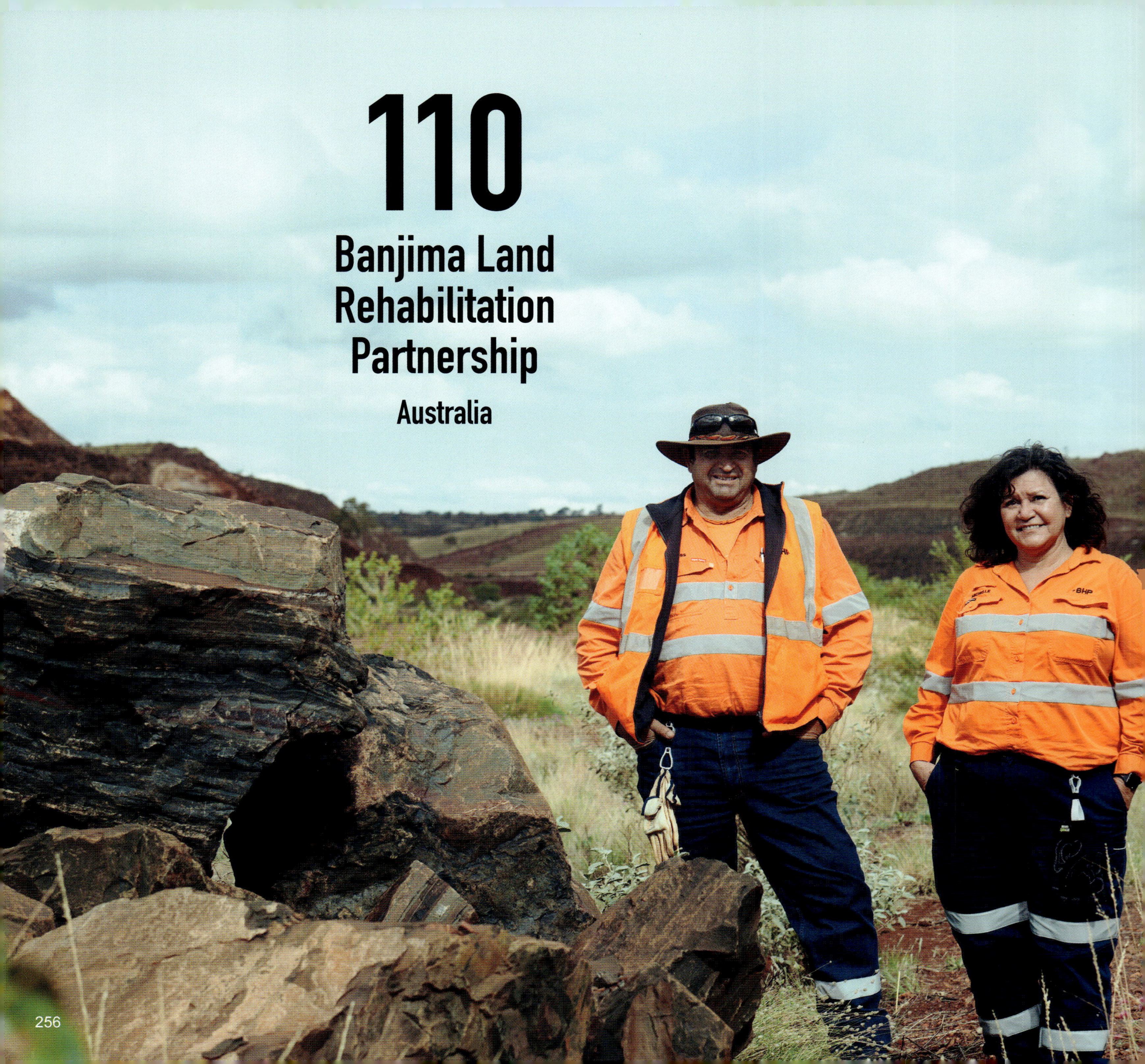

110

Banjima Land Rehabilitation Partnership

Australia

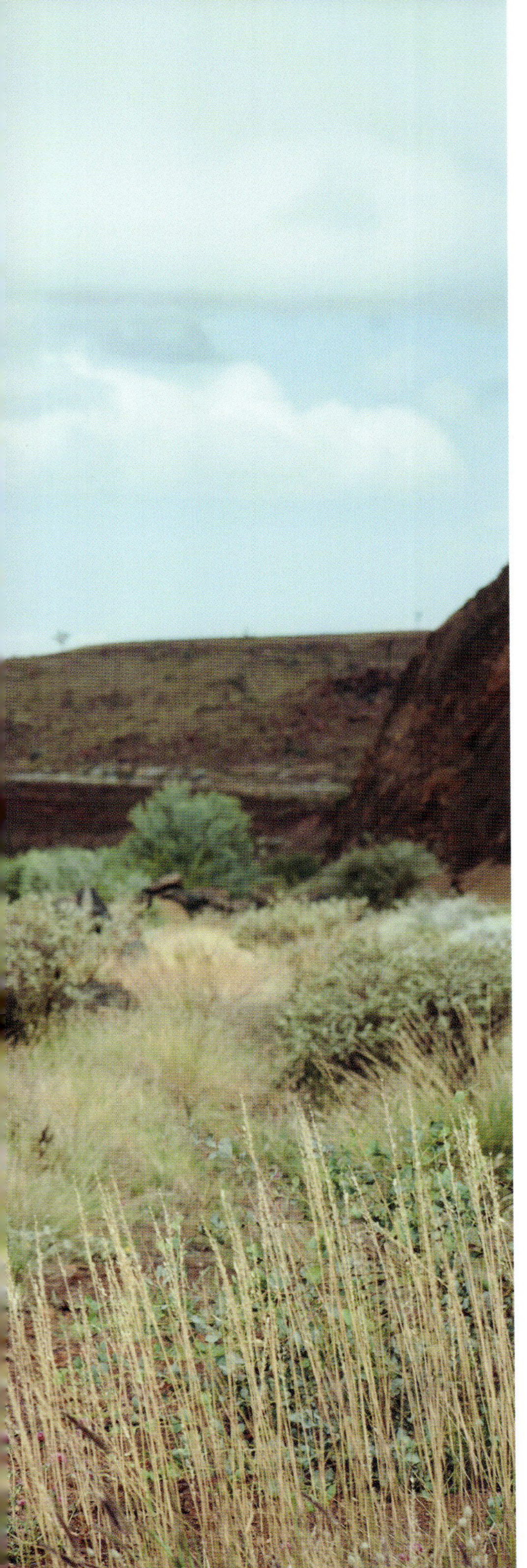

BHP's Yandi iron mine sprawls across the parched, rust-coloured country of Western Australia's Pilbara. 'Country' is also a term used by Traditional Owners to describe the terrain to which they have been physically, spiritually and culturally connected through countless generations, in this case, the Banjima people. Progressive ecological restoration is underway as the mine ramps down to closure, prompting two Aboriginal BHP employees, Ross West and Michelle Adams, to develop a programme to train and employ Banjima people in land rehabilitation, offering a pathway to transferable skills and future employment prospects in a remote, sparsely populated region where mining is big business.

Led by the Banjima Native Title Aboriginal Corporation in partnership with BHP, which provides dedicated personnel and infrastructure, the Banjima Land Rehabilitation Partnership is a unique ground-up collaboration. It aims to grow the regional labour pool, build regional business capacity and increase indigenous representation in mining and land management activities.

Funded by the Yandi closure provision, the programme combines classroom learning facilitated by TAFE (a vocational training provider) to achieve a Certificate II in Conservation and Ecosystems Management, with self-development and on-the-job training. Rehabilitation activities include a tree nursery to raise 20,000 native woody plants for restoration projects, seed collection on Country, seed storage, native plant propagation and care, and participation in R&D seed production trials at Perth's Kings Park and Botanic Garden nursery. Since starting in late 2021, the programme is a success with the majority of trainees transiting straight into work or obtaining the Certificate II.

The success of the partnership has been widely acknowledged, and deployment across other BHP sites in Australia is under consideration. Recent recognition includes the Indigenous and Community Engagement Award at the Australian Mining Prospect Awards and BHP's in-house HSEC Award for Excellence in Social Value.

The collaboration's success is due to its progressive understanding of community development – it is not project management or a short-term proposition. It links community and business aspirations through land and takes advantage of a growing desire and increased funding to deliver social value (especially for indigenous stakeholders) and close the emerging gap in resources for land rehabilitation. The evolving restoration economy will require better-informed stakeholders and a larger workforce with improved skillsets and rehabilitation approaches, and the principles of this project are applicable globally.

111
Charlie Bigham's
UK

Dulcote Quarry contains a lot of life for what was, until fairly recently, a worked-out limestone quarry.

Charlie Bigham, the founder of the Charlie Bigham's meal range, was hunting for a new HQ for his business when he stumbled across the disused 30-hectare quarry in the Mendips. Aside from being largely flat and close to the road, there was little infrastructure on the site, but Bigham saw its potential as a commercial site and a way to embody his company's values.

Charlie Bigham's is not a conventional oven-ready meals company. It doesn't produce ultra-processed food. Fresh ingredients are cooked in small batches and finished by hand – it's more like a large hotel kitchen than a factory. Its approach has resulted in numerous awards and made it one of the fastest-growing grocery brands in the UK, with sales of over GBP 120 million per year. Yet, as a B Corp, it measures success in terms of its social and environmental performance, not just profitability, and these values have helped shape the post-mining development of Dulcote Quarry.

Quality of life was high on the list of priorities when the company was looking for new premises. Located just outside of Wells in the county of Somerset, Dulcote offers a very different pace of life to

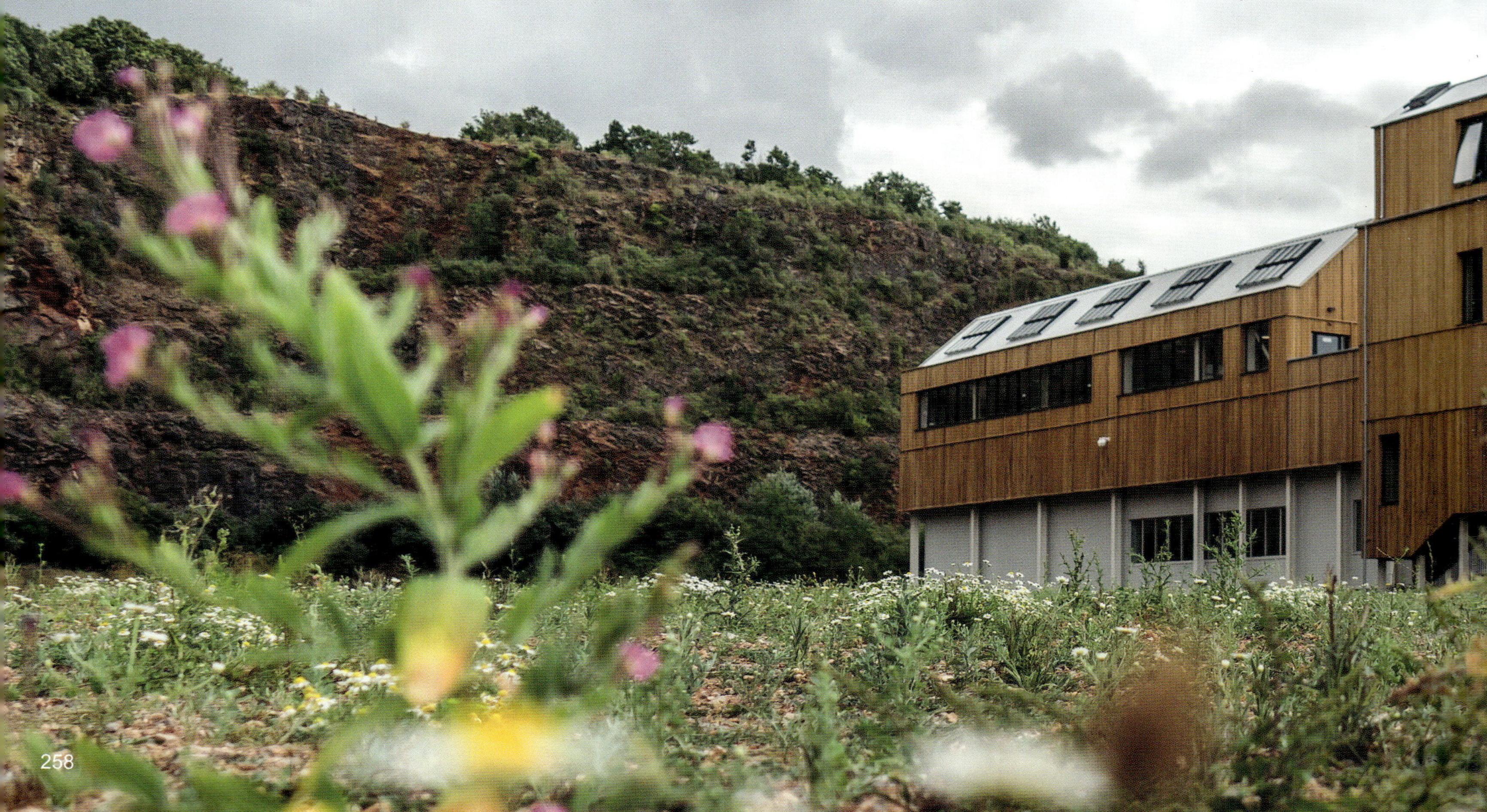

London and better value for money when it comes to housing. Its smaller London premises are still used for developing new lines, but many staff chose to relocate. Four hundred people now work at the Dulcote Quarry, which produces the majority of Charlie Bigham's dishes. While planning permission was in place for light industrial and office use, Bigham was guided by his principle that 'high-quality food can only be made in a high-quality environment.'

Designed by architects Feilden Fowles, the buildings comprise kitchens, offices and a staff café on the southern side of the quarry. The kitchens are housed in the more industrial-looking part of the building, whose sawtooth profile and rusty red paintwork echo those of the original quarry sheds. There is just under 400 KWh of solar PV on its jagged roofline, with a further 2.2 MW planned for an adjacent former quarry waste tip. Wastewater is collected and purified on-site, and ponds have been created for drainage that also support wildlife.

Over a quarter of the eight-hectare bare quarry floor has been turned into a wildflower meadow, and hundreds of trees have been planted. The wildlife that colonised the quarry when it closed has stuck around. The original population of 100 great crested newts is now 500; the greater horseshoe bats that had moved into the remains of the old weighbridge now have a purpose-built bat house, and families of peregrine falcons live on the cliffs behind the factory.

This is just the first phase of a 20-year master plan for the site. And, although Charlie Bigham's plans to increase the site's production capacity, it also wants to create more of a campus feel that would include a café and possibly a visitor centre. None of this will be at the expense of the wildlife – a wetland area has been proposed, and almost half of the site is reserved for nature – ensuring that the environment prospers alongside the company.

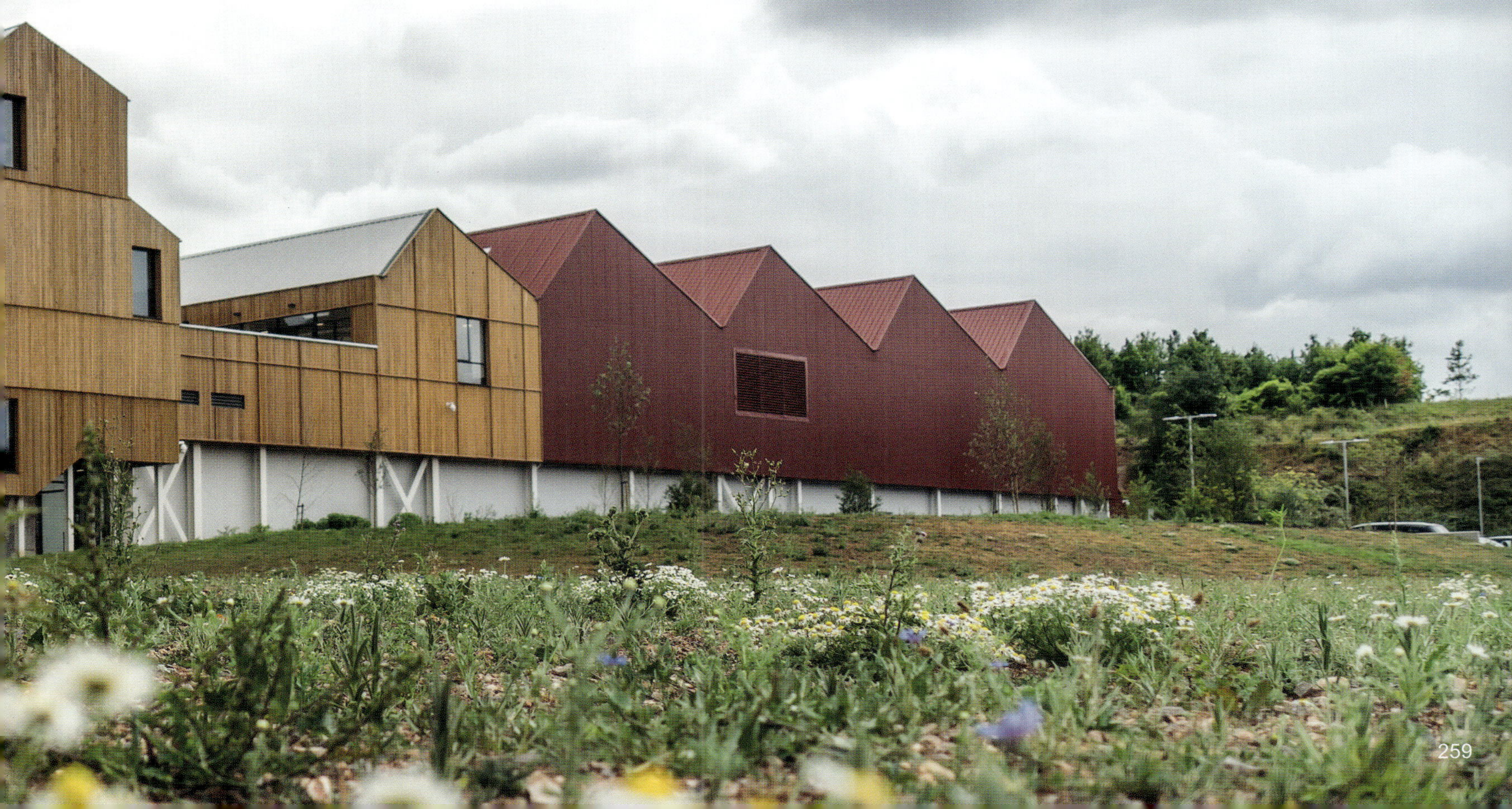

Without whom...

Many people went above and beyond to help make *102 Things* happen.
Please accept our apologies for any inadvertent omissions.

Bill Adams (formerly Rio Tinto), Michelle Adams (BHP), James Alexander (De Beers),Virginia Alexander (Rio Tinto), Celestina Allotey (Gold Fields), Patrick Angel (formerly Appalachian Regional Reforestation Initiative), Matt Baida (VAST Landscape Architecture), Chris Barton (Green Forests Work), Peter Beckett (Laurentian University), Niall Benson (formerly Durham County Council), Debbie Berthelot (BHP), René Betemps (Cooperativa Produttori Latte e Fontina s.c.a.r.l.), Stephen Bird (Queensland Government), Mike Boon (Diamond Coast Aquaculture), Rudolph Botha (Anglo American), Stephen Bourn (Rio Tinto), Anna Maria Branduzzi (Green Forests Work), Jen Brereton (Mine Land Rehabilitation Authority), James Breslin (National Trust), Dawn Brock (Anglo American, formerly ICMM), Carolyn Burns (Resolve), Hunter Burrow (Mettiki Coal), Chris Cafe (Blue Derby Foundation), James Cameron (BHP) Alan Carter (The Land Trust), Harry Catherall (Newcastle University), Wayne Chapman (Stawell Gold Mines), Carole Charron (NORCAT), Preston Chiaro (Death Valley Conservancy), Troy Cole (Stawell Gold Mines), Nick Cotts (Newmont), Mark Couch (Short Run Press), Max Cromie, Karen Daglish (formerly Durham County Council), Richard Davis (BHP), Ettienne de Jager (Namdeb), Steve D'Esposito (Resolve/Regeneration), Stuart Devenish (Shire of Collie), Caroline Digby (formerly Eden Project), Greg Dipple (Arca), Ralf Donat (Heinz Sielmann Stiftung), Blair Douglas (BHP), Eibhlin Doyle (Government of Ireland),

José Francisco Martín Duque (Complutense University of Madrid), Don Duval (NORCAT), Wesley Earl (Zip World), Paul Eldon (Elliot Lake XC Ski & Bike Club), Kevin Erwin (North Grampians Shire Council), Emily Evans (Premier Coal), Anna Farrell (Collie Motorplex), Kim Ferguson (WSP, formerly BHP), Karsten Feucht (Karsten Feucht Architektur), Leo Fincher-Johnson (University of Melbourne), Tony Ford (Genex Power), Chloe Ford-Welman (Haller Foundation), Ben Foster (Eden Project), Nathan Francis (Rio Tinto),Elizabeth Freele (Sympact), Michael French (Green Forests Work), Emma Gagen (ICMM), Rebecca Getty (BHP), Jim Gillon (Gateshead Energy Company), Henry Golas (Death Valley Conservancy), Blanco Racionero Gómez (Levin Sources), Jess Gosling (DeepStore), Emma Grimster (The Land Trust), Rhona Guertin (Elliot Lake Retirement Living), Euan Hall (formerly The Land Trust), René Haller (Haller Foundation), Mark Harman (Maryland Department of Natural Resources), Peter Harvey (Rio Tinto), Fiona Haslam-McKenzie (University of Western Australia & CRC-TiME), Timo Hauge (Regionalverband Ruhr), Trevor Heaton (Rio Tinto), Bill Hickman (Pikeville-Pike County Airport Board), Shane Hirst (USP Big Sandy), Kathleen Hofmann-Mitzschke (LMBV), Frida Holst (VAST Landscape Architecture), Cameron Hope (Stawell Gold Mines), Greg Howard (Dorset Council), Mat Howlett (EnergyAustralia), Derek Jang (Britannia Mine Museum), Andrew Johnson

(GCRE), Deron Johnston (Britannia Mine Museum), Petrus Jordan (De Beers), Romy Kaltschmidt (LMBV), Daniel Karlsson (AdventureMine), John Kearney (Government of Western Australia), Hans Kgasago (De Beers), Nigel Kieser (formerly Rio Tinto), Graeme King (Gateshead Council), Colby Kirk (One East Kentucky), Michal Kozikowski (BHP), Manuel Lapp (Saxon State Office for Environment, Agriculture and Geology), David Lawson (MountainRose Vineyard), Alexandra Le Van (Rio Tinto), Nikisi Lesufi (Minerals Council South Africa), Estelle Levin-Nally (Levin Sources), Hanna Lohmann (Stiftung Zollverein), Martin Lokanc (World Bank), Nick Loubser (Viking Aquaculture), Zhihai Luo (EBP), Rae Mackay (Mine Land Rehabilitation Authority), Jones Mantey (Gold Fields), Ian Marsh (National Trust), Tina McCaffrey (City of Greater Sudbury), Sarah McConnell (BHP), David McGavin (Engie), Tom Measham (University of Queensland & CRC-TiME), Nigel Mercer (Dorset Tasmania History Society), Russell Merz (METS Ignited), Aaron Miller (Mettiki Coal), Kim Mintern-Lane (University of Melbourne), Jon Missen (Engie), Maxwell Morapeli (De Beers), Maxime Morin (BHP), Emma Ekua Akyineba Morrison (Gold Fields), Allan Morton (enviroMETS Qld), Miles Moulding (Go Below), Stephinah Mudau (Minerals Council South Africa), Ritva Muhlbauer (formerly Thungela), Kado Muir (Ngalia Heritage Research Council Aboriginal Corporation), Rhonda Norrish (BHP), Tim O'Connor (BHP), Erin Parham (De Beers), Dave Payne (Anglo American), Georgina Pender (née Pearman, formerly Eden Project), Ryan Perry (Rio Tinto), Ryan Peterson (Teck), Greg Radford (Intergovernmental Forum on Mining, Minerals, Metals and Sustainable Development), Neil Rein (Thungela), Scott Rennie (City of Greater Sudbury), Mike Reynolds (Death Valley National Park), Piper Rhodes (Rio Tinto), Carol Richards (King Edward Mine), Vivienne Robertson (Reclaim the Void), Victoria Ruane (The Land Trust), Wendy Russell (BHP), John Sabraw (University of Ohio), Luis Salazar (Crops Trust), Steven Sannoh (Resolve), Janine Scharf (Ferropolis), Thies Schröder (Ferropolis), Peter Sheppard (King Edward Mine), A. Lena Sieler (Landschaftspark Duisburg-Nord), Sonal Singh (Haller Foundation), Neeltje Slingerland (WSP), Jill Smillie (The Butchart Gardens), Lee Smith (Veolia), Elmarie Snyman (Namdeb), Veronika Sochorova (Euromines), Gerry Stanley (Government of Ireland), Uwe Steinhuber (LMBV), Heather Sugden (Newcastle University), Phil Tanner, Sean Taylor (Zip World), Cassandra Tolsma (EnergyAustralia), Peter Toth (Newmont, formerly Rio Tinto), Laura Tyler (formerly BHP), Natalia Valderrama, Eben van Heerden (De Beers), Ronell Van Staden (Futureworld), Juanjo Villanueva (1270 Craft Beer), Werner Voigt (Anglo American), Friedrich von Bismarck (formerly StuBA), Gero von Daniels (StuBA), Georgina Voß (Stiftung Zollverein), Lance Wallace (EnergyAustralia), Weina Wang (DnA_Design and Architecture), Heather Warman (SUP Kentucky), Jack Wasserfall (Namdeb), Barbara Wernick (WSP), Adam Webb (Charlie Bigham), Stephen Wheston (Tembusu), Katie Whitbread-Abrutat, Maxine Whitbread-Abrutat (Future Terrains International), Carleigh Whitman (Teck), Damien Wieland (North Grampians Shire Council), Glenn Willmott (Government of Western Australia), Abby Wines (Death Valley National Park), Nikki Wright (Queensland Government), Charlene Wrigley (Gold Fields), Roman Wunderlich (Wunderlich Backstuben), Wang Yaqiong (China Three Gorges Renewables Group), Elena Zabudskaya (City of Greater Sudbury).

Indexes

The case studies are grouped by mine site feature – features found on or in association
with mine sites and after-use – generic groupings of similar post-mining uses.

Mine Site Feature	Rationale	Case Study Numbers
Voids	Open pits at the surface; underground workings.	1, 2, 7, 9, 11, 14, 16, 18, 22, 25, 28, 29, 30, 31, 34, 37, 38, 40, 41, 43, 47, 48, 51, 52, 56, 58, 61, 62, 63, 65, 66, 68, 69, 77, 78, 79, 84, 86, 87, 88, 90, 91, 95, 96, 97, 109, 111
Mineral Wastes	Includes waste rock, processing wastes, and overburden.	5, 12, 15, 19, 20, 21, 22, 26, 35, 39, 44, 57, 60, 67, 71, 74, 81, 85, 89, 100, 106, 107, 108, 109
Mine Waters	Water emanating from mine sites, including below-ground.	3, 24, 65, 78, 82, 85, 86, 93, 104, 107, 108
Infrastructure, Buildings & Equipment	Mine buildings, roads, railways, power infrastructure, water treatment plants, etc.	4, 6, 10, 33, 53, 54, 55, 72, 73, 75, 80, 81, 84, 85, 87, 88, 91, 92, 94, 97, 98, 100, 103, 105, 108
Communities	Settlements and people closely associated with mine sites.	3, 6, 8, 10, 22, 23, 28, 32, 33, 34, 37, 45, 50, 52, 58, 70, 75, 76, 79, 81, 83, 85, 91, 93, 96, 97, 98, 99, 100, 102, 106, 107, 108, 110
Land & Landscapes	Land within a mine's license area and how it connects to the surrounding landscape.	4, 6, 7, 12, 13, 17, 22, 26, 27, 33, 36, 42, 44, 45, 46, 49, 54, 59, 60, 61, 64, 67, 68, 72, 74, 75, 76, 79, 81, 83, 85, 88, 89, 91, 92, 94, 97, 99, 100, 101, 102, 106, 107, 109, 110, 111

After-use	Rationale	Case Study Numbers
Green Energy & Climate Change	Producing or storing green energy & mitigating climate change.	1, 20, 22, 39, 48, 51, 56, 72, 81, 85, 88, 89, 93, 102, 105, 107, 111
Active Recreation & Health	Active recreational activities, including sports.	6, 12, 16, 18, 19, 20, 22, 26, 33, 37, 42, 44, 45, 47, 55, 58, 61, 70, 71, 73, 75, 78, 79, 80, 81, 82, 84, 85, 86, 87, 92, 98, 100, 102, 103, 107, 109
Art, Design & Performance	Projects and places featuring art and design, including gardens, music, performance and visual arts.	1, 6, 7, 8, 15, 19, 22, 23, 25, 26, 28, 29, 34, 37, 38, 47, 50, 60, 63, 67, 73, 75, 79, 81, 83, 87, 95, 97, 100, 102, 103, 104, 107, 108, 109, 111
Industrial Heritage	Specifically industrial heritage, as opposed to broader cultural heritage.	6, 10, 12, 16, 22, 23, 25, 28, 32, 33, 45, 49, 50, 52, 53, 55, 58, 60, 70, 73, 75, 78, 79, 81, 83, 85, 86, 87, 91, 92, 93, 95, 97, 98, 100, 103, 108
Visitor Destination	Tourist destinations with natural, cultural or historic value and/or beauty.	1, 2, 6, 7, 8, 10, 12, 14, 15, 16, 18, 19, 20, 22, 25, 26, 28, 29, 32, 33, 35, 37, 38, 41, 42, 43, 45. 47, 49, 50, 52, 55, 58, 59, 60, 61, 63, 64, 69, 70, 71, 73, 75, 78, 79, 80, 81, 83, 84, 85, 86, 87, 91, 92, 93, 95, 96, 97, 98, 100, 101, 102, 103, 108, 109
Spirituality	Places of worship.	28, 34, 87, 91, 94, 96
Indigenous Heritage	Projects relating to indigenous peoples.	34, 57, 79, 83, 91, 92, 108, 110
Research, Technology & Formal Learning	Research, education and training, and new technology development.	1, 5, 14, 17, 25, 29, 39, 51, 53, 54, 61, 62, 66, 89, 93, 104, 105, 106, 108, 110
Food & Drink	Food and drink production.	3, 4, 5, 9, 24, 30, 35, 36, 40, 52, 65, 69, 77, 99, 100, 106, 111
Re-mining	Using old mining and processing wastes as new mineral resources.	21, 57, 104
Nature	Ecological restoration, landscape restoration, rewilding and conservation.	1, 6, 12, 13, 22, 26, 27, 33, 37, 44, 49, 57, 59, 60, 61, 64, 67, 71, 74, 75, 76, 79, 81, 82, 85, 89, 92, 100, 101, 102, 104, 107, 108, 109, 110, 111
Commercial & Property Development	Projects for commercial, property and business development.	4, 6, 14, 18, 20, 21, 22, 24, 25, 29, 35, 41, 42, 44, 45, 46, 50, 52, 54, 56, 58, 64, 66, 68, 69, 70, 72, 73, 75, 79, 80, 81, 84, 85, 87, 88, 91, 97, 100, 102, 104, 105, 106, 107, 109, 111
Storage & Refuge	Storing information, energy and products of all kinds and sheltering people.	1, 2, 9, 11, 30, 31, 40, 43, 48, 51, 69, 72, 77, 88, 90, 91, 93

Photography credits

Credits are organised 01–111 by case study number. Specific page numbers given where more than one photographer is featured. All images are copyright © of the photographers and/or organisations listed below, but we have retained © in credits where requested.

Cover image David Folland / Eden Project
1 Main image: Hufton + Crow; p.11 Kathryn Nichols; p.12, James Street; p.13 (top) Big Lunch, (bottom) Ben Foster
2 Albin Marciniak/stock.adobe.com
3 Peter Whitbread-Abrutat
4 Peter Whitbread-Abrutat
5 Peter Whitbread-Abrutat
6 All images © Jochen Tack / Zollverein Foundation except p.22 Simon Bierwald, © Red Dot Design
7 Peter Whitbread-Abrutat
8 BJ Warnick / Alamy Stock Photo
9 Photo 12 / Getty
10 motive56 / Alamy Stock Photo
11 Bloomberg / Getty
12 Main image: United Archives / Getty; p.39 Peter Whitbread-Abrutat
13 Ainsley Hatt
14 Nataliia Sokolovska / Shutterstock
15 Mark Asthoff/Unsplash.com
16 Go Below Underground Adventures Ltd
17 Marius Claassen
18 © Zip World
19 Peter Whitbread-Abrutat
20 Alpincenter
21 DRDGOLD
22 Main image: Peter Radke / LMBV; p.57 Peter Whitbread-Abrutat; p.58 Peter Radke / LMBV; p.59 all by Peter Whitbread-Abrutat
23 Michael Klug
24 Peter Whitbread-Abrutat
25 DnA_Design and Architecture
26 Main image: Tony McLaughlin / The Land Trust; p.67 Over and Above / The Land Trust
27 Flament / Andia / Alamy Stock Photo
28 AntonyM / Alamy Stock Photo
29 Fariz Abasov / Shutterstock
30 James Padolsey / Unsplash
31 As above
32 David Kettles / Alamy Stock Photo
33 Main image: Rainbow Joe / iStock; p.78 Kevin Britland / Alamy Stock Photo
34 Nic Duncan
35 Peter Whitbread-Abrutat
36 Esin Deniz / iStock
37 Peter Whitbread-Abrutat

38 Dalhalla
39 Acra
40 Roman Wunderlich, Wunderlichs Backstuben
41 Anton Vierietin / Shutterstock
42 Peter Whitbread-Abrutat
43 Kristof Bellens / Alamy Stock Photo
44 chris276644 / Shutterstock
45 Main image: Joshua Firth; p.100 FiledIMAGE / Shutterstock
46 USP Big Sandy
47 Peter Whitbread-Abrutat
48 Bloomberg / Getty
49 Peter Whitbread-Abrutat
50 Peter Whitbread-Abrutat
51 Genex Power
52 Doce Setenta
53 Ralph Menz
54 Matthew Nichol Photography / GCRE
55 Max Willcock
56 China Three Gorges Renewables (Group) Co.
57 Thomas C. Kline, Jr. / Alamy Stock Photo
58 Adventuremine
59 Peter Whitbread-Abrutat
60 Peter Whitbread-Abrutat
61 Peter Whitbread-Abrutat
62 Virginia Kilborn
63 Vinicius Bacarin / Alamy Stock Photo
64 David Davies / Alamy Stock Photo
65 Edwin Orrillo
66 Peter Whitbread-Abrutat
67 Miguel Ángel Langa
68 Pike County Airport Board
69 Calin Stan / Shutterstock
70 Peter Whitbread-Abrutat
71 Peter Whitbread-Abrutat
72 Image supplied courtesy of Engie Hazelwood
73 All images Ferropolis except p.161 Peter Whitbread-Abrutat
74 Rio Tinto
75 Main image: Wirestock / iStock; p.166 Oberhaeuser / Alamy Stock Photo; p.167 (top), Erik / Adobe Stock, (bottom) Jochen Tack / Alamy Stock Photo
76 Peter Whitbread-Abrutat
77 Cooperativa Produttori Latte e Fontina s.c.a r.l.
78 © Jenn Coleman / Coleman Concierge

79 Peter Whitbread-Abrutat
80 Peter Whitbread-Abrutat
81 Brett Schreckengost
82 Stuart Wilson / Biosphoto / Alamy Stock Photo
83 Peter Whitbread-Abrutat
84 © Zip World
85 PhotoBliss / Alamy Stock Photo; p.194 Peter Whitbread-Abrutat
86 SUP Kentucky
87 p.197 Eva Pruchova / Alamy Stock Photo; p.198 Edwin Remsberg / Alamy Stock Photo; p.199 David Lyon / Alamy Stock Photo
88 Peter Morris Photography
89 Peter Whitbread-Abrutat
90 DeepStore
91 Peter Whitbread-Abrutat
92 Peter Whitbread-Abrutat
93 Gateshead Council
94 Main image: © Death Valley Conservancy; p.218 Peter Whitbread-Abrutat
95 Elenaphotos / Alamy Stock Photo
96 José Luis Holguín
97 blickwinkel/M. Woike / Alamy Stock Photo
98 Kay Roxby / Alamy Stock Photo
99 Ezekiel Kargbo / Freetown Media Centre
100 Main image: Flament / Andia / Alamy Stock Photo; p.230 Denis Charlet / AFP / Getty; p.231 ProCip / Alamy Stock Photo; p.232 LECLERCQ Olivier / hemis.fr / Alamy; p.233 Ludovic Maillard / Alamy Stock Photo
101 Main image: WorldFoto / Alamy Stock Photo; p.236 Peter Whitbread-Abrutat; p.237 Greg Taylor / Alamy Stock Photo
102 Peter Whitbread-Abrutat
103 Jochen Eckel/Süddeutsche Zeitung Photo / Alamy Stock Photo
104 Ben Siegel / Ohio University
105 Peter Whitbread-Abrutat
106 Peter Whitbread-Abrutat
107 Peter Whitbread Abrutat
108 Main image: Edgar Bullon / Alamy Stock Photo; p.254 Britannia Mine Museum
109 Phil Tanner
110 BHP
111 © Charlie Bigham